Groundwater in the Environment

For Keith Anderson:

Exegetically erudite
Ebulliently enquiring
Exhilaratingly eccentric
Entertainingly extreme

Cover photograph: This fountain of water is emerging of its own accord from a borehole penetrating the Chalk Aquifer in the UK (Tancred Pit Observation Borehole, Broachdale, near Kilham, East Yorkshire). Normally, the water level in this borehole lies 7–10 meters below ground level. At the time this photograph was taken (in the winter of 2000/01), rainfall in preceding months had exceeded all previous records in this region, and groundwater levels had risen to unprecedented heights. This borehole is located in what is normally a naturally dry valley. The standing water visible in the background of the photograph is due to natural emergence of the water table above ground level: this is an example of a groundwater flood (as discussed in Section 8.2.4). The interaction between groundwater levels, topography, and surface water flows illustrated by this case underlines many of the principles discussed in Chapters 1 and 5 of this book.

Groundwater in the Environment: an introduction

Paul L. Younger

HSBC Chair in Environmental Technologies
Institute for Research on Environment and Sustainability
University of Newcastle
Newcastle Upon Tyne
United Kingdom

BLACKWELL PUBLISHING
350 Main Street, Malden, MA 02148-5020, USA
9600 Garsington Road, Oxford OX4 2DQ, UK
550 Swanston Street, Carlton, Victoria 3053, Australia

First published 2007 by Blackwell Publishing Ltd

1 2007

Library of Congress Cataloging-in-Publication Data

Younger, Paul L.
Groundwater in the environment: an introduction/Paul L. Younger.
p. cm.
Includes bibliograhical references and index.
ISBN-13: 978-1-4051-2143-9 (pbk.: alk. paper)
ISBN-10: 1-4051-2143-2 (pbk. : alk. paper) 1. Groundwater. I. Title.

GB1003.2.Y68 2007
551.49–dc22

2005034231

A catalogue record for this title is available from the British Library.

Set in 10.5/12.5pt Goudy
by Graphicraft Ltd., Hong Kong
Printed and bound in United Kingdom
by TJ International, Padstow

The publisher's policy is to use permanent paper from mills that operate a sustainable forestry policy, and which has been manufactured from pulp processed using acid-free and elementary chlorine-free practices. Furthermore, the publisher ensures that the text paper and cover board used have met acceptable environmental accreditation standards.

For further information on
Blackwell Publishing, visit our website:
www.blackwellpublishing.com

Contents

Preface

"Groundwater for beginners" would have been an appropriate alternative title for this text. Using simple language with a conversational style, this book provides a thorough overview of the occurrence, nature, behavior, and environmental role of natural underground waters. It does this while carefully avoiding the use of scary-looking mathematical notation.

What's wrong with presenting the principles of groundwater science using formal technical language and partial differential equations? Absolutely nothing. If that's what you feel you need right now, there are plenty of existing books which will better meet your requirements (of which the most up to date is the highly commended text by Hiscock: *Hydrogeology: principles and practice*. Blackwell Publishing, 2005).

Rather, the reason for writing this book is to serve the needs of newcomers to the topic of groundwater, especially those who would prefer to acquire a clear grasp of key principles and concepts, but who would be intimidated to find sparse text wrapped around dazzling collations of symbols such as f, ω, ψ, and ∂. Frankly, as an author I would have found it easier to resort to mathematical shorthand. However, from 15 years' experience of teaching university students across the full range of study years (from pre-freshman "access" classes to final-year engineering PhD students), I know that there are many people who find it difficult to learn new concepts if these are expressed principally in equation form, whereas they can grasp them readily when expressed in lucid English prose, using analogies to everyday experiences. While it would be satisfying to think that at least some of the readers of this book will draw from this book the inspiration to go on and become fluent in the mathematical expression of a secure conceptual understanding of hydrogeology, this text aims solely to furnish that understanding.

This is not the first book to pursue this goal. I recall with affection how much I learned from Mike Price's book *Introducing Groundwater* (first edition 1985, George Allen & Unwin, London), and with admiration the manner in which Frank Chapelle chronicled the broad sweep of human–groundwater interactions in *The Hidden Sea* (Geoscience Press, Tucson, 1997). What makes this book different from these two impressive predecessors?

This is the first groundwater textbook *at any level* to discuss in detail the interdependence of groundwater and freshwater ecosystems (Chapter 6). For instance, it contains substantial original material on wetlands (Section 6.2), including a discussion of their vulnerability to overpumping of aquifers (Box 6.1). As such, this book will likely prove useful to many practicing hydrogeologists who left the classroom many years ago, but find themselves increasingly called upon to deal with ecological issues. Where older textbooks concentrate on issues of groundwater resource development, this book adopts a more holistic perspective: whilst not neglecting to describe the crucial role of groundwater as a resource (Chapter 7), this book provides far more detailed coverage of groundwater quality than previous introductory-level texts (Chapter 4), and also describes the threats posed to groundwater and dependent ecosystems by pollution, changes in the use of overlying land, and human-induced climate change (Chapter 9). The coverage of the use of groundwater as an energy source (Chapter 7) and of remediation technologies for polluted groundwaters (Chapter 11) are also novelties for an introductory text, but they are novelties which very much reflect current major trends in groundwater engineering practice.

In contrast to its benefits to humans and ecosystems, groundwater is often a nuisance, if not a mortal hazard, to engineers working in the mining and construction industries, as explained in particular detail in Chapter 8. This book also boasts the first ever nonmathematical explanation of the principles and practice of groundwater modeling: thus even this most mathematical of hydrogeological topics is explained without resort to the Greek alphabet!

Finally, the book is exhaustively referenced throughout, so that users are equipped to pursue topics which interest them in the classic sources, without having to first wade through a higher-level textbook for guidance on the key literature.

Who should read this book? This book has been written with junior-level bachelor's degree program students in mind. The nonmathematical treatment will make it especially appealing to many students on non-numerate science programs in environmental science, geography, and geology. However, the text is sufficiently rigorous and detailed that it can also be used as an introductory treatment of the subject for students on engineering and numerate science programs who are new to the subject of groundwater, but who may well go on later to study the subject in greater mathematical detail. It is even envisaged that newcomers to specialist master's programs in hydrogeology, groundwater engineering, and geotechnical engineering will find this book a very helpful place to begin their studies, albeit they will soon "outgrow" the level of coverage offered here.

In addition to a potential student readership, the book is also intended to serve as a readable introductory text for professionals already working in fields such as ecology, environmental science, environmental engineering, agriculture, forestry, surveying, land reclamation, planning, development control, environmental regulation, public health management, etc., who come across groundwater in their work and would benefit from learning more about it, but are disinclined to take another master's course or delve into highly mathematical tomes on the subject. Finally, anyone with an interest in contemporary environmental issues is likely to find much accessible material of interest in these pages.

How to use this book. This book contains sufficient material to support a full semester course (of around 20 contact hours) for newcomers to groundwater science. Alternatively, for students who have already received some formation in the topic, individual chapters can be used as the basis for seminar work and group discussions; this is especially the case for the later chapters (7, 8, 9, and 11) which include much material on which alternative viewpoints might be debated profitably in class.

The following notes on the organizational logic of the text may prove helpful to users. Essentially, the book can be divided into three distinct sections:

1. An introduction to the scientific foundations of hydrogeology (Chapters 1 through 5).
2. An exposé of the main areas of applied interest related to groundwaters (Chapters 6 through 9).
3. An explanation of the principles and present-day practices of analysis and management of groundwater systems (Chapters 10 and 11).

In relation to the scientific essentials of hydrogeology, Chapters 1 through 5 provide coverage of the following:

- The relative abundance of groundwaters amongst the freshwater resources of Earth (Chapter 1).
- How water actually occurs underground (Chapter 1), and the fact that almost all of it comes from surface precipitation (Chapter 2).
- How rainfall becomes groundwater, clearly noting the increasing importance of indirect pathways with increasing aridity (Chapter 2).
- The phenomenon of head in groundwater systems, and a robust explanation of Darcy's Law (Chapter 3).

- How groundwater flows, both in response to natural head gradients and in response to pumping (Chapter 3).
- The nature and origins of groundwater quality parameters, presented in the form of a guide to reading water analyses (Chapter 4).
- Natural groundwater discharge processes, via springs and in the form of baseflow diffusely entering rivers (Chapter 5).

While the emphasis throughout these "scientific basis" chapters is on conceptual understanding, essential analytical tools are described in a simple (but still usable) manner. For instance, besides presenting Darcy's Law (Section 3.2.1), the manner in which it can be used to analyze flows in real aquifers as represented on flow nets is also explained (Section 3.3.2). In relation to hydrochemical interpretations, the necessary complication of converting mass/volume units to equivalents/volume units is handled using a simple conversion table (Table 4.2). A few of the more advanced approaches are described in discrete text boxes, such as the practical use of the Jacob method to determine transmissivity and storativity (Box 3.3), and the preparation of flow duration curves for the analysis of groundwater discharge to rivers (Box 5.2). Boxes are also used for more extended examples of particular principles or phenomena at levels of detail beyond those strictly necessary for the acquisition of a basic understanding.

Having established the principles of groundwater occurrence, movement, and quality, Chapters 6 through 9 cover the main areas of applied interest with which groundwater specialists are grappling worldwide in these opening years of the twenty first century. Pride of place had to be given to the interrelatedness of groundwater and freshwater ecosystems (Chapter 6), not least because the management of all human interference with groundwater systems is increasingly being judged on the basis of its implications for ecosystem health. This perspective is increasingly manifest in national and international laws, such as those relating to assessment of environmental impacts, management of water resources (e.g. the European Water Framework Directive, and the South African Water Act), and waste management regulations.

The more traditional applied hydrogeology topic of water resources management is covered in Chapter 7. A novel and simple classification of water resource use categories (A (for Agriculture), B (for Big industrial uses), C (for Cooling in power stations), and D (for Domestic)) is introduced here, and used as a framework for:

- Presenting current levels of groundwater use worldwide (Table 7.1).
- Exploring constraints on the usability of groundwaters, and
- Pointing out the undesirable side-effects of groundwater use

The organizational logic of the book is clearly manifest in the application of lessons from the first five chapters to practical issues. For instance, it is at this point that the "scientific" presentation of water quality (Chapter 4) is applied to the very important issue of the limitations which dissolved substances impose on the practical usefulness of waters for agriculture and human consumption (Tables 7.3 and 7.4). The same approach of applying tools from Chapters 1 to 5 to real-world problems is also followed in the discussions of groundwater geohazards (Chapter 8) and the threats posed to aquifers (Chapter 9) by over-pumping, human-induced climate change, and pollution from point and diffuse sources as diverse as landfill leachates and natural saline waters.

The final "analysis and management" section of the book (Chapters 10 and 11) first presents the modern approach to integrated analysis of aquifer systems by means of conceptual and (where appropriate) mathematical modeling (Chapter 10), and then succinctly presents techniques for management of the issues raised in each of Chapters 7 through 9.

For teachers planning a full semester course, it will likely prove most beneficial to follow the organizational logic of the text (science → issues → analysis and management). For teachers of more advanced courses wishing to use some of the later chapters as the basis for seminars and class discussions, the "issues" chapters (6–9) could well be used independently of the earlier and later material in the book.

Given its emphasis on transmitting concepts rather than techniques, the book does not include problems or exercises. However, these might easily be developed where appropriate using material from the various boxes as the basis for simple exercises in literature searching, description, or basic calculations. Indeed, the lack of a formal mathematical presentation makes this book potentially valuable to teachers of advanced courses, as they can set challenging exercises in which numerate students are asked to develop mathematical expressions to represent processes such as those illustrated on Figures 2.2, 3.2, 5.5, 6.1, 6.2, 7.4, 9.3, 9.4 and 10.2, and/or described in Sections 3.1.2, 3.3.3, 4.4.4, 5.3.1, 6.3.2, 7.4.2, 10.3.4, and 10.5.2.

Paul Younger
University of Newcastle,
Newcastle Upon Tyne, UK

Artwork from the book is available to instructors at **www.blackwellpublishing.com/younger** and by request on CD-ROM.

Acknowledgments

The idea of writing this book was first mooted by Dr Ian Francis at Blackwell Publishing. His enthusiasm and professionalism brought the project to life, and it has subsequently been sustained by his colleagues Delia Sandford and Rosie Hayden. Pat Croucher managed the desk-editing process, and Oxford Designers and Illustrators expertly prepared the final drafts of the figures from my rough sketches. Careful and constructive critical reviews of the draft manuscript, provided by Kevin Hiscock (University of East Anglia) and three anonymous referees, helped to sharpen the text considerably.

The contents of this book are essentially the fruits of some 15 years' experience in the teaching of hydrogeology to students, mainly at the University of Newcastle and mainly at master's level. My former students merit not only my gratitude but also my apologies, to the extent that they were the unwitting guinea-pigs of my experiments in explanation. Hopefully they will agree that only the better analogies and examples used in class made it to these pages.

It's now 20 years since I was regularly on the receiving end in class, but my intellectual inheritance from my own teachers remains enormous. Those who have the privilege of knowing the work of those who first initiated me into the mysteries of hydrogeology at Oklahoma State University in the mid 1980s will be able to detect the powerful influence of Professor Wayne Pettyjohn (especially in Chapters 1, 3, and 5) and Professor Arthur Hounslow (Chapters 4, 5, 9, and 11). Many subsequent mentors challenged and changed my hydrogeological world-view. Especial thanks go to Professor Enda O'Connell FREng (Chapters 5, 10, and 11), Mr Brian Connorton MSc C.Geol. (Chapters 10 and 11), Professor Steven A Banwart (Chapters 4 and 11), Professor David Lerner FREng (Chapter 2), Professor Emilio Custodio (Chapter 9), Professor Rafael Fernández-Rubio (Chapters 8 and 11), and Dr Alexander Klimchouk (Chapter 8, and Figure 1.3c).

The text is, I hope, enriched by implicit and explicit allusions to many aspects of my international professional experience. None of this experience was gained in isolation. I am particularly grateful to numerous former colleagues and current collaborators in the Environment Agency (England and Wales), Centro YUNTA (Bolivia), the House of Water and Environment (West Bank), and the European Commission. Special mention must be made of David Gowans and David Wright of Project Dewatering Ltd (www.projectdewatering.co.uk), who unerringly present me with the most challenging of groundwater management problems against which I can continually hone my interpretational skills. It's fun, lads; thanks!

The finalization of this book was made possible by a timely sabbatical period, sponsored by HSBC Bank as part of their inspirational "Partnership for Environmental Innovation." I am especially grateful to Francis Sullivan, Environment Advisor to the Board of HSBC Holdings plc, for his enthusiastic support for this project. Equally important was the willingness of work colleagues to field problems in my absence: thanks go especially to Dr Adam Jarvis and Dr Jaime Amezaga. I am also very grateful to Janet Giles for helping to compile the Glossary.

It's a bit of cliché to close acknowledgments with a vote of thanks to one's immediate family. In this case, my debt of gratitude is extraordinary: not only did my wife Louise and my sons Thomas, Callum, and Dominic fully grant me the time and space to write this book, they never once complained about the hassle this caused them; and all of this against a backdrop of turmoil and bereavement which greatly strained the emotional resources of us all. *Is gràdh Dia; agus an ti a tha a fantainn ann an gràdh tha e na chòmhnaidh ann an Dia, agus Dia annsan.*

1

Occurrence of water underground

How silently in former ages all this water had found its way, perhaps drop by drop, into the stony reservoirs! How silently it had lain there, under solid strata, no one suspecting its existence! But now at length, man must trouble the peaceful waters . . . the fountains of the deep in the hollow places of the earth have been broken up by rude hands.

(James R Leifchild, 1853)

Key questions

- What is groundwater?
- Where does it fit in the wider world of natural waters?
- How does water actually occur in soils and rocks?
- What is the water table?
- What is an aquifer?
- What is an aquitard?
- What governs the storage and transmission of groundwater in aquifers?
- What is porosity, and does it matter?
- What is meant by "hydrostratigraphy"?
- Which kinds of rocks make the best aquifers?
- How do geological structures such as folds and faults affect the occurrence of groundwater?

1.1 Groundwater and the global water cycle

Appearances are often misleading. Nowhere is this more so than in the natural water environment. Viewed from outer space, the two main bodies of water on our planet are immediately obvious: the oceans and the polar ice caps. From the more down-to-earth perspective of land mammals, these impressive stores of water are of little avail; the oceans are too salty for our use, and we cannot afford the economic or environmental costs of actively melting the ice caps. Fortunately, potentially useful freshwater is almost always within sight, in the form of rain, rivers, or lakes. Freshwaters adorning the land surface are often strikingly beautiful: a majestic waterfall, a still lake surface mirroring hills and clouds, the sparkle of tap water in sunlight. It is easy to get entranced by the beauty and to consider these manifestations of freshwater as representing its very essence. But Nature is more bountiful than our eyes can tell us: by far the most abundant usable freshwaters are largely hidden from our gaze, in the ground beneath our feet. Subsurface water accounts for just under 99% of the total volume of freshwater presently circulating on our planet (Figure 1.1). Surface water, that is all of the readily-visible water present in rivers, lakes or wetlands, amounts to less than 1% of the total, with the balance (a puny 0.16%) being present in the form of atmospheric moisture (Herschy 1998). Given the overwhelming dominance of subsurface waters in the global freshwater budget, enlightened self-interest alone would thus suggest that it is wise to specialize in studying groundwater.

So what is "groundwater"? The simplest definition would propose to equate "groundwater" with "subsurface water," that is any and all water beneath the ground surface. While such a simple definition *does* find use in some legal codes and nonspecialist literature, in technical circles more complicated definitions are generally preferred. This is not because of scientific conceit, but rather arises from utterly practical considerations, reflecting the long human experience of exploiting subsurface water as a resource. From the earliest times, well diggers observed that water will only flow spontaneously into a well below a certain horizon, usually termed "the water table."[i] The depth to this water table varies from one locality to another, though it is generally found to form a relatively flat horizon over short distances between two neighboring wells. For early well diggers, the discovery of the water table led to the impression that water was *only* present underground below this horizon. Although we now know that moisture is invariably present between

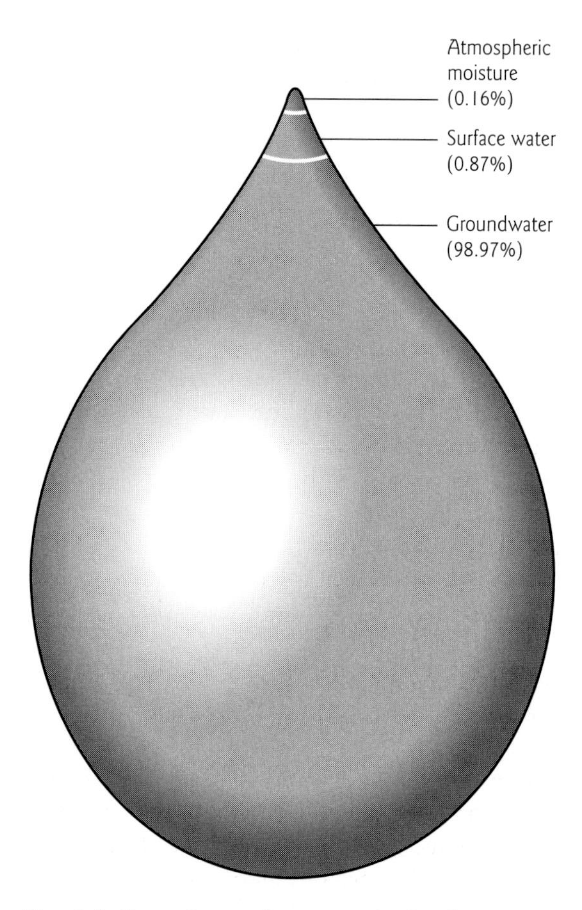

Fig. 1.1 Groundwater dominates the Earth! The droplet represents all liquid freshwater on our planet, showing the proportion in each of the three principal compartments at any one time. Water is of course constantly moving between the three compartments; this diagram shows the average net amount of water undergoing turnover in each compartment at any one time (Herschy 1998). "Surface water" includes all rivers, lakes, and freshwater wetlands. The surprisingly puny proportions of water in temporary storage in the atmosphere and surface waters reflect the very rapid rates of water movement through these compartments – groundwaters move much more slowly. (Of course very large quantities of freshwater are presently (but for how much longer?) trapped in polar ice and other glaciers; and salty marine waters, though of little use to mammals, far dominate the overall global water budget.)

the ground surface and the water table, it remains true to this day that we can only economically pump those subsurface waters that will flow freely into wells, i.e. those that occur below the "water table." Hence the term "groundwater" came to be defined pragmatically by water engineers to mean "subsurface water below the water table." Therefore **subsurface water** is the sum of **soil moisture** (above the water table) and **groundwater** (below it).

While these pragmatic distinctions serve us well for most purposes of groundwater resource evaluation (see Chapters 3 and 7) and analyses of the contribution of groundwater to wider catchment processes (see Chapters 5, 6), it is always crucial to bear in mind that a significant proportion of "soil moisture" is on its way to becoming "groundwater" (see Chapter 2), and as such we cannot protect groundwater from contamination without first making sure we are also protecting the quality of soil moisture (see Chapters 9, 11). When all's said and done, water is water no matter where it is currently located: groundwater forms part of an unbreakable continuum not only with soil moisture, but also with atmospheric moisture, surface waters, marine waters, and ice. It is well known that all of these forms of water are dynamically linked in space and time via the hydrological cycle (Figure 1.2). Despite the widespread appreciation of the interconnectedness of ground and surface waters, the tendency for professionals to specialize in studying only one of the two is surprisingly persistent. Groundwater tends to be the specialist province of the "hydrogeologist," while surface waters are the particular domain of the "hydrologist" (or, more correctly, the "surface water hydrologist"). Often, the groundwater specialists have Bachelor's degrees in geology, while the surface water specialists hold degrees in civil engineering or physical geography: thus the topics are often segregated between different university departments whose students seldom meet! This disciplinary boundary has unreasonably hindered coherent thinking about the water environment. As a friend once quipped: "To judge from the behavior of many water scientists, you could be forgiven for thinking that groundwater and surface water were immiscible liquids!"

It is high time that we did away with maintaining a discipline boundary between groundwaters and surface waters. We should do this not only to improve scientific communication, but also for the far more important reason that, with very few exceptions, it is not possible to disturb a groundwater system without also affecting a surface water system, and *vice versa.* This much should be evident from the particular representation of the hydrological cycle presented in Figure 1.2, in which groundwater discharge to rivers and the ocean is clearly depicted: one cannot artificially remove water from a groundwater system without removing the amount available to discharge naturally to the surface environment. In extreme cases, artificial removal of groundwater can lower underground water levels to such an extent that surface waters can be induced to flow into the subsurface. Of course the magnitude of the collateral effects of using ground or surface waters varies greatly from one situation to another. Nevertheless, acknowledgment that ground and surface waters constitute a single resource is the first step towards developing a mature approach to their management. We must recognize that *any* use of a natural water resource, surface or subsurface, is a *choice*, and a choice which inevitably limits the scope for making other possible choices concerning the same overall water resource system. Of course on short time-scales the limitations imposed by making one choice may not be very obvious, and may well be reversible. But cumulatively, drop-by-drop so to speak, we gradually drain the cup of our water resource options with each additional use.

Having accepted that resource use involves making a choice, a plethora of questions immediately arise: Who makes the choice? For whom is the choice made? Is the choice we've made the best option from an economic, ecological or social perspective? This book provides an introduction to the scientific insights and techniques needed to develop well-informed responses to these and related questions. It is hoped that the reader will be inspired to pursue the subject in greater depth, in which case they will be able to

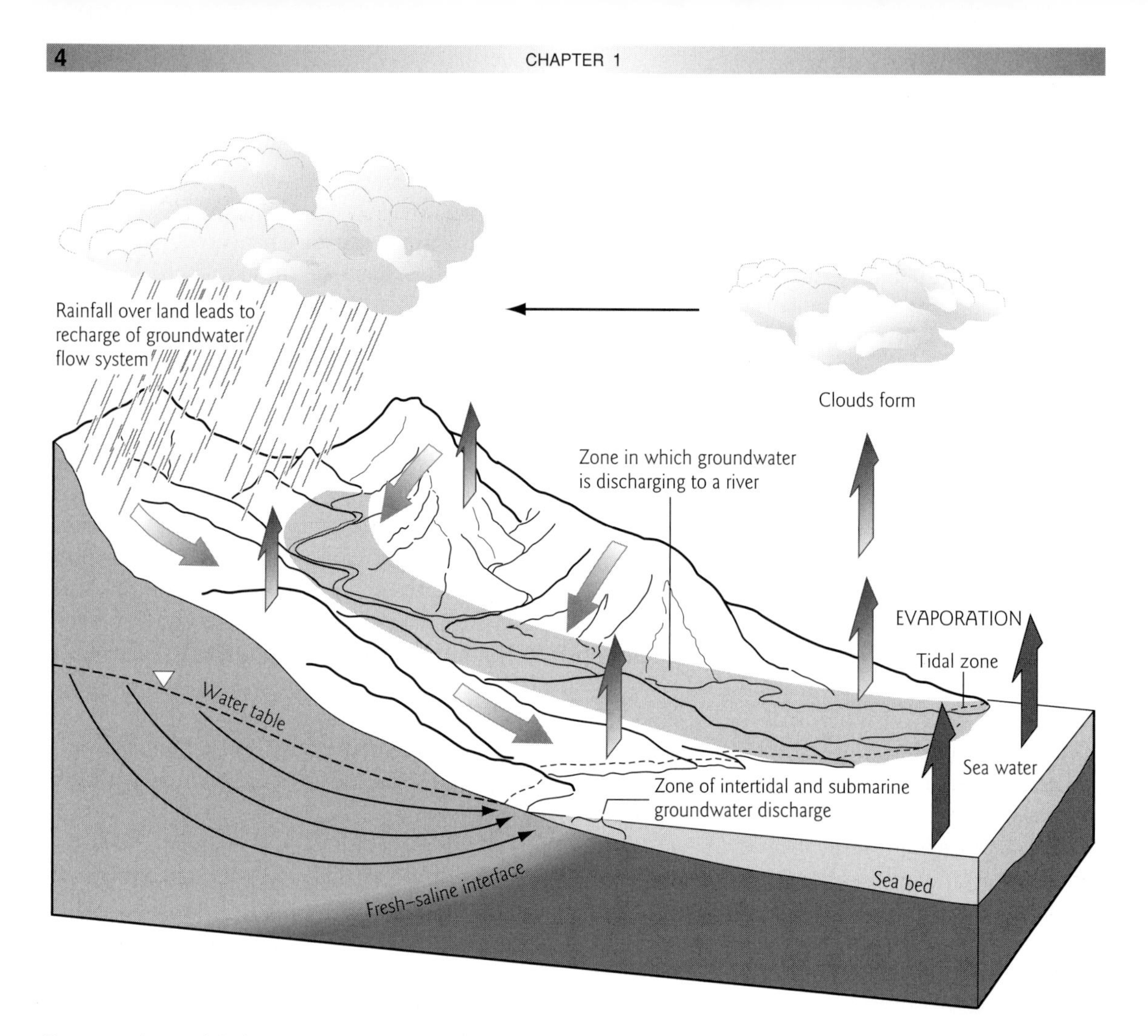

Fig. 1.2 A simplified representation of the hydrological cycle, emphasizing (though not exaggerating) the importance of groundwater. The water table is denoted, as is conventional, by a line surmounted by an inverted triangle, and the curved black arrows illustrate the approximate manner in which groundwater flows towards areas of low elevation. Broader areas on the ground surface illustrate the general directions of surface runoff (overland flow and in channels). Above the ground surface, the vertical arrows represent both evaporation from the ocean and evapotranspiration (i.e. the combined effects of evaporation and transpiration) from inland watercourses, lakes, and vegetated soils. Note the major zones of groundwater discharge: to inland rivers and directly to the sea.

avail themselves of more formal and detailed treatments of topics introduced here, in texts such as that of Hiscock (2005).

Having just argued that we ought to remove the disciplinary boundary between the study of groundwaters and the study of surface waters, the reader would be justified in questioning why this book focuses on groundwater. The choice of this particular "bottom-up" view point reflects the overwhelming volumetric dominance of groundwaters over surface waters within the hydrological cycle (cf. Figure 1.1). Many perennial rivers can best be understood as the visible fringes of water resource systems which exist predominantly as groundwaters (cf. Figure 1.2). Just as the study of human history so often benefits by considering the lives of the majority of the population rather than just those of the kings and social elites, hydrology benefits by taking the "majority" groundwater perspective from time to time.

1.2 The natural zonation of water underground

Appearances can indeed be very misleading. For many people, if they think about groundwater at all they think of the quasi-mystical figure of the water-diviner, striding out confidently across open country, forked twig to the fore, until some minor convulsions of the wrist prompt them to announce that they are now standing above an "underground stream." A driller or well-digger is then instructed to perforate the ground at this precise point until, lo and behold, water is indeed encountered, flowing readily into the newly sunk hole. The water-diviner's reputation is bolstered and the existence of "underground streams" is apparently demonstrated beyond doubt. Or is it? If it's a large enough excavation, and you have the courage to climb to the bottom, or if you take advantage of modern technology and send a closed-circuit television camera down the well, you will almost always find that the water is not entering the hole from anything that looks even remotely like a stream. Rather, you are far more likely to see a damp horizon of soil beneath which water is oozing out of a multitude of small cracks or pores. In fact, underground streams are very rare; most groundwater occurs in the small openings in soil or rock known as pores, which may either correspond to the gaps between sediment grains (Figure 1.3a), or else be due to the presence of fractures (Figure 1.3b). We will discuss pores in further detail in Section 1.4 below.

Pore space may be partially or completely filled with water. Thanks to gravity, the distribution of water underground is not random, but tends to be vertically zoned (Figure 1.4). The zone in which pores are completely filled with water is termed the **saturated zone** (or **phreatic** zone), and its upper surface is the **water table**. Above the water table the pores are only partly filled with water; this is generally called the **unsaturated zone** (or **vadose** zone), and it is the zone in which water is referred to as **soil moisture** (recalling Section 1.1) or, less commonly, "vadose water". It is possible to further subdivide the unsaturated zone into the soil and sub-soil zones. The **soil zone** is the uppermost layer of earth, which typically supports plant life. In most climatic zones, the soil zone will tend to remain unsaturated for most of the year, as the root systems of the plants remove moisture from the soil for use in metabolic processes. Excess water taken up by plants is not usually returned to the soil zone, but rather is lost to the atmosphere from the leaves, by a process termed **transpiration**. The base of the soil zone is defined by the maximum depth from which water can be removed by root suction; we can therefore term it the **root-suction base**. Between the root-suction base and the water table is the **sub-soil zone**, comprising unsaturated soils and/or rocks in which the soil moisture is slowly seeping downwards, destined eventually to replenish the store of groundwater below the water table (Chapter 2).

A slight complication arises at the interface between the sub-soil zone and the saturated zone. While we have already defined the uppermost surface of the saturated zone to be the **water table** (which as we saw in Section 1.1 is simply the level to which water will settle in a well dug into the saturated zone), in reality the pores tend to be completely filled with water for a short height above that level. In effect, the water table is surmounted by a thin mantle of fully saturated pores, which is usually referred to as the **capillary fringe**. The height of this fringe depends on the size of the pores; while it may be only a fraction of a millimeter in coarse gravel, it may reach several meters in clays, silts, and rocks with small narrow pores (Box 1.1). Some authors consider the capillary fringe to form part of the saturated zone, on the reasonable grounds that the pores within it are certainly saturated. However, as will be explained in greater detail in Section 1.3 and Chapter 2, the hydraulic behavior of the capillary fringe is essentially the same as that of the sub-soil zone, which means that for all practical purposes it is more convenient to assign the capillary fringe to the unsaturated zone.

Box 1.1 Fringe activities: capillarity can be fun.

A formal explanation of the phenomenon of capillarity, and thus the development of the capillary fringe above the water table, is beyond the scope of this book; if you are really keen to get your head around the soil physics involved, I'd suggest you start by reading the recent account by Hiscock (2005, pp. 32–33, 151–154). For most of us lesser mortals, a better starting point on the road to understanding can be gained at the kitchen sink.

You need some transparent tubes (open at both ends) of different diameters. For the largest one, try cutting the ends off an old soft drink bottle (e.g. 500 ml capacity plastic cola or lemonade bottle). For the medium-sized tube, the outer plastic casing of a common ballpoint pen is ideal. If your ballpoint pen has a refill that's nearly spent – brilliant! Snip off the part with the remaining ink in (i.e. the bit closest to the nib) and the ink-free length will serve for your smallest size tube. Keep on raiding the grocery cupboard and the toiletry cupboard to get as wide a variety as possible of different see-through tubes. Once you've got as many tubes as you can snaffle, put some water in a dish. If you've got some food colouring, add it to the water – it will make some effects easier to see. Now, starting with the largest tube, inserts its tip into the water and hold it steady. Apart from slight adhesion of the water to the tube walls like so:

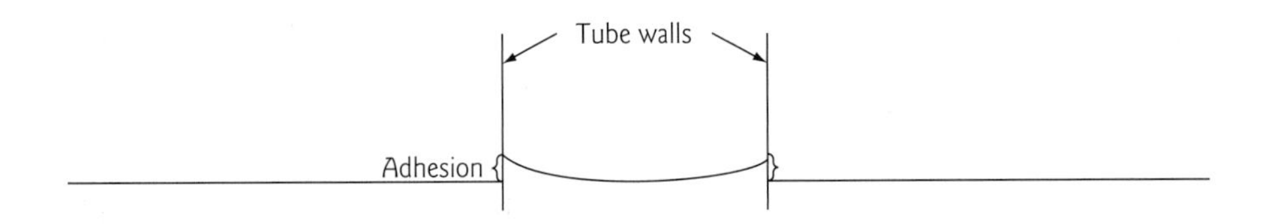

you're unlikely to be able to see any difference between the water level inside and outside the tube. Now do the same with the next largest tube. Note any difference? Is the adhesion higher than before? Is the water level basically the same inside and outside the tube? Keep going, testing ever smaller tubes. Eventually, and certainly by the time you get to the ballpoint pen refill, you'll find that the water level inside the tube tends to rise significantly above the water level in the dish. If you've been observant (or better still, if you've had the dexterity to actually measure how high within each tube the water surface rises above the dish water level, then you'll be able to demonstrate that **THE NARROWER THE TUBE, THE HIGHER THE WATER LEVEL RISES**:

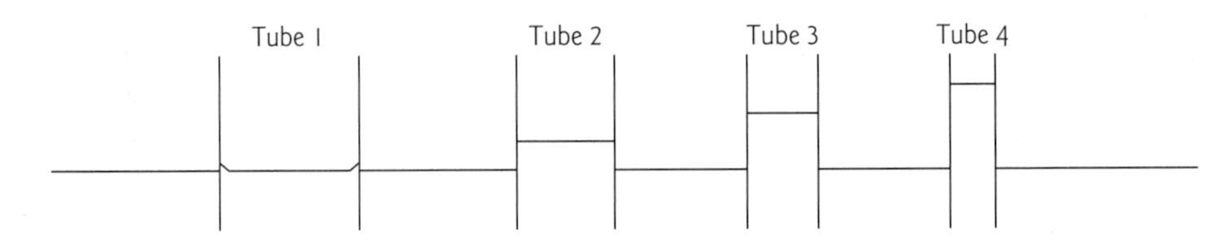

What's going on? The attraction between the water molecules and the solid material of the tube walls is greater than that between the molecules themselves. That much is evident from the adhesion at the edge of the water surface in even the largest diameter tube. (Similar adhesion is seen in all other cases, by the way, but is not easy to draw at small scales.) In the very smallest tubes, the water molecules are all so close to the tube walls that the water–solid attraction overwhelms the attraction between the molecules themselves, and the water level in the tube rises ("capillary rise") until as many molecules as possible are in direct contact with the solid surface.

In natural soils and rocks, the same principles apply. In a gravel (grains >4 mm), the pores are wide and only slight capillary rise occurs; at the other extreme, between grains of silt (between 0.03 and 0.004 mm) or mud (<0.004 mm diameter), the pores are very narrow, leading to high capillary rises.

So what about the underground streams so beloved of the water-diviner? The prosaic truth is that true underground streams are very rare features, being almost wholly restricted to terrains composed of three types of rock:

- Rocks composed of readily soluble minerals in which cave systems (Figure 1.3c) have developed over geological time (Ford and Williams 1989). This is most notably the case in relation to calcite ($CaCO_3$), the main constituent of limestones, and gypsum ($CaSO_4 \cdot 2H_2O$), which forms thick and extensive beds in some sedimentary sequences.
- Volcanic rocks, especially basaltic lavas containing caves (see Section 1.5.2).
- Sandy or silty soils and rocks which have been subjected to particular modes of weathering known as "piping" and "sapping" (Higgins and Coates 1990), which give rise to cave systems broadly comparable to (if usually smaller than) those commonly found in limestone country.

Even where stream caves do occur, the general principles of subsurface water zonation (Figure 1.4) still apply in modified form, as all underground streams tend to drain down to some common level of ubiquitous flooding, corresponding to the water table. Whether our pore space is cavernous or microscopic, therefore, the ubiquity of the water table means that when a water diviner professes to detect the *presence* of groundwater they address a nonissue: in almost all geological settings, groundwater will be present, if we drill deeply enough. Far more pertinent questions than presence or absence of groundwater are:

- How far below ground level does the water table lie?
- In what kind of soil or rock is the groundwater present?
- How readily will these soils or rocks yield groundwater to a spring or a well?
- How high a yield can we depend upon in the long term?
- Is the groundwater of good or poor quality?

While one might find a water-diviner willing to express an opinion on one or two of these points (e.g. Applegate 2002), I have yet to meet one able to comment with credible confidence on all five. The right person for that job is a hydrogeologist.

The simplicity of this brief outline of the natural zonation of subsurface waters between the ground surface and the water table (Figure 1.4) rather belies the extraordinary complexity of the processes of water movement in the unsaturated zone (see Chapter 2). Nevertheless, it will be easier to appreciate some of that complexity after we have further reflected on the occurrence of water in the saturated zone.

1.3 Water pressure, the saturated zone, aquifers, and aquitards

Imagine hand-digging a well. Better still: try it some time. As we dig down through the unsaturated zone, we are working our way through ground that contains much water in the form of soil moisture: the dirt we remove from the hole will likely feel rather moist. So what is stopping this soil moisture from flowing into our hole as we dig? Provided we have not inserted artificial barriers, it would at first seem that *nothing* should be preventing the soil moisture from entering the hole. Take a deep breath and think again: there *is* something in the hole stopping the water

Fig. 1.3 The three principal types of pores commonly occupied by groundwater. (a) *Pores between sediment grains*. This is a photomicrograph of a thin section of a specimen of the Fell Sandstone Aquifer of northernmost England. Width of field of view: 3 mm. Most of the mineral grains are quartz, which appear white here; the darker, irregularly shaped grain near the bottom-center of the view is a potassium feldspar crystal. Pore space is consistently shaded, and visible between the grains. (b) *Fractured rock*. Groundwater emerging from a discrete fracture intersected by a tunnel 430 m underground in the Scottish Highlands. The groundwater contains chemically reduced manganese, which precipitates to form the dark coating on the granite as it encounters oxygen in the tunnel atmosphere. (c) *Cave*. A photograph from the unsaturated zone of a major karst aquifer in Ethiopia, showing an underground stream flowing though a large cave *en route* to the water table (courtesy of Alex Klimchouk).

entering: air. Although atmospheric pressure is routinely discussed on TV weather forecasts, we somehow seem to forget the importance of air pressure as a powerful force affecting our daily lives. Yet the atmosphere is actually quite highly pressurized, at around 10 tonnes per square meter. Unless the pressure of water in pores is greater than this value, the water will never flow laterally into our hole. Eventually, after much digging, we reach the water table. By definition, any water present in the ground below this level will flow freely into our hole. As it does so, it displaces the air, which moves up and out of the hole. There is only one possible explanation for this behavior: the pressure of water in the pores of the saturated zone is greater than atmospheric pressure. Taking the reasoning a small step further, we can offer the following physical definition of the water table: it is that subsurface horizon upon which (at any given point in time) the pore water pressure exactly equals atmospheric pressure. As we know from the weather forecast, the atmospheric pressure at any one place varies continuously as masses of warm and cool air move over the surface of our planet. Small fluctuations (a few millimeters at most) in the precise elevation of the water table

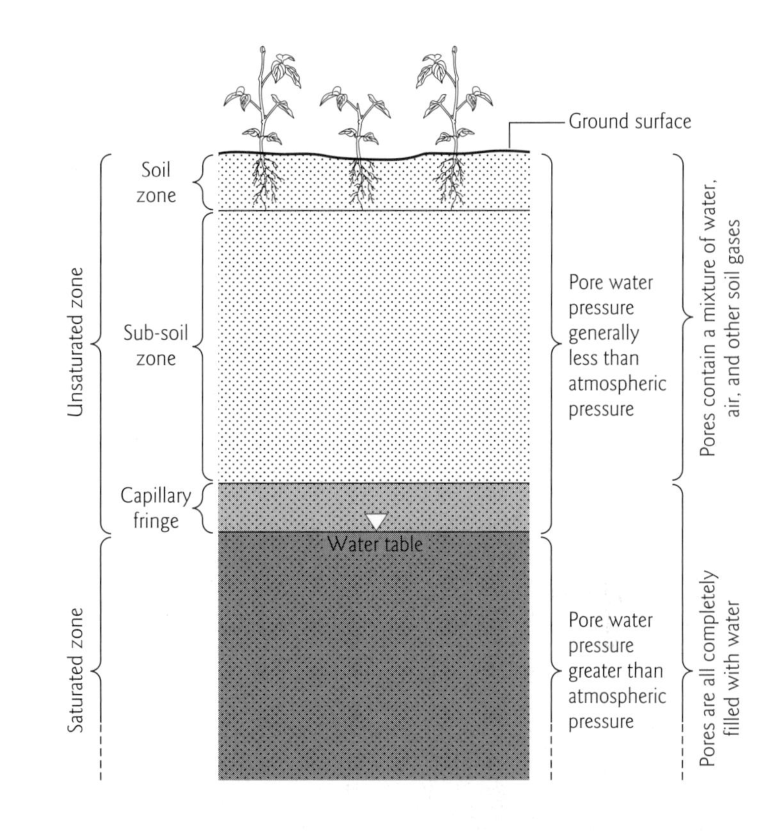

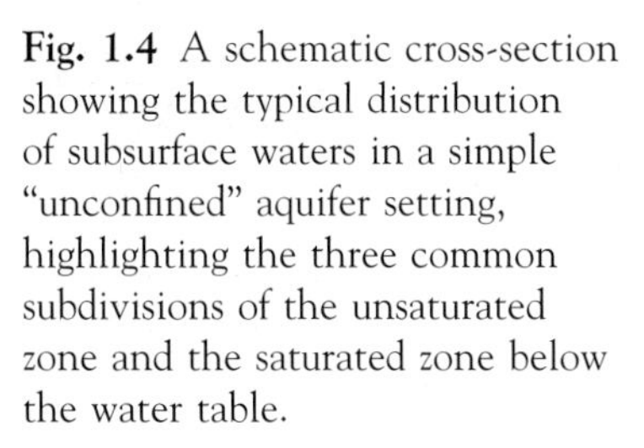
Fig. 1.4 A schematic cross-section showing the typical distribution of subsurface waters in a simple "unconfined" aquifer setting, highlighting the three common subdivisions of the unsaturated zone and the saturated zone below the water table.

can therefore be expected as atmospheric pressure changes; this is exactly what we see whenever we monitor water table levels at sufficiently high precision. The capillary fringe is always at hand to "lend" water to, or "borrow" it from, the true saturated zone.

So, the tendency for groundwater to flood an excavation is down to the fact that its pressure exceeds that of the air. However, for any one value of atmospheric pressure, experience teaches us that the tendency for groundwater to flow into a well or borehole varies dramatically from one piece of ground to another. Indeed, in many practical cases it is possible to distinguish between those soils and rocks which release copious quantities of groundwater very rapidly and those which release it so slowly it may be unnoticeable over time-scales of interest to humanity. A number of terms have been coined to assist us in making these distinctions:

- An **aquifer** is a body of saturated rock that both stores and transmits important quantities of groundwater.
- An **aquitard** is a saturated body of rock that impedes the movement of groundwater.

As these terms are used throughout this book, it is important to clearly understand their significance.

Definitions of the term "aquifer" given in earlier textbooks (e.g. Tolman 1937; Todd 1980; Marsily 1986; Price 1996; Domenico and Schwartz 1997; Fetter 2001) included a requirement that an aquifer must be capable of yielding "economic quantities of water to wells or springs." In former times, when most hydrogeologists were concerned with developing water supply wells, this requirement made sense. However, now that groundwater specialists are engaged in a much wider range of activities, ranging from waste disposal to ecological conservation, it makes little sense any more to retain this element in the definition. The formulation given above has been purposely left less specific, referring only to "important quantities" of water. What constitutes "important" depends utterly on the focus of a particular investigation: a bed of silt might be regarded as an aquitard if

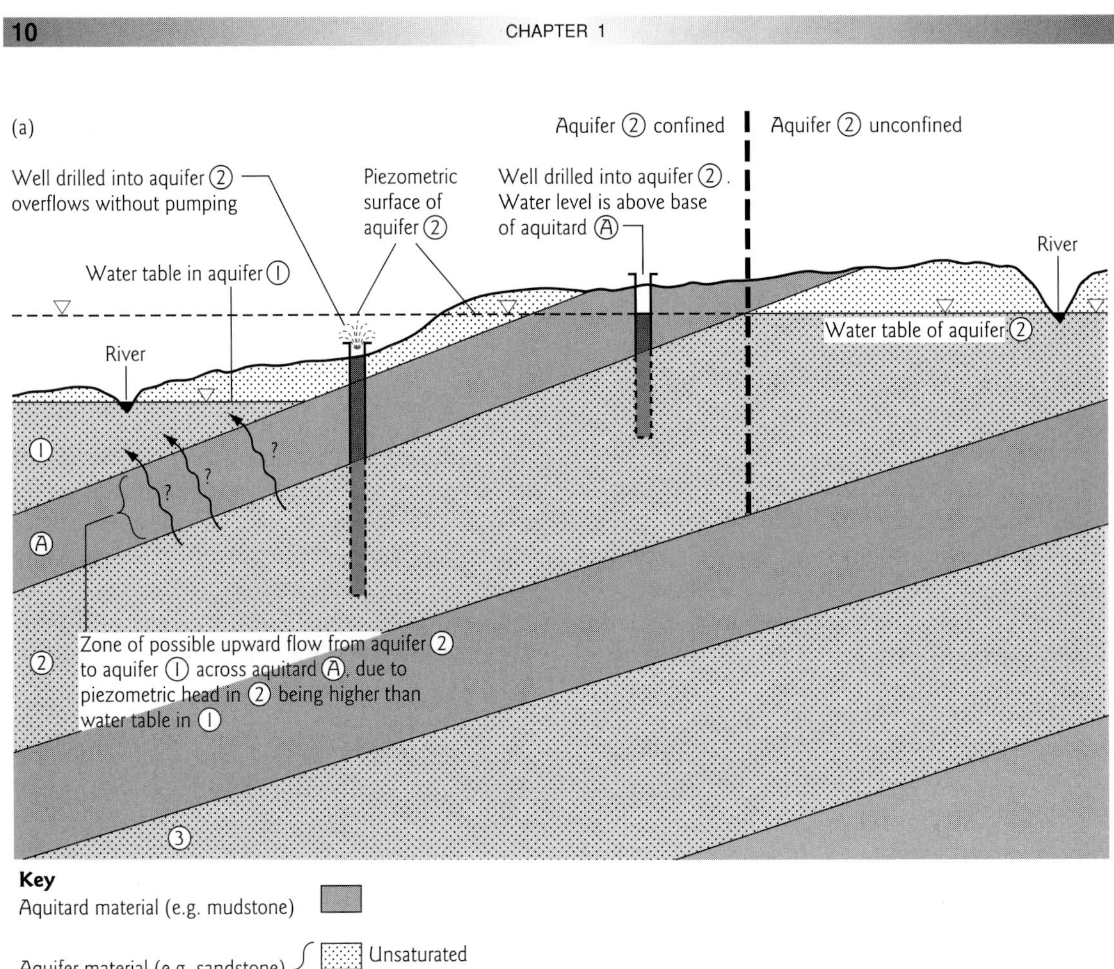
(a)
Aquifer ② confined
Aquifer ② unconfined
Well drilled into aquifer ② overflows without pumping
Piezometric surface of aquifer ②
Well drilled into aquifer ②. Water level is above base of aquitard Ⓐ
River
Water table in aquifer ①
River
Water table of aquifer ②
①
Ⓐ
②
③
?
Zone of possible upward flow from aquifer ② to aquifer ① across aquitard Ⓐ, due to piezometric head in ② being higher than water table in ①
Key
Aquitard material (e.g. mudstone)
Aquifer material (e.g. sandstone)
Unsaturated
Saturated

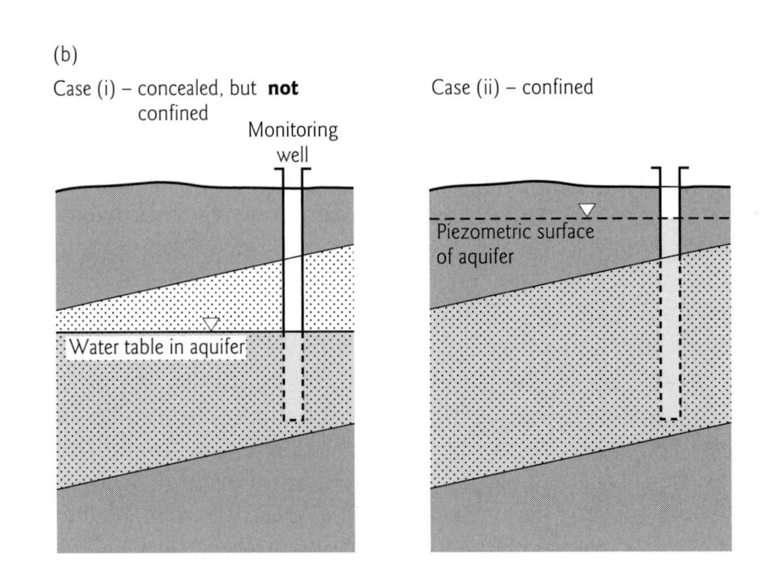
(b)
Case (i) – concealed, but **not** confined
Case (ii) – confined
Monitoring well
Water table in aquifer
Piezometric surface of aquifer

it separates two highly permeable gravel beds, whereas if it lies between two mud layers it may be deemed to be an aquifer. Consider a thin bed of sand which would never be a reasonable prospect for a public supply well, even if it contained water of the finest quality; if such a bed provides the pathway by which grossly polluted waters are migrating to an ecologically sensitive pond or wetland it is likely to be identified as the most important aquifer in the local geological succession.

The term "aquitard" is probably the most widely used descriptor internationally for low-permeability rocks. However, alternative terms have found favor in particular circles (Todd 1980; Seaber 1988). For instance, the US Geological Survey prefer the term "confining bed" (Meinzer 1923a,b; Lohman 1972); however, many so-called "confining beds" are neither "beds" in stratigraphic terms, nor "confining" in hydraulic terms (see below). (Other alternative terms, including "aquifuge" (Davis 1930) and "aquiclude" (Tolman 1937), are now essentially obsolete.)

It is very important to note that aquitards *impede* groundwater flow, rather than stopping it altogether. Over long time-scales, significant quantities of water do flow across aquitards. Cross-aquitard flows are sometimes crucial contributors to regional-scale groundwater flow (e.g. Bredehoeft et al. 1983) and have recently been found to play a crucial role in the evolution of the world's largest-known cave systems (Klimchouk et al. 2000).

The relative positions of aquifers and aquitards at any one site are of great practical importance. Figure 1.4 illustrates what is perhaps the simplest case, in which aquifer material immediately underlies the soil zone. By contrast, it is very often found in practice that the soil zone is immediately underlain by an aquitard, with the shallowest aquifer horizon lying below this aquitard. Such aquifers often contain groundwater that is under sufficient pressure that, when a well is sunk through the overlying aquitard, water from the aquifer will rapidly rise within the well to a rest position substantially higher than the base of the aquitard (Figure 1.5). Indeed in some cases, the water will rise so far within the well that it will overflow from the top of the well. The origins of such excess pressure in aquifers lying beneath aquitards lie in the regional interplay of geology and surface hydrology.

These two principal modes of aquifer occurrence are classified using the following terms:

- **Unconfined aquifer:**[ii] i.e. an aquifer in which the upper limit of saturation (neglecting the capillary fringe) is the water table, so that unsaturated soil or sub-soil lies between the upper boundary of the aquifer and the ground surface (Figure 1.4).
- **Confined aquifer:**[iii] i.e. an aquifer lying below an aquitard, in which there is no unsaturated zone between the base of the aquitard and the groundwater within the aquifer (Figure 1.5a). In most cases, groundwater within a confined aquifer is under sufficient pressure that, in a well penetrating to the confined aquifer, it will settle to a level higher than the base of the overlying aquitard. The horizon formed by the levels to which groundwater in a confined aquifer would rise were it to be penetrated by a number of wells is called its "piezometric surface" (Figure 1.5).

Although the unconfined/confined distinction might at first seem pedantic, it turns out to have

Fig. 1.5 (*Opposite*) Confined aquifer conditions. (a) A schematic cross-section through a system of three aquifers and two intervening aquitards. While aquifer 1 is unconfined throughout the field of view, aquifer 2 is only unconfined in the right-hand third of the diagram (to the right of the vertical dashed line), and confined elsewhere beneath aquitard A (which is itself the basal aquitard to aquifer 1). Lying at great depth, aquifer 3 is everywhere confined, beneath aquitard B. (b) Concealment versus confinement. In case (i) we have an aquifer which is actually unconfined (i.e. there is a saturated zone in the aquifer material, above the water table), though concealed beneath aquitard material which extends to the ground surface. In case (ii), with exactly the same geology as the previous case, the aquifer is completely filled with water and is therefore truly confined.

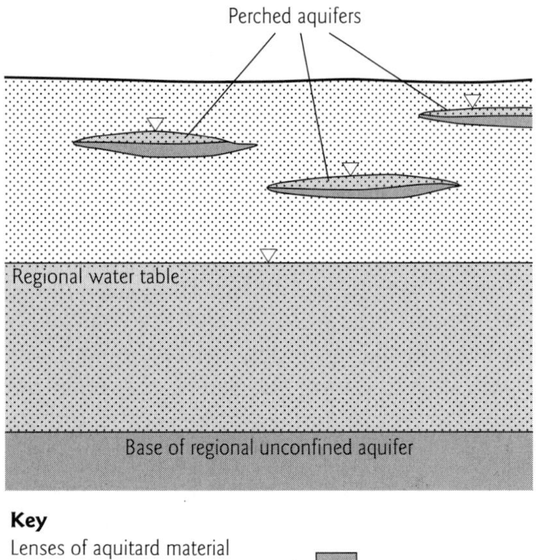

Key

Lenses of aquitard material (e.g. mudstone) and basal aquitard

Aquifer material (e.g. sandstone)

Fig. 1.6 Perched aquifer conditions, which arise where lenses of low-permeability aquitard material intercept recharge and "pond it" above the regional water table in the form of localized lenses of saturation (perched aquifers).

tremendous importance for the ways in which changes in groundwater storage occur within different aquifers (see Section 1.4).

In certain geological settings (Figure 1.5), aquifers can switch between the confined and unconfined states over time. Confinement can be lost if the water pressure in the aquifer drops over time (due to pumping, for instance) such that the piezometric surface drops below the contact between the aquifer and its overlying aquitard; at this point the piezometric surface becomes, by definition, a water table and the aquifer becomes unconfined. Although the unconfined aquifer is still concealed beneath its overlying aquitard it is no longer hydraulically confined (Figure 1.5b).

The possible combinations of aquifers and aquitards under field conditions are almost infinite. However, in most cases, only the shallowest two or three aquifers (and any intervening aquitards) will be of practical interest from a water resources perspective. Nevertheless one particular combination of aquifers and aquitards does merit special mention in this context. This is the combination which gives rise to what are known as "perched aquifer" conditions. Figure 1.6 illustrates how unconfined aquifers of limited lateral extent can develop above the regional water table, lying on top of localized pockets of aquitard material. Great care must be taken during groundwater investigations not to confuse such perched aquifers with the regional unconfined aquifer below: such a mistake can lead to rapid exhaustion of perched groundwater resources, whereas supplies would have been available indefinitely had the wells been sunk all the way to the regional unconfined aquifer.

1.4 Aquifer properties: effective porosity, permeability, storage

1.4.1 Aquifer properties

In defining an "aquifer" as a body of saturated rock that both *stores* and *transmits* important quantities of groundwater, the two verbs were chosen with care. This is because thorough analysis of groundwater systems depends more than anything else on quantifying the factors that govern the ability of the aquifer to *store* and *transmit* groundwater. Indeed measurement (or estimation) of the storage and transmission properties of aquifers is a major routine task for hydrogeologists.[iv] Both storage and transmission properties are controlled fundamentally by geological factors which for any given rock mass determine: (i) the volume and sizes of the pores it contains; and (ii) the strength of the rock mass when subjected to compression by the weight of overlying ground.

1.4.2 Pores and effective porosity

Characterization of pore space is an important activity in many areas of science and engineering, and many specialist laboratory techniques exist for measuring the dimensions and volumes of pores in rock samples. Less accurate field estimation

...haracteristics also exist, which ...sophisticated geophysical tools ...density of the surrounding rocks ...ered down boreholes. The most ...re of pore occurrence is **porosity**, ...oportion of a given volume of rock ...ed by pores. Hydrogeologists tend ...ted only in the *interconnected* pores ...ch water can flow. Indeed, most of ...ques hydrogeologists use to measure ...nly measure the interconnected pore ...nce hydrogeologists commonly talk in terms ... **effective porosity**, which is the ratio of the volume of interconnected pores to the total rock volume.

Effective porosity arises from a range of rock properties. In unconsolidated sands and gravels, much of the effective porosity will be intergranular in nature (e.g. Figure 1.3a). It is important to note that grain *size* does not in itself correlate with effective porosity: a skip full of ball-bearings will have the same effective porosity as a skip full of ten-pin bowling balls. Rather, effective porosity tends to correlate to other aspects of the sediment fabric, such as:

- **Grain shape:** the more platy or more angular the grains, the closer they can pack together, and therefore the lower the effective porosity will be.
- **Grain sorting:** sediments composed of grains with a relatively uniform grain size tend to be more porous than those composed grains of a wide range of sizes; in the latter case, the small grains tend to occupy spaces that would be open pores in the uniform sediment.
- **Grain packing:** where depositional processes have tended to align the long axes of grains parallel to one another, the effective porosity will be lower than if the same sediment were dumped with grains orientated chaotically.

As unconsolidated sediments undergo burial, the weight of overlying strata tends to increase the packing density of grains. Various geochemical and mineralogical changes collectively referred to as **diagenesis** can result in changes in effective porosity, be this destruction (e.g. by precipitation of mineral "cements" in pores) or creation (dissolution of soluble minerals to create new pores). A vast literature exists on diagenetic controls on effective porosity in sandy sediments (e.g. Davis 1988), much of it collected by the petroleum industry (Wilson 1994).

Thoroughly cemented sandstones will retain little *primary* effective porosity, i.e. effective porosity which was acquired when the rock was first formed. The same is true of limestones, which are often pervasively recrystallized during diagenesis (see Section 1.5.2). A lack of primary effective porosity is also common in many igneous and metamorphic rocks. However, various forms of *secondary* effective porosity (i.e. effective porosity introduced to the rock mass later in its history) often assume great importance in such rocks. Secondary, intergranular effective porosity (formed by the dissolution of cements or sediment grains) can be important in petroleum reservoirs; in aquifers, however, the most important form of secondary effective porosity is undoubtedly **fracture porosity** (e.g. Figure 1.3b), which arises from a range of geological processes (see Section 1.5.3).

1.4.3 Transmission of water: from effective porosity to permeability

It often comes as a shock to newcomers in hydrogeology when they learn that there is no general correlation between the effective porosity of a given rock and its **permeability** (i.e. its ability to transmit water, to use the term informally). Before this lesson hits home, many students blithely use the two terms as if they were synonymous! Why is there no general correlation? Effective porosity tells us the proportion of a given rock mass that is occupied by interconnected pores. However, beyond satisfying the condition that at least *some* effective porosity must exist if there is to be any permeability, the *proportion* of pores is unimportant. The real control on permeability is the *size of the pore necks*, i.e. the sizes of the openings which connect each pore to its neighbours. A cursory glance at Figure 1.3a will suffice to show that pore necks are generally far smaller than the pores which they interconnect. Thus while pore diameters are themselves modest in most rocks (e.g. intergranular pores diameters rarely exceed a few hundred micrometers (μm), and fracture apertures in excess of one millimeter are relatively uncommon), pore necks are more slender still. One of the most crucial consequences of the modest dimensions of pore necks is that (in contrast to the rapid, turbulent flow so characteristic of surface waters) most groundwater *seeps* very slowly, in a gentle, laminar fashion.

The relationship between pore neck size and permeability is well illustrated by the fact that, for a range of samples of well-sorted sandstones which have the same effective porosity but differ in grain size, those with the largest grain size will also tend to exhibit the highest permeability. The bigger the grains, the broader will be the pore necks that remain after the grains are packed together (e.g. Bloomfield et al. 2001). It is further possible to conceive of a range of samples of sandstones which differ in effective porosity but are all of roughly the same mineralogical composition and grain size. In such a case, one would expect to find that the more porous sandstones will also be the most permeable. However, any observed proportional relationship between effective porosity and permeability would be expected to break down if rocks other than sandstone were added to the sample suite, as these are likely to have different grain shape and packing characteristics.

So beyond the requirement for at least *some* effective porosity if a rock is to transmit water, no general relationship between effective porosity and permeability exists. Indeed some of the most permeable rocks in the world have very low porosities, as low as 1% in the case of many cavernous limestones for instance (e.g. Figure 1.3c). At the other extreme, some of the most porous rocks in the world are also the least permeable: this is so, for instance, in the case of many mudstones of very even grain size, which may have porosities of 50% or more, yet transmit water extremely slowly. Of course, we could also list examples of rocks which combine high porosities with high permeabilities, and others which combine low porosities with low permeabilities. The point is, however, that effective porosity is only a *prerequisite* for permeability; its magnitude is determined by pore neck size.

1.4.4 Storage properties of aquifers

In contrast to the rather weak correlation between effective porosity and permeability, the storage properties of unconfined aquifers *are* directly related to their porosities. To understand why this is so, we must first consider what is meant by a change in storage in an unconfined aquifer. Let's say we have a very simple unconfined aquifer such as that shown in Figure 1.7, which is entirely surrounded by impermeable bedrock. Although most aquifers are not this simple, such a scenario can occur in a desert setting, for instance, where wind-blown sand has accumulated

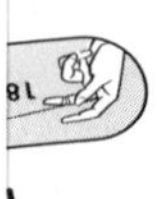

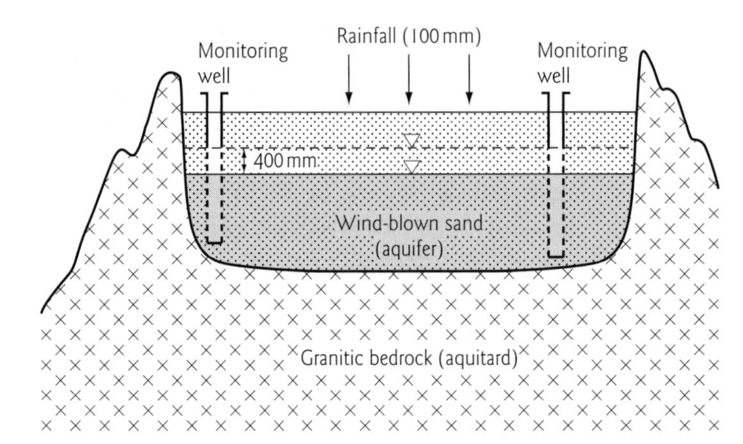

Fig. 1.7 Sketch cross-section of a small hollow in granite bedrock (assumed to be virtually impermeable) deeply filled with highly permeable sand. Only 100 mm of rainfall soaking into the sand causes a 400 mm rise in the water table, due to the fact that the sand has a "fillable porosity" (specific yield) of around 25%. See Section 1.4.4 of the text for further explanation.

to a considerable depth within a hollow in granite bedrock. Consider the following sequence of events:

1 We measure the elevation of the water table relative to sea level, by making measurements in a number of monitoring wells in the basin (as will be explained later in Box 3.1). The elevation of the water table (Figure 1.7) is conventionally denoted by a solid line surmounted by the symbol "∇".
2 The night after we make our measurement of the water table level, a thunderstorm rolls in over our little desert mountain basin. Rain-gauges record that the storm delivers 100 mm of precipitation in the space of a few hours. As this is all falling on clean, wind-blown sand, all of the rain quickly soaks into the ground and seeps down to the water table, so that the upper meter of sand has already drained completely before first light.
3 A few days later, we return to the desert basin and measure the elevation of the water table once more. Not surprisingly, we find it has risen to a new level, marked with a dashed line on Figure 1.7. This new water table elevation turns out to be 400 mm higher than the level we measured before the storm.

If the 100 mm of rain had fallen into a swimming pool with vertical sides, we would expect it to cause the water level in the pool to rise by 100 mm. (This is, of course, exactly how the 100 mm of rainfall was measured in the rain gauge.) Yet in our aquifer the water level has risen by four times as much. Why? It is simply because the water table is present within a porous sand body, so that when the infiltrating rain water reaches the water table, water fills the overlying pores, causing the water table to rise. For 100 mm of rain to lead to a 400 mm rise in water table elevation, it is clear that the available pore space is only one quarter of the total volume of the rock. In other words, the "fillable effective porosity" in this case must be about 25%.

In the simple example just given, the "fillable effective porosity" will correspond very closely to the *total* effective porosity. However, in most soils and rocks some of the pore space is already occupied by adhered water (cf. Box 1.1) before the water table ever rises, so that water can only fill those portions of the pore space which were not previously wet. In most cases, therefore, the "fillable effective porosity" is rather less than the total effective porosity. It is intuitively obvious that the converse applies during periods of water table lowering: pores do not drain completely, but a certain amount of water is left behind, adhered to the grains. We might equally talk, therefore, of "drainable effective porosity," which corresponds exactly to the "fillable effective porosity" for the same body of rock, and thus will also be somewhat smaller than the total effective porosity.

The water which is held in pores after they have been allowed to drain under the influence of gravity is held in place by forces of electrostatic attraction between the water molecules and the surfaces of mineral grains (cf. Box 1.1). The strength of the electrostatic attraction varies from one type of mineral to another. Some minerals commonly found in aquifers exhibit relatively weak electrostatic attraction to water: this is true, for instance, of quartz and calcite, the principal components of sandstones and limestones respectively. However, other types of mineral exhibit strong electrostatic attraction to water, such as the hydrated oxides of iron and manganese and various clay minerals (some of which absorb water into their very structure, swelling in the process). Sedimentary organic matter also tends to retain considerably more moisture than, say, clean quartz sands. Furthermore, just as pore neck size controls permeability, so narrow pore necks favor the retention of moisture in the unsaturated zone. Pores will not drain under the influence of gravity alone if they have neck diameters less than about 10 μm. (Such small pores can, however, be drained by the influence of plant roots exerting high suction pressures.)

Rather than talk clumsily about "fillable" or "drainable" effective porosity, the hydrogeological literature deploys rather more elegant terms: the amount of water which drains freely from a unit volume of initially saturated rock per unit decline in water table elevation is termed the **specific yield**. This property is expressed in exactly the

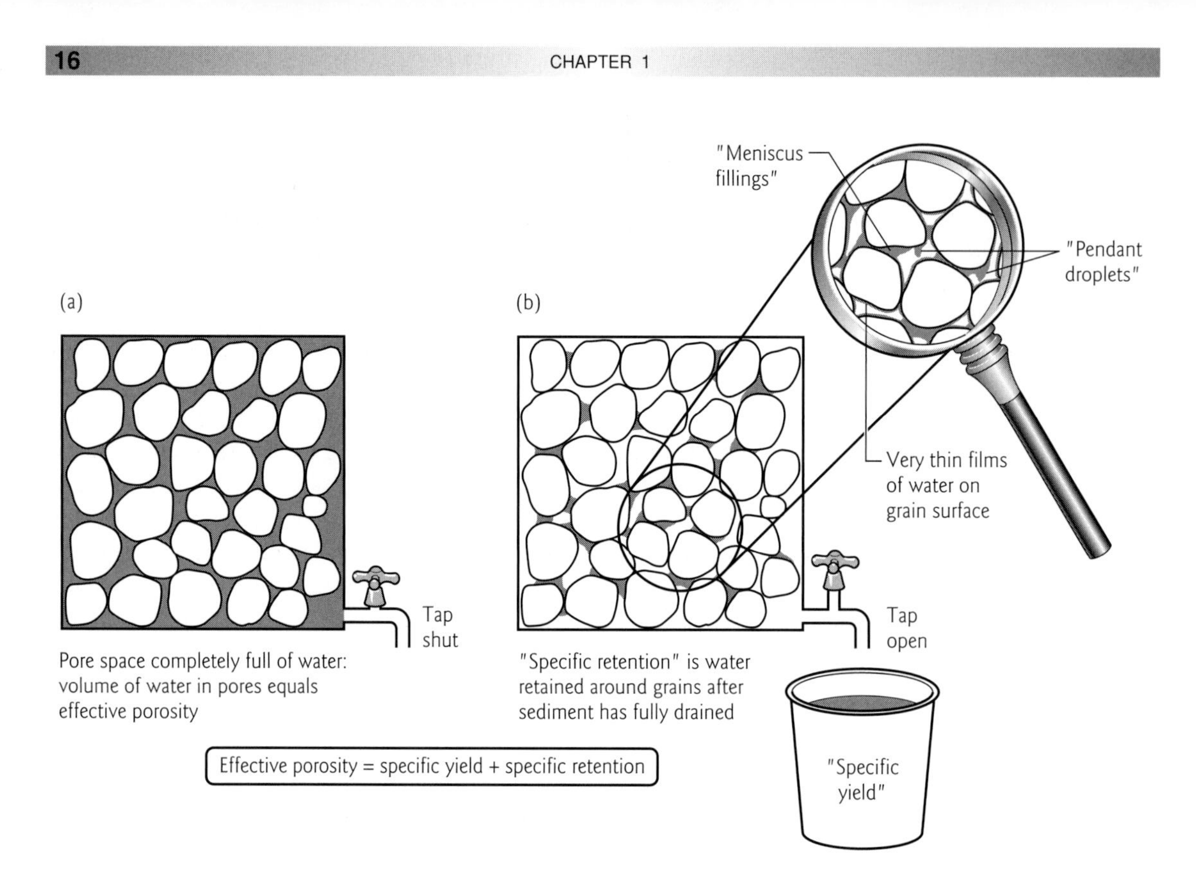

Fig. 1.8 A conceptual sketch to explain the storage properties of unconfined aquifers. (a) A cubic Perspex container fitted with a tap, packed to the top with pebbles and completely filled with water. (b) After opening the tap and waiting until all water has drained from the cubic container, the amount of water in the bucket must correspond to the drainable portion of the effective porosity (which is called the "specific yield"). If we look closely through the sides of the container (using our magnifying glass), we'll see that some water has not drained out (the "specific retention").

same way as effective porosity, either as a percentage or as a decimal fraction of the total unit volume of initially saturated rock. The water retained in the unsaturated zone after drainage under the influence of gravity is termed the **specific retention**, and it corresponds numerically to the difference between the specific yield and the total effective porosity (Figure 1.8).

It is essential to remember that our discussion of aquifer storage up to this point has been strictly limited to *unconfined aquifers only*. The factors controlling storage behavior in confined aquifers are radically different. This is simply because when the water level drops in a well penetrating a confined aquifer, none of the pores within the aquifer have actually been drained: as long as the aquifer remains confined, it will always be completely full with water all the way to the base of the overlying aquitard (cf. Figure 1.5). However, given that the water level in the well has dropped, there is no doubt that there has been a reduction in the amount of water stored in the aquifer. How can we explain this apparent paradox?

Water in a confined aquifer is under pressure, and as such it will press with equal force against all of its enclosing walls. Just think of a balloon: as it inflates, it expands evenly all over, so that the surface is equally taut at all points on its surface. (If we filled the balloon with water rather than air, the same would be true, albeit the weight of the water would make the job of inflating the balloon rather more tedious.) Returning to the pores within a confined aquifer, the pressurized water within each of these will be pressing evenly against the outer pore boundaries, along the surfaces of individual mineral grains (as in

Figure 1.3a) or the rock surfaces which comprise the walls of a fracture (cf. Figure 1.3b). Because of this pressure, the width of each pore tends to be slightly greater when the water is under high pressure than when the water is at low pressure. Even more subtly, under very high pressures the water itself will compress slightly, such that the same mass of water will occupy a smaller volume than it will at lower pressures. As the water pressure in a confined aquifer drops (due to pumping perhaps) water levels in wells penetrating the aquifer will decline, the pores will contract, and the water itself will expand slightly. Thus even though all of the pores remain completely filled with water, some water has been removed from storage due to compaction of the aquifer framework (i.e. the solid "mesh" of mineral grains and fracture surfaces which enclose the pore space) and, to a lesser extent, slight expansion of the formerly pressurized water. Returning to our balloon for a moment: after we have let some of the air escape, the balloon remains completely full of air. However, the air remaining in the balloon is at a far lower pressure, and the walls of the balloon have accordingly moved closer together. This is the phenomenon of **elastic storage**, and it is just as real in the pores of a confined aquifer as it is in a party balloon.

In fact elastic storage occurs in *all* aquifers, both confined and unconfined. However, in unconfined aquifers the contribution which it makes to the overall aquifer storage capacity is miniscule in comparison to the far more bountiful processes of filling/draining pores at the water table. In confined aquifers, elastic storage is the only form of storage available. One does not need to know very much about rock mechanics to appreciate that pressing the walls of pores apart, or squeezing water until it compresses, are far less prolific sources of storage than simply filling vacant pore space. For this reason, confined aquifers have far less storage capacity per unit volume than unconfined aquifers. If we define a storage parameter for confined aquifers along similar lines to the "specific yield" property which was defined above for unconfined aquifers, we come up with the following definition: **storativity** is the amount of water which can be removed from a unit volume of confined aquifer per unit decline in water level (measured in wells penetrating that aquifer). By convention, we express this amount of water as a decimal fraction of a unit volume of the aquifer.

In most hydrogeological literature, storativity is represented by the symbol S. The same symbol is often also used to denote specific yield, for despite their utterly distinct physical origins, the two quantities are treated identically in calculations. However it is desirable to make clear the distinction between the two, so specific yield is usually denoted by S_y. S and S_y are so different in magnitude that the distinction is usually obvious in practice: the S of a confined aquifer will never exceed a value of 0.001 (i.e. 10^{-3}), and in aquifers of limited extent and very high rock strength it can take on values as low as 10^{-12}. By contrast, the values of S_y exhibited by unconfined aquifers rarely drop below 0.01 (i.e. 10^{-2}) and are often much higher than this, occasionally reaching 0.30 in well-sorted sandy aquifers.

1.5 The geology of groundwater occurrence

1.5.1 Hydrostratigraphy

It is becoming difficult to discuss groundwater occurrence in any detail without referring to its manifestations in different types of rock. Indeed, it is impossible to talk in any detail about the interrelations between aquifers and aquitards without specifying the kinds of rocks involved. As soon as we begin to discuss the various types of rocks and their spatial relationships with one another we are well and truly in the realms of the classical geological discipline of "stratigraphy" (e.g. Rawson et al. 2002). In modern parlance, stratigraphy is defined as the study of the "formation, composition, sequence and correlation of the stratified rocks of the earth's crust" (Oldroyd 1996). Stratigraphy is divided into subdisciplines according to the method used to classify rock sequences (e.g. Rawson et al. 2002). Where rock sequences are subdivided solely on the basis of rock type (i.e. "lithology"), then we talk of "lithostratigraphy." Where the subdivision is achieved by means of recognizing distinctive fossils then we talk of "biostratigraphy." Where the stratigraphic sequence is constrained by reliable estimates of absolute age of the constituent rocks, we talk about "chronostratigraphy." Logically, therefore, we can claim to be pursuing **hydrostratigraphy** when we subdivide sequences on the basis of the ability of the rocks to store and transmit water. Although this practice has a very long pedigree, having been practised by the "Father of Stratigraphy" himself, William Smith (1769–1839),[v] the term hydrostratigraphy was only coined in the 1960s (Maxey 1964). Its use was given considerable impetus by influential work undertaken by the US Geological Survey in the late 1980s (Seaber 1988), and by the mid 1990s the concept was beginning to be employed beyond North America (e.g. Al-Aswad and Al-Bassam 1997; Younger and Milne 1997; Al-Bassam et al. 2000).

In essence, hydrostratigraphy amounts to nothing more than identifying, naming, and specifying the extents and properties of the aquifers and aquitards in a given geographical area (e.g. Stone 1999). Where a detailed geological

(a)

Chronostratigraphic units			Lithostratigraphic units	Hydrostratigraphic units	Rock types
System	Age (Ma)	Series	Formation/ member	Aquifer or aquitard?	
Neogene		Miocene and Pliocene	Hofuf	Hofuf aquitard	Mudstone
			Dam	Hasa aquifer	Limestone
			Hadruk		
Paleogene		Eocene	Dammam		
			Rus		Siltstone
		Paleocene	Umm Er Radhuma	Umm Er Radhuma aquifer	Limestone
	67				
Cretaceous		Campanian	Aruma	Aruma aquitarad	Mudstone
		Cenomanian	Wasia	Wasia aquifer	Sandstone
					Mudstone
		Alpian	Biyadh	Biyadh aquifer	Sandstone
		Aptian	Buwaib	Buwaid aquitard	Mudstone
		Neocomian	Yamama		
			Sulaiy	Layla aquifer	Limestone
	140				
Jurassic		Upper	Hith		
			Arab		
			Jubaila		
			Hanifa		
			Tuwayq Mountain	Tuwayq aquitard	Mudstone
		Middle	Dhruma	Az Zulfi aquifer	Limestone
		Lower	Marat	Marat aquitard	Mudstone
	204				
Triassic		Upper	Minjur	Al Suwaidi aquifer	Sandstone
					Mudstone
		Middle	Jilh	Jalah aquifer	Sandstone
					Mudstone
				Shamasiyah aquifer	Sandstone
	245	Lower	Sudair	Sudair aquitard	Mudstone

Fig. 1.9 Hydrostratigraphic classifications of real rock sequences. (a) The previously established chrono- and litho-stratigraphic units of the Mesozoic and Cenozoic sequences of Saudi Arabia, relabeled to identify major hydrostratigraphic units (after Al-Bassam et al. 2000).

map already exists for the area in question, the task of the hydrostratigrapher is often relatively easy. Usually all that will be required is to use groundwater data from the local area, such as yield records for wells and springs, to identify which elements of the local stratigraphic sequence act as aquifers and which as aquitards. The local stratigraphic column, and therefore the local geological map, can then be re-labeled to provide a useful guide to the relative positions of aquifers and aquitards in the landscape. For example, Figure 1.9a shows a stratigraphic column which has been re-labeled to identify the principal aquifers and aquitards in Saudi Arabia.

Occasionally, the hydrogeologist will have to work in an area which is either devoid of a reliable geological map, or else has been mapped only at very low resolution (say at 1 : 250,000 scale or greater). In such circumstances it will not be possible to develop a hydrostratigraphic classification by simply re-labeling an existing stratigraphic column, and primary geological surveying will be necessary. Although this is likely to prove far more time-consuming than working from an existing stratigraphic classification, it does have its benefits. Most importantly, if a primary geological survey is conducted by a team which includes a hydrogeologist, it is possible to produce a lithostratigraphic classification which takes hydrogeological observations into account. For instance, water level records from periods of pumping can provide compelling evidence of stratigraphic continuity between poorly exposed sandstone units which would be difficult to correlate otherwise (e.g. Turner et al. 1993). Figure 1.9b gives an example of a stratigraphic column for an area in west-central Bolivia which was mapped to Formation level (the previous

(b)

Series	Group	Formation	Thickness (m)	Lithologies	Hydrostratigraphic classification	Notes on groundwater occurrence/quality
Pliocene	Umala	Khari	30000 m	Sandstones, mudstones, and acidic ashflow tuffs	Aquifers	Mainly thin (20 m) sandstone aquifers, mostly confined by mudstones or tuffs. Best yields in sub-tuff aquifers. Water quality very good.
		Khuchiquiña	500–1000 m	Laminated mudstones, rare thin distal acid tuffs and tuffaceous sandstones	Aquitards	Very little groundwater present. The few sandstones are too thin and too heavily-cemented to yield much water.
		Mekha	700–1800 m	Laminated mudstones with gypsum beds, acid tuffs and rare sandstones	Aquitards	Low-permeability mudstones predominate; gypsum beds are thin and non karstified. The few wells in these strata contain highly mineralized water.
		Unconformity				
Miocene	Corocoro	Totora	4000 m	Muddy sandstones and sandy mudstones with acid tuffs	Aquitards	Little evidence of groundwater in these strata; sandstones too poorly sorted/too cemented to store much groundwater.
		Huayllamarca	2000–5000 m	Arkosic sandstones with lenses of cupriferous mudstone	Aquifers	Thick, fractured sandstone units contain significant quantities of groundwater, but pollution associated with copper mines renders this of dubious qualtiy.
		Chuquichambi	500 m	Gypsum and gypsiferous mudstones and sandstones	Aquitards	Severely deformed (décollemeitzone); Low-permeability mélange of mudstone and gypsum. Any water likely of poor quality.

Fig. 1.9 (*continued*) (b) The Miocene and Pliocene sequence of the central Altiplano of Bolivia, which was surveyed by the author simultaneously for litho- and hydro-stratigraphic purposes.

mapping having been restricted to Group level) during a hydrogeological survey. Reflecting the influence of hydrogeological evidence, the boundaries of the various designated Formations correspond closely to those of the principal hydrostratigraphic units. This contrasts significantly with the lack of one-for-one correspondence between the lithologically and hydrogeologically defined boundaries in a typical "re-labeled" stratigraphic sequence (e.g. Figure 1.9a).

1.5.2 Which rocks make the best aquifers?

Whenever an experienced hydrogeologist approaches an area which is new to them, they inevitably bring to bear a store of knowledge based on other projects elsewhere in the world. Experiences gained in various geological settings invariably predispose the hydrogeologist to expect certain kinds of rocks to behave largely as aquifers, and other kinds of rocks to behave largely as aquitards. For better or for worse, when I travel to a new destination and begin to examine the local rock sequence for the presence or absence of aquifers, I instinctively turn my attention first to any of the following four rock types in the area: (i) unconsolidated sands and gravels; (ii) sandstones; (iii) limestones; and (iv) basaltic lava flows. Of course I am well aware that it is possible to quote examples of sandstone, limestone, and basalt *aquitards*; yet I am also aware that more than 80% of all the aquifers I have encountered in my career to date have corresponded to one or other of these four rock types. Similarly, I normally expect any mudstones, siltstones, metamorphic rocks, and plutonic rocks to behave as aquitards, and I am rarely proved wrong. Again exceptions exist, but they are still greatly outnumbered by the many aquitards of these lithologies. Finally, my mental map includes a third category of rock types, including volcanic tuffs, coals, and many evaporites, which refuse to occur predominantly as aquitards or aquifers; they may be either, depending on their specific nature and the vagaries of local geological history. I suspect there are few hydrogeologists who would come up with three categories markedly different from my own.

Why should sands and gravels, sandstones, limestones, and lava flows predominate in the unwritten inventory of the world's greatest aquifers? Clearly unconsolidated sands and gravels are archetypal aquifer materials (cf. Figure 1.3b; see also Sharp 1988). No child can spend a happy morning on the beach without coming to realize how readily water soaks into these materials. However, as sand and gravel deposits undergo burial and diagenesis they inevitably lose some of their original permeability (e.g. Davis 1988), due to reductions in pore neck apertures by compaction and cementation (cf. Section 1.4.2). Indeed many ancient sandstones are so heavily cemented that they no longer retain any appreciable primary effective porosity. Nevertheless, some of the world's most important aquifers are consolidated sandstones: for instance the Dakota Sandstone of the northern Great Plains region of the USA (e.g. Fetter 2001, pp. 268–272), the Guaraní Sandstone of Brazil, Paraguay, Uruguay, and Argentina (e.g. Kemper et al. 2003), the "Buntsandstein" of northwest Europe (e.g. Hinderer and Einsele 1997; Kalin and Roberts 1997; Bloomfield et al. 2001), and the Nubian Sandstone of North Africa (e.g. Abd El Samie and Sadek 2001; Ebraheem et al. 2002) and the Middle East (e.g. Rosenthal et al. 1992). Why should old, cemented sandstones continue to function as good aquifers? The principal reason is that when a cemented sandstone is subjected to deformation, which will often amount to nothing more than gentle "extension" (i.e. stretching of the Earth's crust) during a period of uplift, it tends to develop "clean" fractures which have wide enough apertures to give rise to significant permeability.

A rather similar story can be told for limestones. Recently formed limestones, such as those which underlie many tropical islands, often have very high primary porosities. This is due to the many hollows arising from the growth habits of bryozoan, coralline and algal colonies which determine the rock fabric of warm-water marine ramp/reef carbonates. The very large pores typical of these deposits give rise to correspondingly high permeabilities. On the other hand, although they also have high primary porosities, the pores in many deep-water marine carbonate deposits are tiny (<10 μm diameter), due to the fine grain size of the microfaunal tests of which they are composed, and therefore have low primary permeability. Over geological time, carbonates are highly prone to recrystallization, to the extent that they often lose virtually all of their primary effective porosity (e.g. Brahana et al. 1988). Like sandstones, however, most limestones will develop clean fractures during extensional tectonic deformation. Under appropriate hydrogeochemical conditions, the apertures of these fractures can become greatly enlarged (e.g. Figure 1.3a), leading eventually to the development of caves, dolines (roughly circular depressions in the land surface, which sometimes connect directly with underlying caves), and other features diagnostic of so-called "karst"[vi] landscapes common to many limestone massifs (e.g. Davis 1930; Ford and Williams 1989).

Lava flows share with sandstones and limestones the tendencies both to clog with mineral precipitates over time (which serves to reduce permeability), and to fracture cleanly in response to extensional deformation (increasing permeability once more). However lava flows differ from most other types of aquifer in the degree to which they can display extremely high permeabilities from the very earliest days of their existence, long before they have been subjected to deformation (Wood and Fernandez 1988). This is because some of the processes which occur during the accumulation and cooling of lava flows inherently produce large, open voids within the rock mass, which provide excellent flow pathways for circulating groundwater once they are submerged beneath the water table. Most dramatic amongst these large open voids are lava tubes (e.g. Larson and Larson 1990). These form during the downhill movement of cooling lava when the uppermost surface of the flow, which is in contact with the atmosphere, chills more rapidly than the interior. A carapace of solidified lava thus tends to develop above the still-molten interior, and the latter will often flow onwards, leaving an empty tube behind. These features are remarkably common in many volcanic terrains. Nevertheless, they are far from being the only features which tend to make lava flows rather permeable. Thermal contraction joints formed during cooling of the lava are typical of both the tops and bottoms of individual flows, within which they typically form thoroughly interconnected hexagonal networks of vertical joints. Rubbly, brecciated zones are also common on the tops of individual flows, along with shrinkage cracks and concentrations of coalescing vesicles (where gas escape was most prolific during cooling, at the lava–atmosphere interface). Deeper within the lava flows, however, isolated pores corresponding to "frozen" gas vesicles are common, and these tend to give rise to a far greater disparity between "total" and "effective" porosities than one would tend to find in other types of rock. This factor needs to be taken carefully into account when interpreting porosity values for lavas derived from geophysical measurements. Overall, many lava flows form very good aquifers, with excellent examples documented in the Columbia Lava Plateau of the northwestern United States (e.g. Lindholm and Vaccaro 1988), on the Hawaiian Islands (Hunt et al. 1988), on Iceland (e.g. Kiernan et al. 2003), the Canary Islands (Custodio and Llamas, 1996, pp. 1472–1481), Sicily (Aiuppa et al. 2003), and in the Deccan Traps of India (Kulkarni et al. 2000). Indeed, lava sequences give rise to some of the world's most prolific springs, including the Fiumefreddo springs on the northeast flank of Mount Etna, Sicily, which have an average yield of 2 m^3/s (Guest et al. 2003, p. 178), and several individual springs in the Snake River Plateau of Idaho which each discharge in excess of 2.8 m^3/s (Lindholm and Vaccaro 1988, p. 44). Of course counter-examples can also be cited, such as the Palaeogene Plateau Basalts of Northern Ireland, which support few springs and provide generally disappointing yields to boreholes (Robins 1996).

1.5.3 Structural factors: folds, fractures, faults

Having established the nature and identity of the aquifers and aquitards in a particular geological succession, there is one further element of the geology which must be evaluated before it is possible to begin analyzing groundwater dynamics in any detail: the structural framework. Structural geology is concerned with the consequences of crustal deformation for the spatial disposition of rock masses. In modern structural geology the emphasis is on deduction of the history of deformation which has affected a given area, including uplift, subsidence, folding, faulting, and other processes. These so-called "tectonic" processes are essentially the regional manifestations of the global-scale movement of oceanic and continental plates of crustal rocks, the interactions of which determine not only the deformation of existing rocks but also the loci of most of the Earth's volcanic activity and the locations and dynamics of areas of active sediment accumulation. In hydrogeological studies, the overall tectonic framework, however interesting, is rarely of primary importance in the development of a clear understanding of the locations and spatial interrelations between one or more specific aquifers and/or aquitards. At most of the scales of investigation relevant to water resources management and pollution remediation, the principal requirement is to understand how the aquifers/aquitards of interest are disposed within the subsurface. To this end, the main focus is usually on understanding the direction and magnitude of any dip displayed by the strata, especially where this varies from one place to another, and any breaches in the continuity of strata arising from the presence of faults.

As has already been implied in Figure 1.5a, the magnitude and orientation of dip is one of the principal controls on the occurrence of confined conditions in multilayered sequences of aquifers and aquitards. The simple cross-section shown in Figure 1.5a effectively shows only one portion of what must logically be a much larger fold structure. By contrast, Figure 1.10a is a cross-section representing a much wider swathe of terrain than Figure 1.5a. Two major folds are discernible in Figure 1.10a, an **antiform** (i.e. an "up-fold", shaped thus: ∩) and a **synform** (i.e. a "down-fold" shaped thus: ∪). These folds obviously have a profound effect on the depths below ground surface at which the various aquifers (i.e. Aquifers 2, 3, and 4) would be encountered when drilling from the surface.

A crucial aspect of the geological history of the area represented in Figure 1.10a is evident in the upper portions of the cross-section in the form of the "plane of angular unconformity" which separates Aquifer 1 (and its two enclosing aquitards) from the three deeper aquifers. Aquifer 1 is not affected by the folds which displace Aquifers 2, 3, and 4. This is because the folding of the deeper rocks was complete, and had been succeeded by an episode of erosion, before deposition of the rocks above the unconformity commenced. The fact that Aquifer 1 extends beyond the subcrop of its underlying aquitard to rest on the plane of unconformity itself is testament to the existence of considerable topographic relief on the eroded surface of the folded rocks before the commencement of the next phase of deposition. Indeed, Aquifer 1 and its associated aquitards display a certain amount of eastward dip themselves, which means that a certain amount of tilting has occurred following the deposition of these younger rocks. If we restore the plane of unconformity to the horizontal, the axes of the antiform and synform in the underlying rocks are seen to become vertical, which is thus deduced to be their original orientation at the end of the pre-Aquifer 1 phase of folding.

From a strictly hydrogeological perspective, supposing the ground surface does not plummet to much lower elevations just outside the frame of the cross-section, it is possible to deduce from Figure 1.10a that:

- Aquifer 1 is in contact with Aquifer 2 in the west of the study area, but does not come into direct contact with the other aquifers in this vicinity. (It is worth noting that we cannot presume that the plane of unconformity is permeable, so that even though these aquifers adjoin each other, they may not be in very good hydraulic connectivity. We could only establish their degree of hydraulic connectivity from additional field evidence, such as groundwater level data and *in situ* measurements of aquifer properties.)
- There is a substantial, synformal segment of Aquifer 2 which is utterly isolated from the

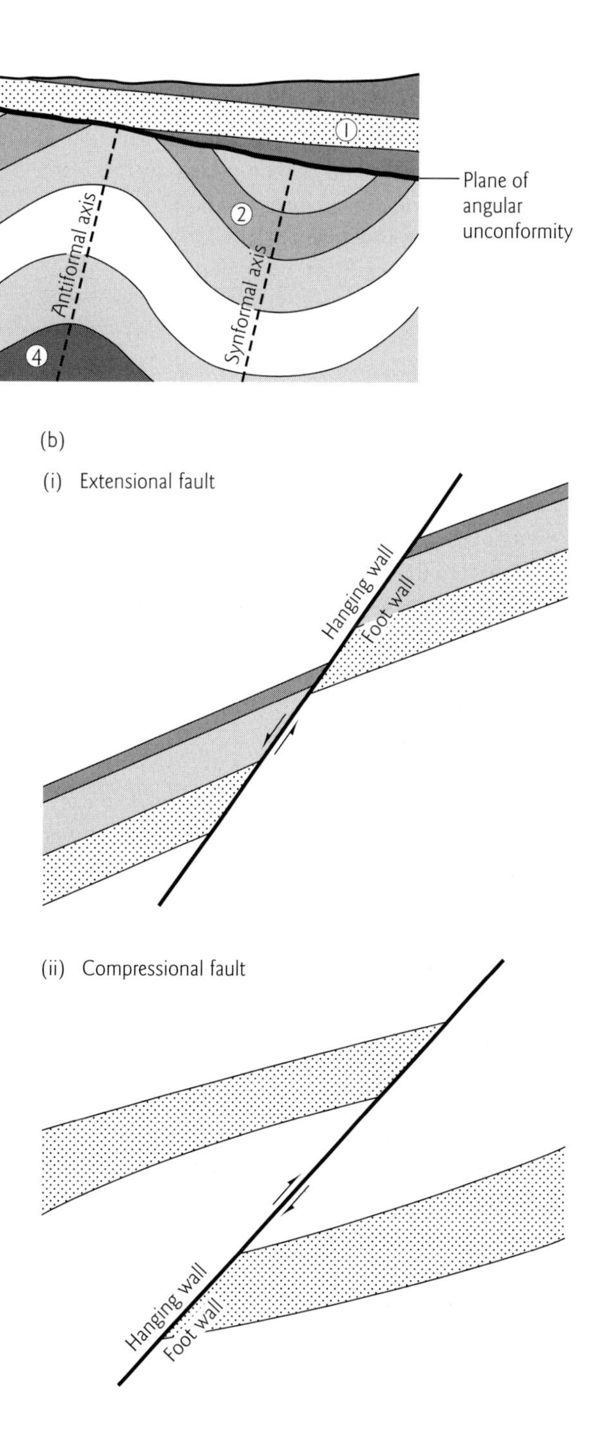

Fig. 1.10 (*Right*) Common structural features which affect the spatial distribution and interconnectivity of aquifers. The horizons numbered 1 through 4 are all aquifers, the unnumbered horizons are aquitards. (a) *Folding and unconformity*. The two principal types of fold are antiforms (upfolds) and synforms (downfold); the centre-lines (axes) of examples of both types are shown to affect aquifers 2 through 4 and their enclosing aquitards. A period of erosion must have followed the episode of folding that affected these aquifers, for an angular uncomformity separates them from the overlying (and evidently younger) aquifer 1, which is unaffected by the folding. (b) *Faults*. The two main types of fault which are commonly found to disrupt the lateral continuity in aquifer horizons: (i) extensional fault; (ii) compressional fault.

westerly segment of the same material due to effective "sealing" by a combination of the folded mass of aquitard mudstones between the synform and the westerly segment, and the cap composed of a younger mudstone lying above the plane of unconformity.

- All portions of Aquifers 2, 3, and 4 are confined, whereas at least the western portion of Aquifer 1 is unconfined.

Similar interpretative exercises are a crucial, albeit routine, part of the work of most hydrogeologists.

Folding and resultant unconformities are of considerable importance in determining the large-scale disposition of aquifers and aquitards. However, the lateral continuity of hydrostratigraphic units can be significantly disrupted by faulting. Faults are planar features across which the elevations of specific rock horizons are displaced (Figure 1.10b). In order to concisely describe the geometry of faults it is helpful to introduce a little jargon. Given that most fault planes are not absolutely vertical, it is normally possible to identify a block of rock which lies above a given fault plane and another which lies below it. The overlying block is called the "hanging wall," whereas the underlying block is called the "foot wall." Two principal types of faults are distinguished on the basis of the relative displacements of the hanging wall and foot wall rocks (Figure 1.10b):

1. Extensional faults (referred to in the older literature as "normal faults") are defined as those in which the hanging wall rocks appear to have been displaced downwards relative to the foot wall rocks. Extensional faults are the most abundant type of fault worldwide, though it is never wise to assume that the faults in a given area will be extensional in the absence of direct evidence.
2. Compressional faults (referred to as "reverse faults" in the older literature) are those in which the the hanging wall rocks appear to have been displaced upwards relative to the foot wall rocks. Where the plane of a compressional fault lies at a low angle (i.e. has a dip of 45 degrees or less), the fault might be referred to as a "thrust" or "overthrust." Compressional faults are especially common in the central districts of most nonvolcanic mountain ranges, and in lowland areas containing rocks which were formerly located in such districts at an earlier stage in geological history.

Although other types of faults do exist (e.g. wrench faults, in which the relative displacement of the rocks is lateral rather than vertical), most faults which exert important influences on groundwater movement are extensional or compressional.

The hydrogeological implications of faulting are not restricted to breaching the continuity of an aquifer or bringing two separate aquifers into contact: as with planes of unconformity, the nature of the fault planes themselves exerts an important control on their own hydraulic behavior. Fault planes can actually be impermeable where they are filled with material of low permeability. This is common in many types of fault systems, where fine-grained material (known as "fault gouge") is smeared throughout the fault plane during the process of fault development. The planes of some large-scale faults are lined with shattered masses of wall rocks (termed "fault breccias") which might also be of low to medium permeability. The temperatures and pressures which develop during the formation of certain types of faults can be so intense that the rocks within the fault plane become metamorphosed, or even melt temporarily, solidifying to form glassy materials ("mylonites") which are typically of very low permeability. On the other hand, faults can form preferentially permeable pathways for groundwater flow in some geological settings. This is especially common in karst terrains, in which the carbonate rocks hosting the faults can retain large apertures, enabling them to function as preferential flow zones below the water table.

It is important to stress that this very brief introduction to geological controls on groundwater occurrence has done no more than scratch the surface of this vast and highly important topic, to which entire books have been devoted. In an attempt to compensate to some degree for the brevity of this introduction, further illustrations of geological controls on groundwater occurrence and movement will be introduced and exemplified throughout this book. There is, however, no real substitute for experience and the old maxim that "a geologist is only as good as the number of different rocks they have seen" can be readily extended to relate the competence of hydrogeologists to the number of different geological settings in which they have plied their trade.

Endnotes

- i. As is shown in Section 1.3, the reason *why* water will only flow into the well below the water table lies in the interrelationship between atmospheric pressure and water pressure, recognition of which leads to a more elegant definition of the water table as the surface on which pore water pressure exactly equals atmospheric pressure.
- ii. Unconfined aquifers are sometimes termed "phreatic aquifer" in older literature.
- iii. The word "confined" is a synonym of "artesian" (a word still found in older/nonspecialist groundwater literature), which is derived from the

name of Artois (Belgium) where confined conditions where first recognized long ago. Given its lack of any intuitive meaning in modern English, and because it is also widely (mis-)used to refer to any well from which water flows without pumping (a phenomenon not restricted to confined aquifers; Freeze and Cherry 1979), its further use is to be discouraged.

iv As is shown in Section 3.4 "aquifer properties" are normally equated mathematically with two specific physical values, transmissivity (T) and storativity (S), values for which are directly yielded by most field test methods.

v William Smith (1769–1839) invented the geological map and was the principal pioneer of stratigraphy (Winchester 2001). For a fascinating account of Smith's water-related work, see Torrens (2004).

vi The term "karst" refers to the assemblage of landscape features such as caves, dolines, dry streams, etc. found in areas underlain by soluble rocks (see Ford and Williams 1989).

2

Sources of Groundwater: Recharge Processes

Now the sun, moving as it does, sets up processes of change, of becoming, and of decay; and by its agency the finest and sweetest water is every day carried up and is dissolved into vapor and rises to the upper region, where it is condensed again by the cold and so returns to the earth.

(Aristotle, 384–322 BCE)

Key questions

- Where does groundwater come from?
- What is the most important source of groundwater?
- What is meant by "recharge"?
- How do recharge processes differ between humid and dry regions of the world?
- How does recharge occur in urban areas?
- How can we estimate recharge in these different settings?
- How does water move through the unsaturated zone to reach the water table?

2.1 Provenance of groundwater

In Chapter 1, we took it for granted that the circulation of fresh water on our planet is driven by the precipitation over land of atmospheric moisture, which is in turn predominantly derived from evaporation of sea water (Figure 1.2). From the opening quotation of this chapter, it is clear that this much was appreciated by Aristotle more than 2300 years ago. It might therefore seem pointless even raising the issue of the provenance of groundwater: it seems obvious that at least some of the precipitation which falls on the land surface must soak into the ground, eventually replenishing the store of groundwater below the water table.

Nevertheless, no matter how obvious this now seems to us, it is sobering to realize that widespread recognition of the atmospheric origin of most groundwaters is actually a rather recent development. For almost two millennia after the death of Aristotle most European intellectuals misinterpreted the implications of his teaching for this topic. Some settled for simple declarations that groundwater had been placed in the subsurface by God during the creation of the Earth, rather tendentiously quoting the book of Genesis (1: 6–7) in support of their views. Others preferred instead to cite the teachings of Plato, who considered groundwaters to be derived from sea water by some mysterious, unobservable process of subsurface distillation within the bowels of the earth, in which waters were drawn inland from the seafloor by some unspecified force, being purified as they passed upwards to reach the Earth's surface as fresh water springs (see Biswas 1970). Even when Aristotle's teaching on the origins of terrestrial freshwater was accepted, it was assumed to explain only the origin of surface waters, with Plato's theory being uncritically accepted to explain groundwater origins. It was not until the late seventeenth century that scientists such as Pierre Perrault and Edme Mariotte in France, and Sir Edmund Halley (he of comet fame) in England, effectively re-invented Aristotle's original concept by means of deductions based on observations of the processes of rainfall, runoff, evaporation, and condensation, going on to unequivocally apply this concept to the origins of groundwaters as well as surface waters (Biswas 1970). Even after these arguments had become well known, many resisted giving up the old Platonic concepts. Thus as late as 1778 we find an eminent Cornish physician and mine adventurer explicitly rubbishing the arguments of Sir Edmund Halley, and re-stating the Platonic concept in the following words:

> *. . . the only true origin of perpetual springs [is] the Ocean . . . our hypothesis is that in the formation of perpetual springs they not only derive their waters from the sea, by ducts and cavities running from thence through the bowels of the earth . . . but that the sea itself acts like a huge forcing engine, or hydraulick [sic] machine to force and protrude its waters from immense and unfathomable depths, through those cavities, to a considerable distance inland . . .* **(Pryce 1778, p. 13)**.

Ultimately, however, such resistance was to prove futile, so that by the mid nineteenth century the atmospheric origin of most groundwaters seems to have been almost universally accepted. Groundwaters originating from the infiltration of atmospheric waters are generally termed "meteoric," a term derived from an archaic adjective for things pertaining to the atmosphere. (The word "meteorology," i.e. the science of weather and climate, has the same root).

While dismissing the mistaken views of Pryce (1778), it is worth pausing a moment to acknowledge that most of the groundwater in the world does indeed lie beneath the beds of the oceans. However the high salinity of marine groundwater renders it of little use to mankind, and it has therefore been little studied in its own right. However, there are a number of motivations for studying the interactions between fresh and saline groundwaters in coastal aquifers. From a water resources management perspective, the high salinity of marine groundwater renders it significantly more dense and more viscous than fresh groundwater, leading to complex flow relations wherever these two meet in coastal aquifers. These complexities mean that very careful management strategies are required to avoid inducing the entry of excessively saline waters into public supply wells (e.g. Robins et al. 1999). Submarine groundwater discharge is also a topic of increasing interest to chemical oceanographers, marine ecologists, and other scientists interested in the nutrient balance of sea waters, and as such has become a fertile area of hydrogeological and biogeochemical research in recent years (e.g. Destouni and Prieto 2003). (Further discussion on fresh–marine groundwater interactions may be found in Section 7.2.4.)

Besides meteoric and marine groundwaters, there are a few other minor sources of groundwater which ought to be noted even though they are very rarely encountered in practice (Fairbridge 1998). Perhaps the most frequently discussed of these minor sources is **connate water**. Strictly speaking, connate water is conceived to be groundwater which has been present in the pores of a sedimentary rock ever since it was deposited. (The term "connate" is derived from a Latin phrase meaning "born with".) Supposedly connate waters are invariably saline, living up to the expectation that they must be fossil seawaters dating back to the time of sediment deposition, and/or that the very long residence times of these waters have allowed them to dissolve substantial quantities of minerals which only react very slowly. Given that most aquifer rocks were deposited millions of years ago, and also given the tendency for groundwater to flow, it is difficult to imagine many circumstances in which an aquifer is likely to retain precisely the same water which was present in its pores at the time of sediment accumulation (see Marty et al. 2003). Nevertheless, saline waters are commonly found at depths exceeding a few hundred to a thousand meters in nearly all bedrock aquifers. Such deep-seated saline groundwaters, which have clearly not fallen as rain water in recent history, are often loosely referred to as connate waters despite the unlikelihood that they have remained utterly static since the sedimentation of their present host rocks. Detailed geochemical studies of such waters tend to confirm that they are of great antiquity, but seldom if ever are they found to be as old as their host rocks. Rather, their salinity tends to reflect various episodes in local geological history, such as prehistoric incursions of the sea (e.g. Elliot et al. 2001), or periods of desertification and associated infiltration of surface waters which have experienced intense evaporation (e.g. Edmunds and Tyler 2002).

If true connate waters are rare, even less common are **juvenile waters**. These are waters which have not previously participated in the hydrological cycle during the entire history of our planet. As such they arise either from deep within the Earth's mantle, or else arrive on Earth from outer space (Fairbridge 1998). As outlandish as these two origins might at first seem, they do both occur in reality. Water arising from the Earth's mantle typically reaches the surface in the form of steam emissions in volcanic areas. It has been estimated that water vapor accounts for more than 90% of the gases released during volcanic eruptions worldwide. However, nearly all of this is actually meteoric water, entrained by the volcanic processes within a few hundred meters of ground surface. Nevertheless, even that fraction of the volcanic steam which originates in the mantle may not be juvenile, for the incorporation of ocean floor rocks into the mantle in subduction zones (e.g. Cervantes and Wallace 2003) inevitably also results in the introduction of meteoric water into the mantle. A further source of juvenile water is so-called "dirty snowballs" that land on the Earth during meteor showers. These mixtures of ice and rock material are fragments of comet tails. It has been estimated that about 40 tonnes of water per annum are recruited to the Earth's hydrological cycle; this rate may well have been substantially greater earlier in the history of our solar system (Fairbridge 1998). However, few hydrogeologists will ever come across juvenile waters in their work, and of those few most will not be able to demonstrate that the waters are truly juvenile.

For all intents and purposes, therefore, the only groundwater which will concern most of us in practice will be meteoric in origin. The remainder of this chapter is therefore devoted to understanding and quantifying the processes by which meteoric waters enter aquifers: so-called recharge processes.

2.2 Recharge processes

2.2.1 Fundamental controls on recharge

Stated most simply, **recharge** is the entry of water into the saturated zone. It is the means by which groundwater storage is replenished, and is thus also one of the principal causes of rises in water table levels. For the most part, recharge occurs from above, by means of downward migration of moisture through an overlying unsaturated zone. However, recharge can also occur by means of lateral flow from a lake or river which is deep enough to at least partially penetrate an adjoining saturated zone. Furthermore, in some confined aquifer settings, recharge can occur by means of saturated flow of groundwater across the bounding aquitard(s).

It is a matter of considerable frustration that, despite the pivotal role which recharge plays in all groundwater systems, it is a rather perplexing phenomenon to deal with. This is principally because, with few exceptions (and none of any practical importance), it is usually impossible to measure recharge directly. When we attempt to quantify recharge, therefore, we have to resort to measuring a range of other phenomena from which we can estimate recharge by subtraction. Any errors in the measurements of these other phenomena translate into uncertainties in the estimated magnitude of recharge. In many groundwater investigations, therefore, the rate of recharge is one of the least certain elements. Some of the reasons why it is so difficult to obtain defensible estimates of recharge will become apparent in the course of this chapter. Suffice it here to note that controversies frequently rage over the rates of recharge assumed in particular investigations, or over the assumptions made in attempting to estimate recharge with some degree of rigor. For many hydrogeologists, recharge is like a guilty secret: it's something they cannot get away from, but would really rather not discuss!

Recharge depends critically on the availability of water. For the most part, the availability of water depends on the interplay between *supply* in the form of **precipitation** (i.e. rain, hail, sleet, or snow) and the *loss* of water back to the atmosphere. Loss occurs by the combined effects of **evaporation** (i.e. the direct vaporization of liquid water) and **transpiration** (the release of water vapor to the atmosphere by plants). In most soil–plant systems these two processes are difficult to distinguish in practice, and their separate quantification is in any case unnecessary for the purposes of quantifying groundwater recharge. Hence it is convenient to lump the two together and refer to them conjointly as **evapotranspiration** (e.g. Oliver 1998). The mechanisms of evapotranspiration are rather complex, and readers desiring comprehensive explanations of the key phenomena are referred to classic works such as those of Thornthwaite and Holzman (1942), Penman (1948, 1949), and Grindley (1967, 1969), together with more recent syntheses such as those of Bras (1990) and Oliver (1998). One concept which must be clearly understood in order to make sense of many recharge estimation techniques is **potential evapotranspiration**. This is the rate at which evapotranspiration would occur, given the ambient conditions of atmospheric temperature, humidity, and solar radiation, if there were no limit to the supply of water to the soil surface and/or to plants. One advantage of specifying values of potential evapotranspiration is that it can be estimated with considerable accuracy from the data routinely collected by automatic weather stations. However, in very many cases the soil will simply be too dry to evapotranspire water at the full potential rate. Under these common circumstances, which hold sway for most of the year in many catchments, the **actual evapotranspiration** will be but a small fraction of the potential rate. In fact the actual and potential rates tend to coincide only during the wet season. Despite the apparent limitations of seeking relationships between actual and potential evapotranspiration rates, a number of reliable recharge estimation techniques do include calculations in which soil moisture availability is compared with the potential evapotranspiration rate (e.g. Penman 1948; Grindley 1969).

Given the dependence of recharge on the magnitudes of both precipitation and evapotranspiration, the rate at which water enters the saturated zone can be expected to vary dramatically from one climate zone to another. Table 2.1 illustrates this point at the coarsest of scales, in terms of long-term annual averages of freshwater availability for the six nonpolar continents. While these continental averages clearly hide great spatial variations in recharge rates *within* each continent, some interesting points emerge

Table 2.1 A simplified annual average fresh water budget for the nonpolar continents of planet Earth, emphasizing the percentage of incoming precipitation which eventually forms groundwater recharge. (Deduced and re-calculated from global water budget data presented by Herschy 1998.)

Element of water budget	Europe	Asia	Africa	North America	South America	Australia
Total precipitation (volume over given continent (km^3))	7,162	32,590	20,780	13,810	29,255	6,405
Total precipitation (% of total global precipitation)	6.51	29.63	18.89	12.55	26.60	5.82
Evapotranspirative loss (% of total precipitation lost to actual evapotranspiration)	57	60	77	69	65	69
Available freshwater (% of total precipitation becoming surface runoff and groundwater recharge)	43	40	23	31	35	31
Surface runoff						
As % of total precipitation	28	30	15	21	22	24
As % of available freshwater	65	75	65	68	63	77
Total annual volume (km^3)	2,005	9,777	3,117	2,900	6,436	1,537
Groundwater recharge						
As % of total precipitation	15	10	8	10	13	7
As % of available freshwater	34	26	35	32	36	24
Total annual volume (km^3)	1,047	3,389	1,673	1,370	3,686	476
Groundwater discharge						
To rivers						
% of available freshwater	21	15	8	18	16	10
Total annual volume (km^3)	646	1,955	382	771	1,638	198
Submarine outflows						
% of available freshwater	13	11	27	14	20	14
Total annual volume (km^3)	401	1,433	1,291	599	2,048	278

which provide us with some valuable starting points for thinking about recharge processes. Firstly, it is remarkable that the percentage of the total available freshwater which becomes groundwater recharge is relatively consistent between all six continents, averaging 31% with a standard deviation of only 5%. Given the marked contrasts in the predominance of arid zones between the different continents, this is a significant finding for it confirms an important general principle, namely that recharge undoubtedly occurs to some extent in even the most arid regions (cf. Lerner et al. 1990, p. 7). In terms of the percentage of total *precipitation* which becomes groundwater recharge, Table 2.1 provides some insights which can help constrain case-specific estimates of recharge. For instance, if we take the figure of 10%, which is applicable to both Asia and North America, it is immediately obvious that this average embraces extreme variations from, say, the arid heart of the Gobi Desert to the temperate rainforest of the Olympic

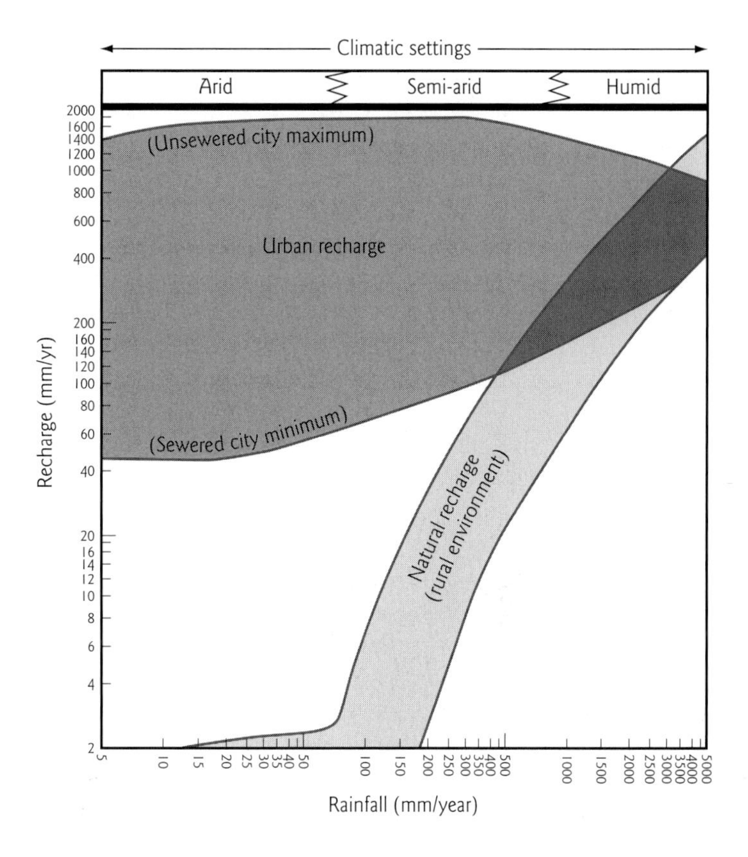

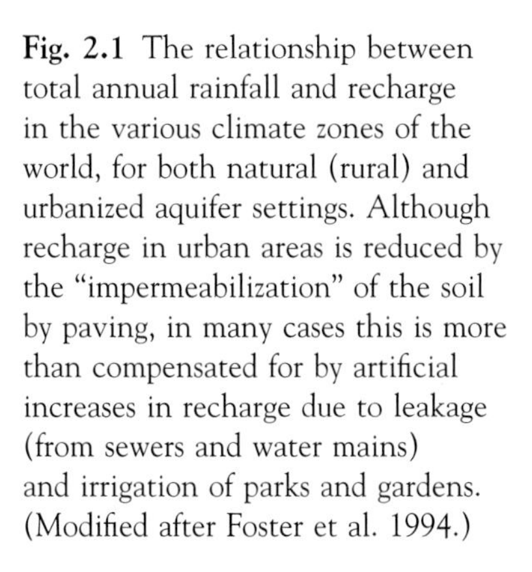

Fig. 2.1 The relationship between total annual rainfall and recharge in the various climate zones of the world, for both natural (rural) and urbanized aquifer settings. Although recharge in urban areas is reduced by the "impermeabilization" of the soil by paving, in many cases this is more than compensated for by artificial increases in recharge due to leakage (from sewers and water mains) and irrigation of parks and gardens. (Modified after Foster et al. 1994.)

Peninsula (Washington State, USA). At first glance, therefore, this average figure is of little utility. However, it does allow us to constrain our recharge estimates by acknowledging that:

- in dry areas of Asia and North America, annual recharge rates are highly unlikely to exceed 10% of the local total annual precipitation, whereas
- in humid areas of these continents, recharge rates are most unlikely to be any less than 10% of the local total annual precipitation

If nothing else, considerations of this nature help to limit our expectations within realistic bounds, prior to making the site-specific evaluations which are essential in practice.

Detailed studies of recharge have been undertaken in many parts of the world; Lerner et al. (1990) have compiled a wide range of examples. From such studies, broad relationships between total annual rainfall and total annual recharge rate have emerged (Figure 2.1). Most estimates of natural recharge rates range between about 5% and 25% of rainfall. In certain types of terrain, values as high as 75% have been recorded: for instance, this is the rate at which recharge occurs in the highly permeable lavas of Mount Etna (Guest et al. 2003, p. 178). However, such exceptionally high rates of conversion from rainfall to recharge are relatively uncommon in practice, and it is normally wise to carefully scrutinize any recharge estimates which exceed about 30% of the total rainfall in the study area.

2.2.2 Modes of recharge: direct vs. indirect

Although Table 2.1 and Figure 2.1 provide interesting empirical evidence for the maximum and minimum ratios of recharge to rainfall, inspection of these sources of summary information does not in itself advance our understanding of the actual processes by which recharge occurs in different climatic settings. For instance, both Table 2.1 and Figure 2.1 provide strong evidence for the occurrence of recharge even in the most arid of regions. However, they do not answer the question which immediately arises from this observation: how can it be that significant recharge occurs in both humid and arid regions, given the vast differences in soils and climatic conditions between the two? Are the processes of recharge the same in all regions, irrespective of their degree of aridity? To answer such questions it is necessary to delve a little more deeply into the processes by which recharge actually occurs (Lerner et al. 1990). Where rainfall landing on the soil surface soaks downwards immediately below its point of impact, passing beyond the root-suction base and continuing all the way to the water table (Figure 1.4), then it is said to constitute **direct recharge**. By contrast, where rainfall fails to soak into the soil surface on which it first lands and becomes surface runoff instead, but then subsequently does enter the subsurface at some distance from its point of initial impact (thereafter soaking on down to the water table), then it is said to constitute **indirect recharge**[i] (Lerner et al. 1990). Two distinct modes of indirect recharge are recognized (e.g. de Vries and Simmers 2002), essentially differing only on the basis of the geomorphological scale at which the surface runoff and subsequent infiltration occur:

- **Localized recharge**, in which surface runoff occurs as overland flow (i.e. not within first- or higher-order channels) and the subsequent infiltration occurs via fractures or "macropores" (as we shall see in Section 2.2.4).
- **Channel leakage**, in which recharge occurs by seepage through the beds and banks of recognizable stream channels (as explained in Section 2.2.5).

The distinction between direct and indirect recharge has far greater importance than mere scientific classification, for there is a distinct contrast in the predominance of the two modes of recharge between humid and (semi-)arid areas (Lerner et al. 1990). While direct recharge is predominant in many humid areas, indirect recharge is often the only form of recharge operative in very dry areas. Indeed the existence of indirect recharge mechanisms in the desiccated landscapes of semi-arid and arid regions explains why the percentage of total available freshwater accounted for by groundwater recharge (Table 2.1) differs relatively little between predominantly arid regions (e.g. 24% in Australia) and predominantly humid regions (e.g. 34% in Europe).

Having outlined the two principal modes of recharge and identified their relative importance in humid and arid areas, it is now appropriate to consider each of these modes in further detail, and to outline methods by which they can be estimated in practice.

2.2.3 Evaluating direct recharge: soil moisture budgeting

An appreciation of seasonal soil moisture dynamics provides the foundation for understanding and predicting direct recharge. Consider a soil which has been baked dry by the summer heat: when a soil is desiccated even the "specific retention" (Figure 1.8) will have been depleted, due to suction by plant roots. Under these conditions, the first rain water to land on the soil will not contribute to recharge, but will be retained within the pores by adhesion to soil particles (cf. Box 1.1), whence much of it will be removed by plant root suction. The sum of the latent water demands of water-deprived plants and parched soil particles are collectively identified by meteorologists using the term **soil moisture deficit** (Grindley 1967, 1969). By the time the first incoming rains have contributed to reducing the magnitude of the soil moisture deficit, there may well be no excess water available to soak on down to the water table as recharge. However, as the rains continue to come, the magnitude of the soil moisture deficit slowly declines, and the soil progressively moistens until the specific retention has been restored and sufficient water is available to meet all current plant demands. At this point the soil is said to be at **field capacity**, and any further rain water entering the soil will be likely to drain on past the root-suction base, and thence on downwards to the water table. At this point in the proceedings, direct recharge is being generated.

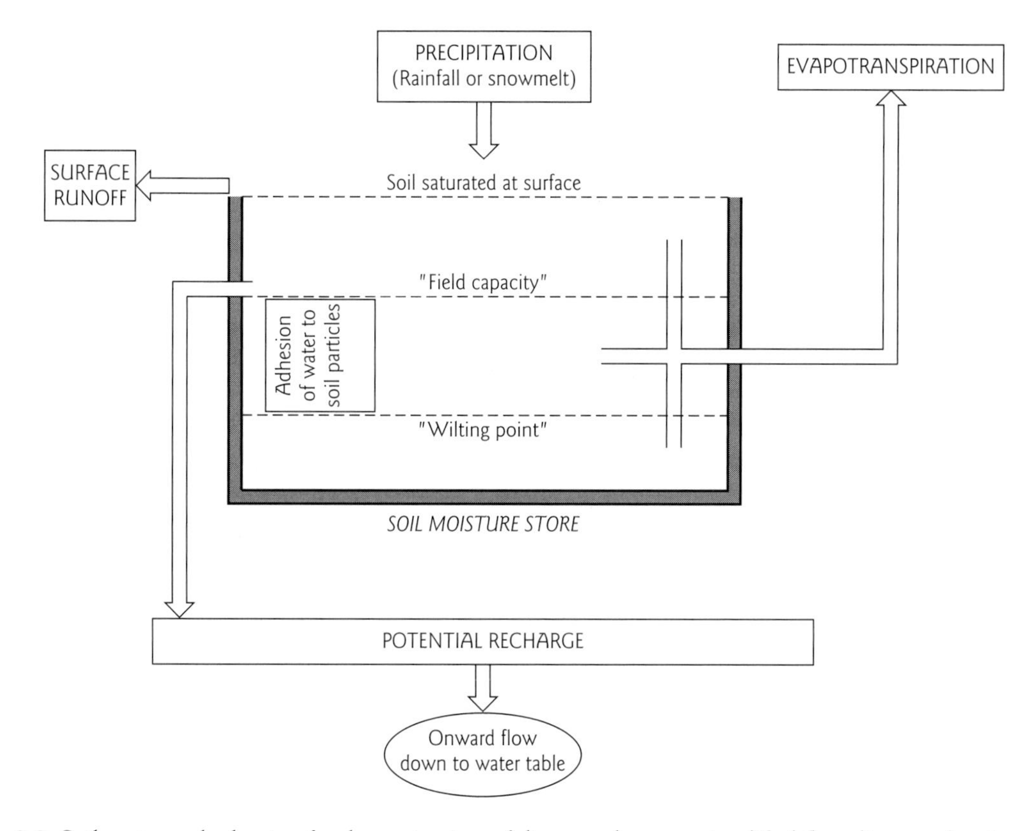

Fig. 2.2 Soil moisture budgeting for determination of direct recharge: a simplified flow diagram showing the principal stores and thresholds which need to be evaluated to evaluate the quantity of "potential recharge" seeping below the root-suction base towards the water table.

Figure 2.2 summarizes diagrammatically the various elements of this soil moisture system. Estimation of direct recharge essentially proceeds by quantification of the volumes within and the fluxes between the various "boxes" shown on Figure 2.2. These calculations yield what is known as a "soil moisture budget," which tracks the evolution of the various water volumes over time. Given the rates at which rainfall and evapotranspiration rates change, it is usual to perform the calculations for daily time intervals. Where weekly or monthly time intervals are used, the resultant recharge estimates are usually found to be unacceptably inaccurate.

Soil moisture budgeting is similar to predicting cashflow through a bank account: by counting up the various payments which are expected to reach the account on certain dates, and deducting any foreseen expenditures, one can predict what the balance of the account is likely to be on any one day. In soil moisture budgeting the "payments" correspond to rainfall, the "expenditures" are evapotranspirative losses, and the "balance" is the soil moisture content. The analogy holds out well up to this point. However, to complete the analogy we would have to imagine that whenever the balance of the account surpasses some pre-set value (threshold) at the end of each day's trading, the bank will automatically transfer any surplus cash into a long-term savings account. If my own banking affairs are anything to go by, this final element of the analogy might well seem far-fetched to the reader. However, suspending disbelief momentarily, the "threshold" in the above analogy corresponds to the moisture content when the soil is at field capacity, the difference between the "threshold" and the "end-of-day balance" corresponds to the volume of water released beneath the root-suction base, which is thus available to become recharge in due course (when it reaches the water table), and the "long-term savings account" is groundwater storage beneath the water table.

Details of the procedures by which such soil moisture budgeting is applied to the calculation of direct recharge in practice are presented by Rushton (2003). The logic of these calculations, follows the sequence shown in Figure 2.2. For each one-day time-step in the recharge estimation period, the budgeting proceeds essentially as follows:

1 Estimation of total daily *actual* evapotranspiration. At least two eventualities need to be considered at this point:
   (a) During periods when no soil moisture deficit exists, actual evapotranspiration will simply equal the rate of potential evapotranspiration, which is calculated from weather station measurements (i.e. air temperature and humidity, net solar radiation, and wind speed), taking into account the water uptake characteristics of the plants in the area under consideration. This in turn requires that the different water demands of the various plant species over their growth seasons be taken into account, and a database of crop water requirements such as that published by the United Nations Food and Agricultural Organization (FAO) (Allen et al. 1998) must be consulted for this purpose.
   (b) During periods when a soil moisture deficit exists, evapotranspiration will occur at less than the full potential rate which we would calculate using weather station data. This is simply because the supply of water to plant root systems is limited as long as the soil water content remains below field capacity.
2 If any rainfall has occurred on the day in question, the daily actual evapotranspiration estimates are subtracted from that day's rainfall total. Where the total rainfall exceeds the actual evapotranspiration, three eventualities need to be considered:
   (a) If a soil moisture deficit currently exists, the excess rainfall will be used up in attempting to reduce this to zero (i.e. to restore the soil to field capacity).
   (b) If no soil moisture deficit exists (i.e. the soil is at or above field capacity) then it is possible for the excess rainfall to seep beyond the root-suction base, and thus to enter the sub-soil zone where it is *en route* to becoming recharge.
   (c) However, in really wet periods the water content of the soil may be so high that all of the pores are completely filled with water: in this case, some fraction of the excess water will be unable to soak into the soil and will tend to remain on the ground surface, where it will give rise to overland flow.
3 On the other hand, where the daily actual evapotranspiration rate exceeds the daily rainfall rate, the excess evapotranspiration will be consumed in removing water from storage in the soil zone, thus drying the soil to some degree. In many cases, this will result in further deepening of the pre-existing soil moisture deficit.

As can be seen, there are many potential obstacles in the way of an individual rain drop before it is definitely on its way to becoming groundwater recharge. Only in step 2(b) above does any soil moisture proceed below the root-suction base and into the sub-soil. Once the moisture gets this far, it is still not certain to reach the regional water table, for it may be intercepted by perched aquifers (see Figure 1.6) which in some cases will hold the moisture back for long periods or even pass it laterally to the surface environment once more. However, infiltrating waters passing the root-suction base can at least be regarded as **potential recharge** (Rushton 2003), the ultimate fate of which depends on unsaturated zone flow processes, as described in Section 2.3.

The processes of evapotranspiration which determine how much rainfall ends up as potential recharge also alter the chemistry of infiltrating waters (Appelo and Postma 1993). Chemical contrasts between rain water and shallow groundwater can therefore be used to infer the magnitude of evapotranspiration. As water molecules leave the water by evapotranspiration, the various dissolved substances are left behind in solution, so that their concentrations inevitably increase. While the concentrations reached by some solutes will be limited by their participation in geochemical reactions (such as the precipitation of new minerals or adsorption to the surfaces of pre-existing minerals), a number of solutes which participate in few such reactions can be used as accurate estimators of evapotranspiration. This is true, for instance, of chloride. Except where the local soils or rocks contain the sodium chloride mineral halite, the increase in chloride concentration between rain water and groundwater will simply reflect the amount of evapotranspiration to which the infiltrating water was subjected. If we compare the chloride contents of rainfall and shallow groundwater we can infer how much of the original rainfall was lost to the atmosphere during its progress through the soil zone. For instance, if the local rainwater contained 3 milligrammes of chloride per liter of water (i.e. 3 mg/L Cl), and the local shallow groundwater contains 20 mg/L Cl, then we can calculate the percentage of the original rainwater which was lost to evapotranspiration as follows:

% of rain water lost to evapotranspiration
$= 100 \times (1 - (3/20)) = 85\%$

or, in other words, of the original rainfall landing at the surface only 15% went to form recharge. Given that we are likely to have measurements of rainfall, such evidence can allow us to calculate average recharge rates without recourse to soil moisture budgeting. However, this approach only yields long-term average recharge rates, and cannot provide the daily estimates afforded by soil moisture budgeting. Nevertheless, this chloride ratio method can provide a useful independent check on the annual average rate calculated by summing daily estimates obtained by other means.

2.2.4 Evaluating localized indirect recharge: bypass routes to the water table

In many semi-arid and arid areas soils may never reach field capacity, and thus direct recharge is unlikely to occur. Even where field capacity is occasionally reached, soil moisture budgeting for sites in semi-arid areas often reveals that direct recharge accounts for only a few percent of the total amount of recharge which is inferred from other evidence to be occurring (for a collation of examples, see de Vries and Simmers 2002). In such cases, one has to assume that much of the recharge is indirect. Where conspicuous ephemeral stream channels are present in the study area, indirect recharge by means of channel leakage is obviously worthy of evaluation (see Section 2.2.5). However, more localized indirect recharge can occur via much smaller features known as **macropores**, which are often widely dispersed in semi-arid/arid landscapes. Macropores are of diverse geometries and origins (e.g. Gee and Hillel 1988), and they include desiccation cracks, animal burrows, root casts (i.e. voids formed where a plant root has decayed), and the interfaces between woody plant roots and the surrounding soil. At slightly larger spatial scales, small topographic hollows may collect surface runoff and provide pathways for entry into outcropping aquifer materials. At the largest scales, major surface depressions such as the dolines characteristic of karst terrains (e.g. Ford and Williams 1989) provide bypass routes for runoff to rapidly flow from the surface to the interiors of carbonate and evaporite aquifers.

Accurate quantification of such localized forms of recharge is very difficult. Flows entering recognizable topographic hollows can sometimes be diverted through flow measurement devices (such as flumes, weirs, or even giant tipping-bucket gauges), providing some evidence of the potential magnitude of recharge occurring via such features. However, unless a very large number of such hollows can be instrumented (which is typically prohibitively expensive) it is difficult to generalize the findings from one or two measuring stations to infer the extent of localized recharge occurring over a wide recharge area. For such purposes, it will generally be more appropriate to determine the localized component of recharge by first obtaining some estimate of *total recharge*, and then subtracting from this the direct component of recharge, determined independently using soil moisture budgeting or some other technique (see Section 2.2.3).

The estimation of total recharge can be approached in a number of ways (de Vries and

Box 2.1 Example calculation of recharge from observed water table rise.

In an aquifer known to have a specific yield (S_y) of 0.15, an annual rise in water table of 425 mm is observed in monitoring wells remote from areas of artificial influence (i.e. away from urban areas and wellfields). We can calculate the equivalent depth of recharge water which produced the observed water table rise by calculating:

Recharge (equivalent depth in mm) = water table rise (mm) × S_y

So in this case:

Recharge = 425 mm × 0.15 = 63.75 mm

(To avoid implying spurious accuracy, given the uncertainties in S_y etc., we would report this value simply as 64 mm.)

Simmers 2002) depending on the temporal resolution of the required estimates. For the purposes of estimating localized recharge, one of the most appropriate techniques (e.g. Healy and Cook 2002) is to examine water table rise in response to individual recharge events (Box 2.1). To estimate total recharge from water table rise, it is necessary to have independently derived estimates of the specific yield of the aquifer under investigation. These are best obtained by analysis of test pumping data (see Section 3.4); where such data are not available, approximate estimates can be obtained using geologically based inference techniques (e.g. Younger 1993). Once we have a credible value of specific yield, we simply multiply this by the water table rise over the period of interest to obtain the "equivalent depth"[ii] of water added to the saturated zone in that time (Box 2.1). Having obtained an estimate for total recharge in this manner, it is possible to estimate the relative contributions of direct and indirect recharge, if the direct recharge element has been independently determined using soil moisture budgeting techniques. (It should be noted that the logic of determining recharge rate from water table rise is simply the inverse application of the principle presented in Figure 1.7, where we calculated specific yield from the observed water level rise corresponding to a *known* amount of recharge.)

Geochemical approaches to recharge estimation can also shed some light on the magnitude of localized inputs. For instance, the occurrence of localized recharge can be demonstrated from the chemistry of pore waters extracted from rock core samples obtained at various depths in the unsaturated zone (e.g. Edmunds and Tyler 2002). As was mentioned in Section 2.2.3, the evapotranspirative loss of water during direct recharge leads us to expect the concentration of chloride (and other solutes) to increase with depth through the soil zone, until some approximately steady concentration is reached, which will persist throughout the unsaturated zone (Figure 2.3a). However, where localized recharge is occurring, marked decreases in chloride concentration at depth may well be detected (Figure 2.3b). This occurs due to the mixing of dilute, rapidly infiltrated localized recharge waters with the more concentrated, directly recharged waters (e.g. de Vries and Simmers 2002). Besides tracing the unsaturated zone profiles of chloride and other major ions, the same approach can also be applied to the isotopes of hydrogen and oxygen present in the water molecules themselves (see Hiscock 2005, pp. 123–126). As elegant as these approaches are, they suffer from the significant drawback that core retrieval is a destructive process, which of its very nature cannot be repeated in precisely the same location. It is therefore difficult to use such approaches to determine temporal variations in localized recharge rates.

Perhaps the most sophisticated approach to estimating the spatial distribution of localized recharge is to sample shallow groundwater chemistry in many boreholes, and then to subject each water analysis to inverse geochemical modeling. This technique involves reconstructing the full

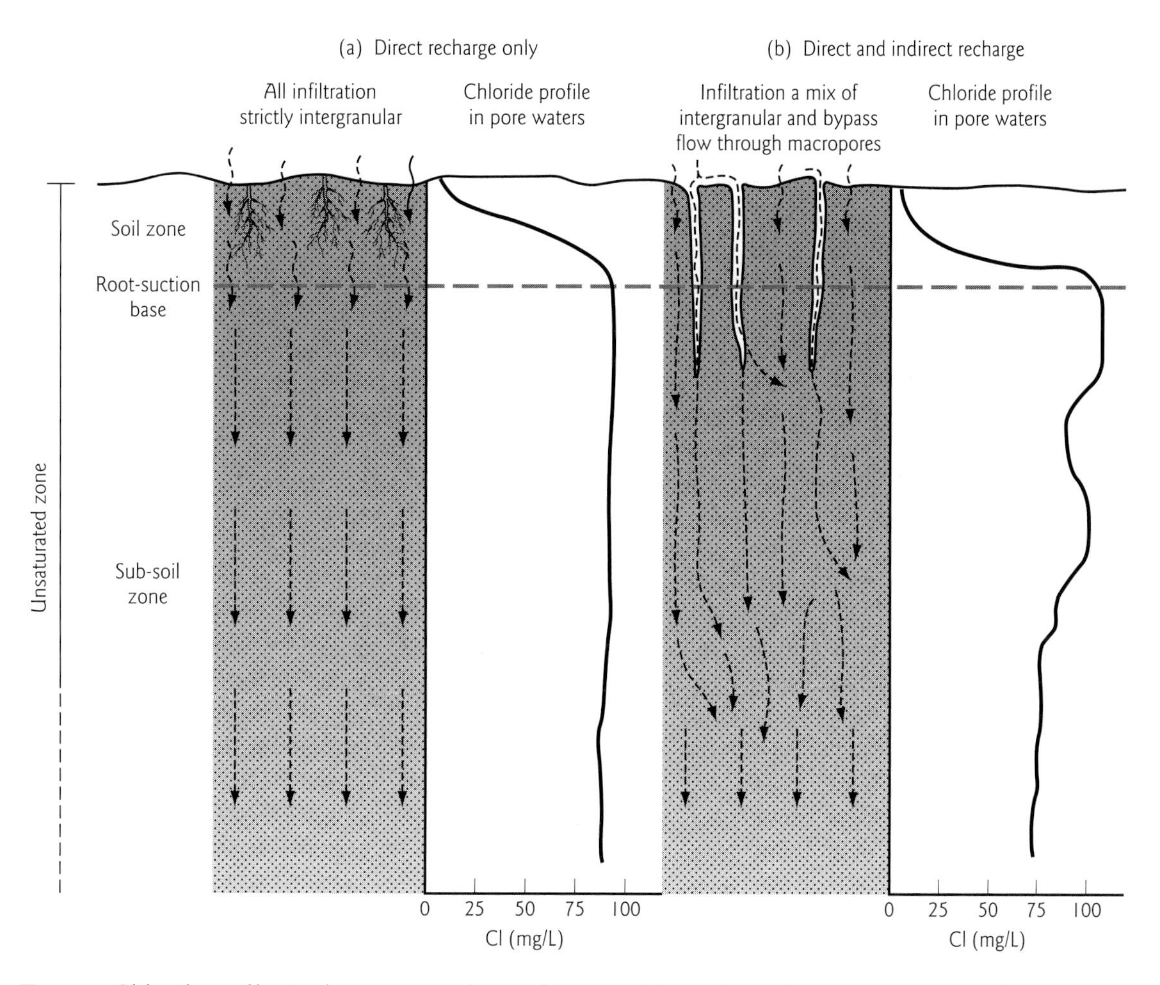

Fig. 2.3 Chloride profiles in the unsaturated zone determined by analysis of pore waters extracted from drilled cores. (a) Typical smooth profile encountered where only direct recharge is operative. (b) Typical "noisy" profile encountered in situations where "bypass" recharge routes introduce indirect recharge to the unsaturated zone at various depths in the sub-soil profile. (Developed after Lerner et al. 1990.)

array of geochemical processes which occur during the flow of water from one location to another, by solving a large number of simultaneous equations which describe the mass balances of individual solutes (e.g. Appelo and Postma 1993; Parkhurst and Appelo 1999). Taking rainwater as the starting point for each individual groundwater analysis, and taking into account the geochemical behavior of minerals known to be present in the local rocks, it is possible to reconstruct the evolution of waters in terms of both rock–water interactions and evapotranspiration (e.g. Chen et al. 1999). Where localized recharge is an important process, wide variations in the influence of the latter will be deduced. The results of the inverse geochemical model simulations will therefore allow delineation of how the relative proportions of direct and localized recharge vary over a given area (e.g. Edmunds et al. 2001).

2.2.5 Evaluating indirect recharge by channel leakage: transmission losses

In many arid and semi-arid regions, the bulk of indirect recharge occurs during periods of storm runoff by means of leakage of water through the

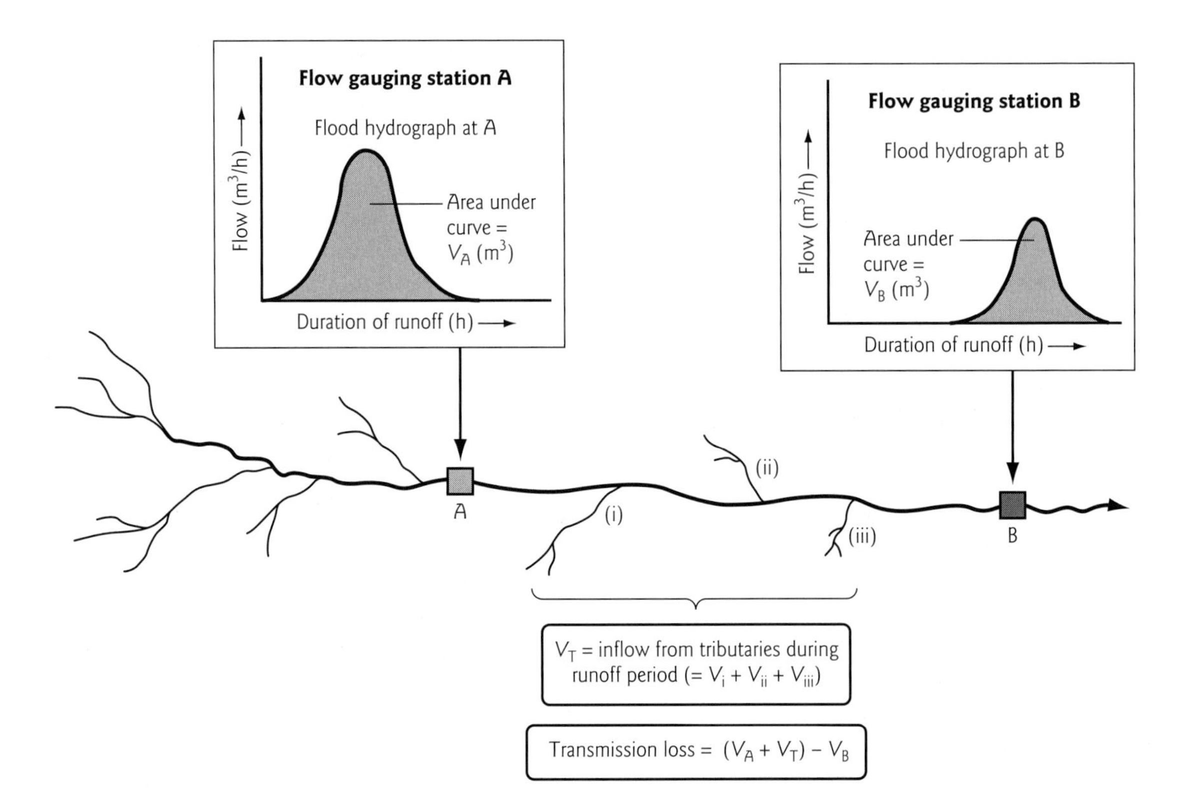

Fig. 2.4 Determining indirect recharge due to channel leakage: evaluation of transmission loss between two successive gauging stations along an ephemeral stream channel. The areas under the two hydrograph curves (V_A and V_B) are total runoff volumes at the corresponding gaging stations for the storm event in question. Taking into account any inflows from tributaries (i), (ii), and (iii) (V_T), and any loss of water to evaporation or abstraction, then the difference between V_A and V_B (i.e. the transmission loss) must equal the volume of recharge generated by leakage through the stream bed.

beds of ephemeral streams (i.e. streams which only carry water during, and immediately after, storms). Where synchronous flow measurements ("gaugings") are made at a number of points along an ephemeral stream channel during a storm runoff period it is possible to directly quantify the amount of water being lost from the channel per unit time and per unit length of channel. The decrease in flow between two successive flow gaging stations is called a **transmission loss**. After accounting for any direct evaporation from the channel (which is usually a relatively minor component of the stream flow budget in any case), and for any artificial abstractions or (in low-gradient river systems) outflows to distributary channels, such transmission losses can be confidently equated with indirect recharge by channel bed leakage (e.g. Abdulrazzak et al. 1989; de Vries and Simmers 2002). Figure 2.4 illustrates the concept, and the way in which quantification of the entire flood volume passing two successive gaging stations can be used to calculate the magnitude of recharge in a given channel reach.

In many mountainous arid regions, channels descending from highland areas carry very high loads of coarse sediment, which they promptly deposit in the form of alluvial fans where they cross the break in slope at the foot of the mountain chain. As they traverse these alluvial fans the stream channels commonly split into numerous distributaries, each of which tends to lose its flow by leakage into the interior of the alluvial sediment pile. This form of indirect recharge by channel leakage is sometimes referred to as **mountain front recharge**. Because this occurs from a large number of small channels, which are highly unstable and prone to avulsion, direct measurement of transmission losses is not feasible. It is therefore common to estimate mountain front recharge using either hydrochemical mass balance techniques (generally using isotopic data or information on dissolved noble gas contents to distinguish different sources of recharge water; e.g. Manning and Solomon 2003) or mathematical modeling of streamflows and infiltration rates (e.g. Chavez et al. 1994a,b). Further discussion on the loss of stream flow to the subsurface (and the converse process) may be found in Section 5.1.6.

2.2.6 Recharge in urban areas

When an area is urbanized, many impermeable surfaces are constructed: roofs, pavements, road surfaces; all are intentionally constructed to be far less permeable than natural soils. It is thus not surprising that surface runoff generally increases markedly as a previously rural area is urbanized. The increase in runoff can be so substantial that it leads to a significant increase in flood risk, with all the implications this has for the security of lives and properties. Some jurisdictions have responded to increasing urban flood risks by demanding that new developments result in no net increase in surface runoff. A number of measures have been devised to comply with the requirement for zero increase in runoff, such as the use of porous pavements, and the routing of roof runoff to underground "soakaways" (i.e. tanks with unlined bases, whence the water is allowed to seep away into the sub-soil; see Section 9.3.3). To date, however, the implementation of such measures around the world has been very patchy. It is therefore generally the case that urbanization tends to substantially *decrease* direct recharge at the expense of increased surface runoff.

Given that direct recharge is almost invariably reduced as an area becomes urbanized, is it appropriate to assume that *total* recharge is less in urban areas than in adjoining rural areas? Perhaps surprisingly, the answer is an emphatic "no" (Lerner 2002). This is because the innumerable sewers and water distribution pipes present below ground in most urban areas are highly prone to leakage. So prolific is leakage from these sources that *indirect* recharge from sewers and water mains more than compensates for any diminution in direct recharge due to "impermeabilization." Consequently recharge is far higher in most urban areas than in nearby rural areas (Lerner 2002). Because the water used for urban water supply often originates well outside the city limits, leakage from water distribution pipes and sewers effectively represents an importation of runoff which would simply never have been present within the city prior to urbanization. The magnitude of the increase in recharge between natural and urbanized conditions is indicated in Figure 2.1. In some cases, the increase in recharge due to urbanization is so extreme that it leads to substantial rises in the water table, giving rise to perennial flows in previously ephemeral stream channels, and even leading to water-logging of the ground and flooding of basements and foundations. Such is the case, for instance, in the city of Riyadh, Saudi Arabia (Rushton and Al-Othman 1994), where rising groundwater levels have been ascribed not only to leakage from water mains and septic tanks, but also to intensive watering of parks and gardens. The presence of a substantial aquitard in the shallow subsurface greatly exacerbates the problems of water table rise caused by an increase in recharge rate, as it leads to ponding of groundwater in the shallow subsurface (Rushton 2003, pp. 327–328).

Given the nature of sewage, and the myriad potential contaminant sources in most urban settings, it is worth bearing in mind that urban recharge is often rather more contaminated than recharge originating in pristine countryside. While this is certainly a management issue for urban aquifers (see Section 9.3), the presence of specific contaminants can be turned to some advantage in that it allows identification of recharge sources and pathways (Barrett et al. 1999).

2.2.7 Other artificial sources of recharge

Besides leakage from sewers and water distribution pipes in urban areas, a number of other human activities, such as irrigation, can lead to substantial increases in recharge over that which would occur naturally. Waters introduced to an agricultural district by irrigation can become recharge either by infiltration through the irrigated soils themselves (mimicking natural direct recharge processes) or else by leakage through the beds of the canals which carry the water to the fields (Lerner et al. 1990; Rushton 2003). Because the rates of supply of irrigation waters are generally well known, it is often possible to estimate these components of recharge relatively accurately, using the same principles as are applied to the quantification of natural recharge.

In some situations, recharge is deliberately introduced to the subsurface, by a number of means including "spreading" of water in unlined basins excavated into permeable material, direct injection into the saturated zone using boreholes, or by inducing recharge from rivers by pumping nearby wells. These practices are known as "artificial recharge" (e.g. Bouwer 2002) and they are an essential component of aquifer storage and recovery schemes, as we shall see in Chapter 7.

2.3 Movement of water through the unsaturated zone

2.3.1 The delay between rainfall and water table rise

Given that the unsaturated zone is often many tens of meters thick, it is reasonable to anticipate that there will often be some time delay between the generation of "potential recharge" at the root-suction base and its conversion to "actual recharge" by arrival at the water table. Such a delay is manifest in the common observation that the peak in annual rainfall often occurs weeks or even months before the annual peak in groundwater levels (Figure 2.5). The delay between rain falling at the surface and recharge arriving at the water table can vary between a few hours (where the water table is shallow and the aquifer is highly permeable) to many months (where the water table is deep and the sub-soil is not very permeable). Where delays are brief, it may not be necessary to take the time lag into account when attempting to quantify groundwater resources. However, where the delays exceed a week or two, it will often be necessary to quantify the delays accurately. There are a number of possible means for doing this.

2.3.2 Simple methods of accounting for unsaturated zone flow

By far the simplest way to account for the delay between rainfall and recharge is to analyze records of rainfall and water table rise and quantify the delay in peak times between the two. It may then be possible to produce a recharge time-series simply by adding a fixed number of days to the time at which rainfall events are known to have occurred. Where the water table is relatively shallow (less than about 15 m) and the soil and sub-soil are fairly permeable, this simplistic approach may well suffice.

However, in many cases the passage of infiltrating waters through the unsaturated zone not only delays the arrival of the peak of a pulse of recharge: it can also cause substantial "smearing" or spreading of the recharge pulse over time (Figure 2.5). For instance, recharge associated with a rainfall event which lasted only 2 days may not only be delayed in reaching the water table by many days, but also the period of time which elapses between the arrival at the water table of the first and last water molecules in the recharge pulse may take 20 days or more. This effect can be effectively accounted for in monthly recharge estimates by assuming that the total recharge calculated to be draining below the root-suction base in any one month (as described in Section 2.2.3) can be divided into a number of fractions, each of which is deemed to arrive at the water table one month after the preceding fraction. For instance, of the total recharge calculated as arising in January, maybe 10% will reach the water table in the same month, 50% will reach it in February, 30% in March, and 10% in April. The total recharge calculated as arising in subsequent months is treated similarly. Hence the recharge arriving at the water table in any one month will

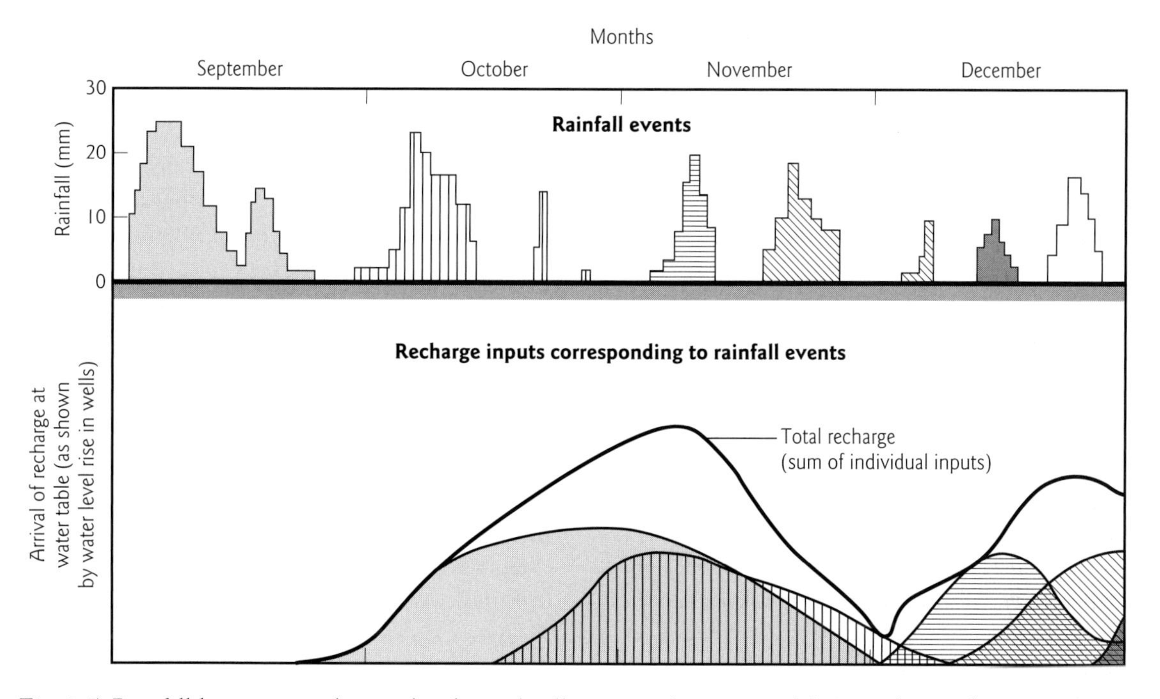

Fig. 2.5 Rainfall becoming recharge: sketch graphs illustrating the nature of the time-lags and attenuation ("smearing") of the rainfall signal which occurs as recharge moves through the unsaturated zone to reach the water table. The shading patterns used in the top portion of the plot (identifying individual rain storms) is maintained in the lower portion, to show when the water introduced by each storm actually arrives at the water table. Note that although rainfall is discontinuous, time-lag and attenuation result in a continuous (if variable) input of recharge from late September onwards. (The time-scales shown here assume a relatively thin unsaturated zone in permeable strata; time-lags are likely to be much greater in other circumstances.)

actually be the sum of the recharge fractions corresponding to rainfall events over the preceding few months. Further details on the practical implementation of this sort of approach, including a number of refinements suited to different hydrogeological settings, are available in the literature (e.g. Senarath and Rushton 1984; Calver 1997; Younger 1998; Rushton 2003). Recharge time-series obtained using this approach have been shown to accurately reproduce observed fluctuations in groundwater levels in a number of major aquifers.

2.3.3 Advanced numerical approaches: elegance and rigor at a cost

From a scientific point of view, the most appealing approach to accounting for the delay between rainfall and water table rise is to apply mathematical models of unsaturated zone flow processes which are based rigorously on what we know of the physics of soil moisture. In comparison with many other mathematical models used in groundwater studies (see Chapter 10), models of unsaturated zone flow have an impressively mature pedigree, for the equation that encapsulates the physics of unsaturated zone flow was first published in 1931 (Richards 1931). The **Richards' Equation** essentially states that, at any and every point in the unsaturated zone:

1. The rate of change of moisture content within the pores equals the difference between the rates of arrival and departure of water at that point.
2. These rates can be calculated by multiplying the permeability of the sub-soil zone (at that specific point) by the gradient of hydraulic head across that

point. (Hydraulic head is essentially a summation of the elevation and pressure of water at any given point; see Section 3.1.2.)

It is important to bear in mind that the permeability of the unsaturated zone varies with moisture content: it is effectively zero at a moisture content of zero, increasing gradually to a maximum value when the pores are completely saturated with water. Hydraulic head also increases proportionately with moisture content. When the Richards' Equation is translated into mathematical notation, it takes the form of a partial differential equation. Only very simple partial differential equations are amenable to exact mathematical solution (see Section 10.4.2). Richards' Equation is unfortunately not so simple, owing to the dependence of both permeability and head on moisture content. This means that exact solutions of the full Richards' Equation do not exist, and approximations must be sought. One approach is to simplify the Richards' Equation itself. For instance, it is possible to neglect the dependency of permeability on moisture content by simply assuming that permeability is invariant, and that it simply equals half of the value which would obtain under saturated conditions. It is also possible to bypass the need to determine the gradient of hydraulic head by deploying the "unit gradient assumption" (i.e. that the hydraulic gradient always equals 1), which may well be close to the truth in situations where the sub-soil is deep and relatively uniform (e.g. Scanlon et al. 2002). With such simplifying assumptions in force, it is possible to solve the resultant equation directly, at least for geometrically simple field conditions (e.g. Lerner et al. 1990). However, if such simplifying assumptions cannot be justified, it will often be preferable to obtain an approximate (though still accurate) solution of the full Richards' Equation using so-called numerical methods (see Section 10.4.3), most notably "finite difference" and "finite element" methods (e.g. Huyakorn and Pinder 1983). While these methods were derived long ago, their implementation was extremely time-consuming prior to the advent of digital computers. Thus it was not until the late 1960s that the first numerical solutions of the Richards' Equation began to be reported in print (Freeze 1969). As computing power has advanced, so has our ability to solve the Richards' Equation at ever-finer scales of resolution. Most recently the Richards' Equation has been re-cast within a formal thermodynamic framework which also facilitates simultaneous accounting for solute transport through the unsaturated zone: the so-called "SAMP model" (**s**ubsystems **a**nd **m**oving **p**ackets) originated by Ewen (1996a,b).

Given that the Richards' Equation is the most scientifically defensible representation of the physics of moisture movement in the unsaturated zone, and given that numerical solutions of it have been available for several decades, one might suppose that it would by now have achieved supremacy amongst the various techniques available for estimating recharge. Yet this is clearly not the case: in practice, far more use is still made of the various approximations discussed earlier in this chapter (see, for instance, Scanlon et al. 2002). The reasons for the lack of practical application of the Richards' Equation are essentially twofold. Firstly, even with modern computer technology, the implementation of numerical solutions to the Richards' Equation is time-consuming and can only be done by highly trained scientific specialists. Secondly, the supply of essential data to the computer simulations requires the repeated collection of site-specific field measurements which can be processed to elucidate the relationships between moisture content, permeability, and head, not only at various depths within the unsaturated zone, but also over time, as they change in response to seasonal patterns of rainfall and evapotranspiration. While techniques for collecting the necessary information are well known (see, for instance, Sammis et al. 1982), their application suffers from two drawbacks:

- They are far more difficult to implement than the techniques which yield equivalent measurements below the water table, and their successful use demands continuous labour by teams of highly skilled technicians.

- They are prohibitively expensive to implement at the scale appropriate to the determination of recharge rates to most aquifers.

Because of these limitations, numerical solutions of the Richards' Equation continue to be of most use as research tools, rather than as day-to-day recharge estimation techniques (Rushton 2003). Their value as research tools is nevertheless enormous, especially in relation to:

- Predicting unsaturated zone behavior under climatic conditions different from those for which recharge has been observed to date (see Section 9.2.3).
- Accounting for the processes affecting the movement of dissolved substances through the unsaturated zone (e.g. Ewen 1996a,b; Robins 1998), which is a key task in the formulation of groundwater protection strategies (see Section 11.4).

Endnotes

i Indirect recharge is also known as "runoff-recharge" in some circles (e.g. Rushton 2003), though this latter term has the disadvantage of a slightly oxymoronic meaning.

ii Recharge volumes can be quoted in the same way as rainfall measurements are reported, as a notional depth of water accumulated over a given area in a specified period of time.

3
Groundwater Movement

The force that drives the water through the rocks drives my red blood;
that dries the mouthing streams turns mine to wax.
And I am dumb to mouth unto my veins
How at the mountain spring the same mouth sucks.

(Dylan Thomas, 1914–1953)

Key questions

- What makes groundwater move?
- How can we quantify the rate of groundwater flow?
- What kinds of flow patterns are exhibited by groundwater systems, and how can we deduce these?
- What features constitute the natural boundaries to groundwater systems?
- How does the pumping of wells affect groundwater flow patterns?
- How can aquifer properties be quantified?

3.1 "The force that drives the water through the rocks"

3.1.1 Gravity and an apparent paradox

What is it that makes groundwater migrate through rocks? At the most fundamental level, the answer is simple: gravity. For just as the Earth's gravitational field causes rain to tumble from the skies and streams to flow to the sea, so it forces groundwater to migrate through rock pores. In general terms, then, groundwater behaves like other natural waters: it tends to flow downwards from high points.

However, there are some subtleties to groundwater flow behavior which need to be clearly understood in order to avoid some serious pitfalls. For instance, at certain locations in many aquifers, groundwater can actually be shown to be flowing vertically *upwards* towards the ground surface. At first glance this might seem strange – after all, streams are never seen to flow uphill, so why should groundwaters be able to apparently defy gravity in this way? In reality, upward-flowing groundwaters are not defying gravity at all; on the contrary, they are slavishly following its diktats within the physical constraints imposed by their surroundings.

To understand how this can be, consider Figure 3.1. In Figure 3.1a, a water tank is connected to a tap by a straight length of pipe. Obviously when we turn on the tap water will flow out, fed by flow down the pipe from the tank. Throughout the length of the pipe, water is flowing downwards, constrained by the pipe walls. In Figure 3.1b, the situation is identical save that the pipe is bent, so that it descends below the level of the tap before finally rising to meet it. Obviously when we turn on the tap in Figure 3.1b, water will flow out as before; however, in this case the water in the pipe will be traveling vertically upwards just before it reaches the tap. In both cases the flow of water proceeds from the highest to the lowest points at which the water is exposed to the atmosphere. This means that, even though the bottom of the U-bend in Figure 3.1b lies lower than the tap, the driving force for water flow is greater at that point than at the mouth of the tap, where the exiting water encounters atmospheric pressure.

From this simple analogy it can be deduced that vertically upward flow in groundwater systems is not so strange as it might at first seem. In groundwater systems, a given packet of water will only be in contact with the atmosphere at the start and end of its subsurface flow path (and nowhere in between these two limits). Further reflection on Figure 3.1 reveals some important truths about the forces that drive water through long subsurface pathways (both in pipes and in systems of interconnected pores in rocks). Clearly the difference in elevation between the water tank and the tap is an important driver for flow. But is elevation in itself enough to explain water movement? Given that the bottom of the U-bend in Figure 3.1b lies lower than the tap (so that water actually flows up from a low point to a higher one), then it is obvious that elevation alone cannot explain subsurface water movement. Rather, the force which ensures that water will flow upwards from the bottom of the U-bend to the tap is the powerful "push" exerted by the greater of the two water columns which lie

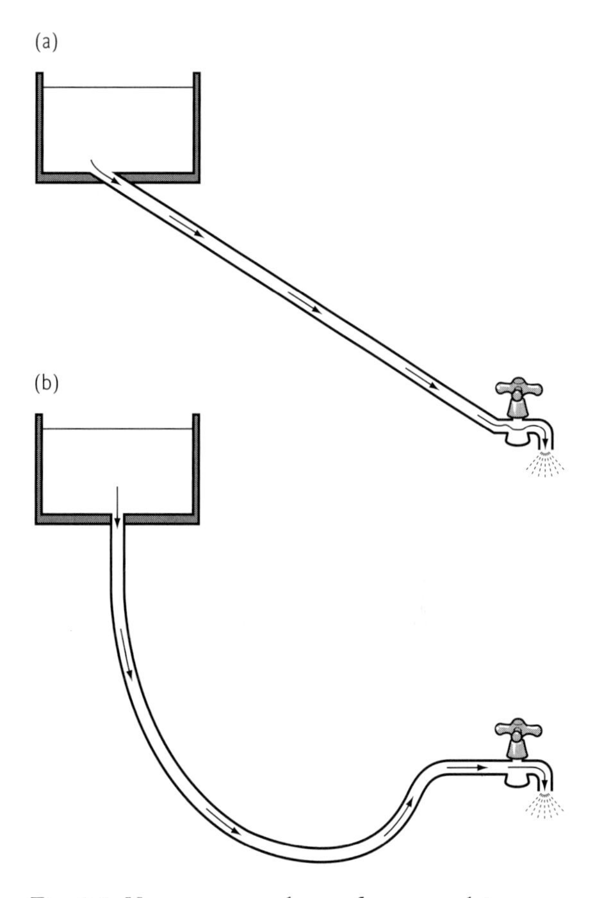

Fig. 3.1 How can groundwater flow upwards? An analogy to a simple header tank and hosepipe. (a) Straight hosepipe, in which flow is always downwards. (b) Hosepipe with downward loop, in which the water must flow upwards to reach the tap. The laminar nature of most groundwater flow means that it tends to behave rather more as if it was in a stack of pipes like this, rather than like water flowing down a stream channel.

either side of the U-bend: obviously there is a far greater weight of water contained in the pipe between the water tank and the bottom of the U-bend than there is between the bottom of the U-bend and the tap. There are therefore two factors at work here to provide the impulse for subsurface water movement:

- The elevation of the water relative to the end-point of the pipe line.
- The pressure exerted at any one point within the pipeline by the overlying water column.

Clearly, both of these factors are manifestations of the force of gravity.

3.1.2 Laminar groundwater flow and the concept of "head"

But is it really reasonable to extend the analogy of a pipe system like that in Figure 3.1b to a real groundwater flow system? After all, the subsurface is not composed of stacks of pipes with U-bends! Rather, as we saw in Chapter 1, aquifers contain myriad pores which are connected to each other in three dimensions. So why should such assemblages of pores behave as if they were pipes? The answer to this question probes deep into the most fundamental differences between the patterns of flow one can observe in a surface stream and those which typify nearly all groundwater flow systems.

As was explained in Section 1.4, the limiting factor on the maximum feasible rate of water movement through rocks is the resistance to flow offered by the pore necks, through which water must pass from one pore to another. In all but the most cavernous of rocks, this resistance is such as to effectively limit the rate of movement of groundwater to very low velocities. Flow rates of less than one millimeter per day are commonplace, while a groundwater moving at more than a few centimeters per day would be regarded as speeding! At such low velocities, water tends to move in a very smooth and orderly manner. There is none of the chaotic mixing and splashing which is so characteristic of turbulent flow in stream channels. Rather, groundwater tends to flow as if it were a stack of separate layers, much like the individual cards in a deck being pushed gently sideways. This quiescent pattern of movement is generally called **laminar flow** (Box 3.1). The tendency for groundwater flow to be laminar has a number of important consequences for various aspects of aquifer behavior and management. Of most immediate importance here is the fact that there tends to be very little mixing between the adjacent laminae in which water moves through an aquifer. It thus happens that gently flowing groundwaters tend to self-organize into lots of parallel flow paths which tend not to cut across one another. Hence, despite the lack of a real physical barrier to the movement of groundwater at angles which are transverse to the principal flow direction, a given packet of groundwater tends to move from its point of recharge to its ultimate point of re-emergence at the Earth's surface just as if it were in a pipe.

Given that laminar flow ensures that the analogy between pipe flow and groundwater flow is a valid description of the manner in which groundwater moves from one place to another, what can we conclude about the driving forces for groundwater? From our discussion of Figure 3.1, it is reasonable to expect that the driving force for groundwater at any one point is likely to reflect both the elevation of that point and the water pressure measured there.

This indeed has been found to be the case. In a landmark paper published in 1940, M.K. Hubbert rigorously analyzed the driving forces of groundwater flow systems and concluded that it is the variations in "the mechanical energy of the fluid per unit mass" which give rise to groundwater flow. Hubbert (1940) proposed the term "potential" as a suitable shorthand for "mechanical energy content per unit mass of water." Although the term "potential" is elegant and physically meaningful, it has largely been replaced in practice with the more earthy term **head**, which is the term we shall use throughout this book. "Head"[i] at any given point on or below the water table equals the sum of:

1. the water pressure measured at that point, and
2. the elevation of the point of measurement relative to a specified datum (usually sea level).

Box 3.1 Measuring groundwater heads in the field.

The measurement of head at a specific point in an aquifer is deceptively simple. All that is required is to measure the depth to water in a borehole, and then convert this value into an elevation (relative to sea level), which is done by subtracting the depth to water from the surveyed elevation of the point at the well head from which depth was measured.

The technology for one-off measurements of head is also pleasingly simple and robust. The most common tool for measuring depth to water is a "dipper," i.e. a graduated tape attached to a probe, in which a simple electrical circuit is completed when it comes into contact with water. With the circuit complete, the user of the dipper is alerted by the sound of a buzzer and/or the illumination of a bulb. The dipper tape is simply lowered down the borehole until a signal is detected and the depth to water recorded to the nearest centimeter.

Where it is desired to monitor changes in head over time, it will often be cost-effective to install an automated digital logger attached to a pressure sensor, which can automatically record changes in water level, either at fixed intervals or whenever the rate of change in head exceeds some user-defined threshold.

It was noted above that these procedures are deceptively simple. This is because it is not always clear precisely which sections of the aquifer are actually sampled by the open-screened section of the borehole (see, for example, Figure 1.5). Boreholes which have long open intervals within their host aquifers will return inherently averaged values of head. Boreholes which have only small open intervals, which are commonly referred to as **piezometers**, will return values of head which represent only that part of the aquifer within which the open interval lies. Great care needs to be taken to keep records during borehole construction if meaningful values of head are to be measured in the completed boreholes at a later date.

Further advice on the installation and use of monitoring boreholes and piezometers can be found in many texts; those by Driscoll (1986), Clark (1988) and Brassington (1998) are particularly recommended.

Fortunately it is rarely necessary to distinguish between these two components of head in practice, and it is possible to make straightforward measurements of total head in the field (Box 3.1).

Groundwater invariably flows from points at which head is high to points where it is low. It is important to realize that it is the gradient of total *head*, rather than of the elevation or water pressure in isolation, which determines the direction and relative velocity of groundwater flow. Figure 3.2 clearly illustrates this point. It is worth running through the four cases shown in a little detail:

- In case 1, the sand in the container has a completely flat water table, and the head below the water table is equal everywhere. Note, however, that this does not mean that water *pressure* is equal everywhere: on the contrary, the further below the water table, the greater is the water pressure, so that the pressure at point B is greater than at point A in this example. However, as the elevation of point B is lower than that of point A, the decrease in the elevation component of head cancels out the increase in pressure so that the total head at A and B is equal.
- In case 2, a clear difference in head (denoted by the symbol Δh) exists across the apparatus from left to right. The total head at point A is greater than at point B, so water flows in that direction. However, because of its greater depth of submergence, the water pressure at B actually exceeds that at point A. We thus have a case here in which water is flowing from a zone of lower water

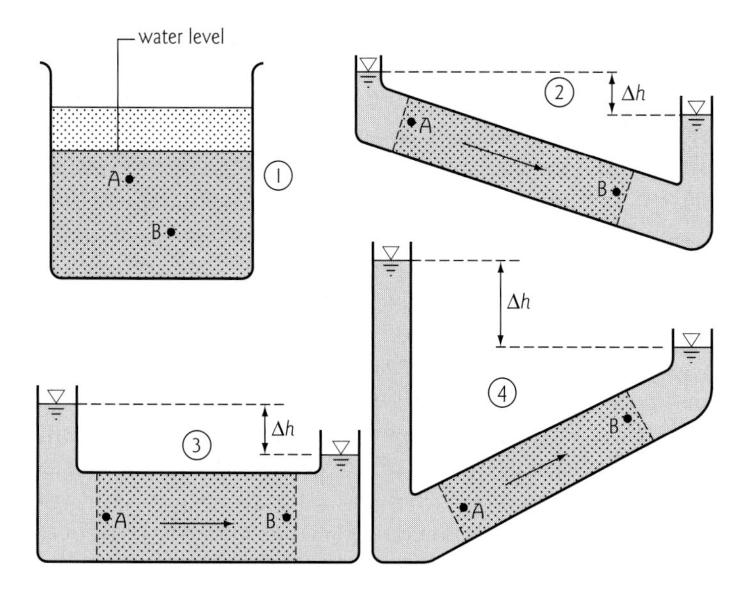

Fig. 3.2 Illustration of the principle that groundwater flows from points of high head to points of low head, and not necessarily from points of high water pressure to points of low water pressure. See text for further explanation. (Adapted after Custodio and Llamas 1996.)

pressure to one of higher water pressure, clearly illustrating that the driving force must be *total head*, not water pressure alone.

- In case 3, the total head at A exceeds that at B, so that the water flows from left to right; the water pressure also declines from left to right.
- In case 4, both the total head and the water pressure are lower at point B than at point A, so that water flows from A to B. However, note that in this case water is flowing from a lower elevation to a higher elevation, making it clear that the driving force is total head, not elevation alone.

Between cases 2 through 4, therefore, we can point to instances in which groundwater flows with and against the gradient in water pressure; however, in no case will water flow counter to the gradient of total head.

In practice, it is common to quantify spatial changes in head relative to the lengths of the flow pathways over which the head decline occurs. The ratio of the change in total head to the distance over which that change occurs is termed the "hydraulic gradient", which is normally represented by the symbol "i" (Figure 3.3). Being a ratio of a length to a distance, hydraulic gradient is dimensionless (i.e. does not have any units).

But why does head decline along a flow path? To answer this question it is worthwhile recalling the principle of conservation of energy, which states that energy can neither be created or destroyed but simply converted from one form into another. Given that head loss amounts to a loss of mechanical energy from the water, where does this energy go? As in most moving physical systems, from a simple pendulum to a space rocket, the loss of mechanical energy in groundwater systems is due to frictional resistance, and the result is the generation of heat. In aquifers, frictional resistance is offered in abundance by the rock matrix, as water is squeezed through tight pore necks and slides across mineral surfaces. Heat is certainly generated in the process, albeit the very high specific heat capacity of water means that increases in ambient temperature are not always easy to demonstrate. Indeed so powerful is the frictional resistance to flow offered by the pore–mineral matrix that it dominates the overall water-bearing behavior of rocks.

3.2 Quantifying flow rates: Darcy's Law and hydraulic conductivity

3.2.1 The unintentional "Father of Hydrogeology": Henry Darcy and his Law

Hydrogeology is a young science, for which the pioneering work dates only to 1856. The term

(a) Unconfined aquifer

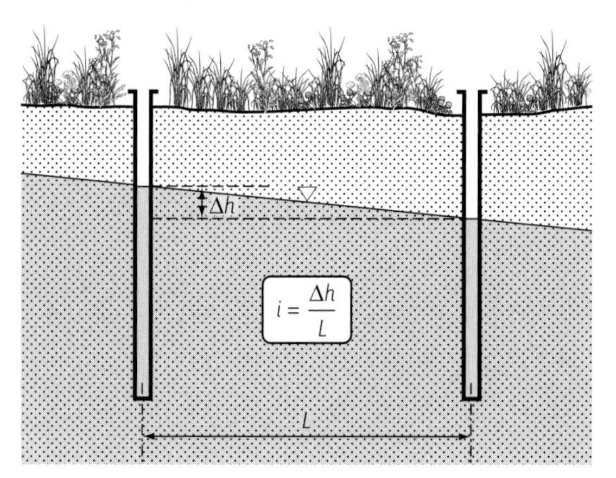

(b) Confined aquifer

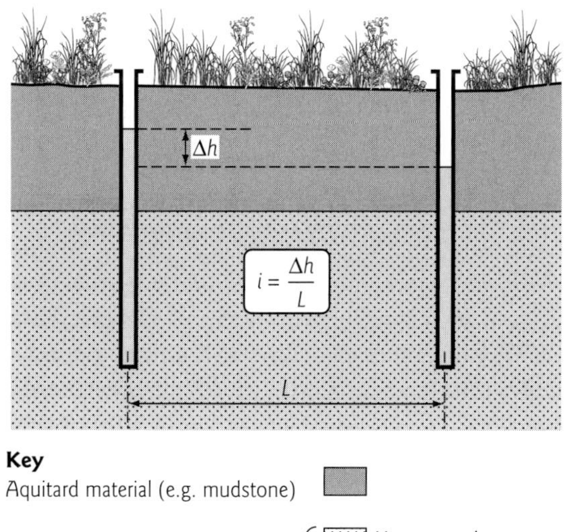

Key
Aquitard material (e.g. mudstone)
Aquifer material (e.g. sandstone) { Unsaturated / Saturated

Fig. 3.3 Cross-section illustrating the concept of hydraulic gradient, i, in an aquifer. (a) In an unconfined aquifer, the hydraulic gradient is simply the slope of the water table. (b) In a confined aquifer, the hydraulic gradient is defined by the slope of the piezometric surface. The measurement of head in practice is described in Box 3.1.

"hydrogeology" (in its presently accepted sense) was coined 17 years later by Joseph Lucas (as described by Mather 2001) and the first complete appraisal of the processes of groundwater flow was published only in 1940 (Hubbert 1940).

Humiliatingly enough for hydrogeologists, the breakthrough which occurred in 1856 did not even arise from any investigation of groundwater flow *per se*. Rather, it emerged from the musings and experiments of a French hydraulic engineer, Henry Darcy (1803–1858), who was one of the leading hydraulic engineers of his age. Besides paving the way for the development of hydrogeology (Simmons 2003), Darcy was instrumental in the derivation of one of a very important formula for describing frictional head losses during flow through pipes (the Darcy–Weisbach Formula, which remains in widespread use today). Without even trying to investigate the movement of natural groundwaters, Henry Darcy managed to derive a fundamental law which has been used ever since for the quantification of flows in aquifers. It came to pass while Darcy was engaged in the development of a reliable public water supply for his native city of Dijon in northeastern France. Having identified a suitable raw water source (a spring located several kilometers from the city) and designed a pipeline to carry the required quantity of water to the city, Darcy turned his attention to the removal of impurities from the water using sand filters. However, the task of designing appropriate sand filters was hindered by the lack of any reliable design criteria. To redress this, Henry Darcy devised and implemented a series of experiments with pilot-scale sand filters, comprising cylinders packed with clean sand. Figure 3.4 illustrates a laboratory sand filter with general characteristics similar to that used by Darcy in his classic experiments (Darcy 1856). (Darcy's original column was oriented vertically, but it is less easy to appreciate the dynamics of flow if a vertical column is sketched.)

By systematically varying flow rates and water levels in the reservoir feeding his experimental filter, by trying out filters of different lengths, and by successively changing the grade of sand used to pack the filters, Darcy soon learned that the flow rate (Q) which could be passed steadily through a sand filter was directly proportional to both the cross-sectional area (A) of the filter and the hydraulic gradient (i) across it. This combined

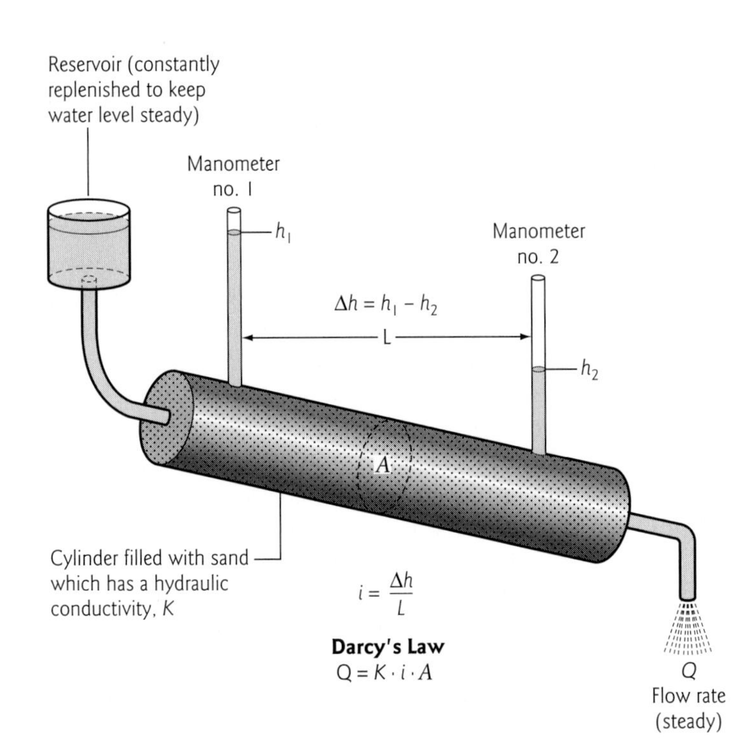

Fig. 3.4 A laboratory apparatus illustrating the terms in Darcy's Law. (The equipment used by Darcy himself was arranged slightly differently, in a manner that makes some of the parameters less obvious to understand.) Similar laboratory apparatus is routinely used to this day for the measurement of hydraulic conductivity values on re-packed soil samples.

proportionality can be written down very simply as follows:

$$Q \propto i \cdot A$$

Given that Darcy had obtained numerous sets of simultaneous measurements of Q, i, and A from his experiments, he was able to move on from this simple expression of proportionality to an "identity" (i.e. a regular-looking formula containing an equals sign). In short he found that he could always calculate Q by multiplying the product of the hydraulic gradient and the cross-sectional area by some factor, a "constant of proportionality." So it is possible to write out the final identity as follows:

$$Q = K \cdot i \cdot A \tag{3.1}$$

where K is the coefficient of proportionality mentioned above, which is nowadays generally referred to as "hydraulic conductivity" (K has units of length per unit time, i.e. the same units as velocity). Equation 3.1 is the most famous formula used in hydrogeological studies, and it is universally known by the anglicized name of **Darcy's Law**. Darcy found that the value of K varied markedly depending on the grain size of the sand in the filter, with high values being found when coarse sand was used, and low values corresponding to the use of fine sand.

3.2.2 Hydraulic conductivity

Hydraulic conductivity is one of the most crucial physical properties studied by hydrogeologists. It is helpful to consider hydraulic conductivity as "the permeability of a given rock with respect to fresh groundwater". This is because the value of K found in Darcy's experiments relates specifically to freshwater, not too dense, saline waters or other fluids. As such, hydraulic conductivity is a function of *both* the properties of the rock *and* the properties of the water (especially its kinematic viscosity and density). As long as we are clearly dealing with freshwater (with, say, less than

10,000 mg/L total dissolved solids), then we can effectively ignore minor variations in water properties and use hydraulic conductivity in exactly the form expressed in Equation 3.1. However, where we want to explicitly consider the movement of saline groundwaters, nonaqueous phase liquids, or gases in the subsurface, we cannot use hydraulic conductivity in our calculations (see Section 3.2.5 below).

Hydraulic conductivity is a remarkable physical property in its own right, for it is known to vary over more than 13 orders of magnitude (from less than 10^{-8} m/day to more than 10^5 m/day), which is a far wider range than is exhibited by most other physical properties (cf. Freeze and Cherry 1979). This high degree of variability has a number of practical consequences. Newcomers to hydrogeology are often perplexed by just how variable hydraulic conductivity measurements in a single aquifer (or even a single interval of a single borehole) prove to be. However, once it is realized that, overall, hydraulic conductivities tend to vary by orders of magnitude rather than by multiples of less than ten, most of the small-scale variations in the parameter begin to fall into perspective. In practice, quantification of hydraulic conductivity to within one order of magnitude of its "true" value is usually sufficiently precise for most purposes of analysis. This in turn implies that it will rarely make sense to express a hydraulic conductivity value to more than two significant figures; much beyond this and we enter the realms of spurious accuracy, which would be like specifying the distance to a particular crater on the moon to the nearest micron!

Besides their utility in allowing quantification of groundwater flow rates using Darcy's Law, hydraulic conductivity values are also very useful as criteria with which to compare the water-bearing capabilities of different rocks. This not only allows us to identify the more and less permeable parts of a given aquifer, but also allows us to compare values obtained for a particular aquifer with those obtained for other aquifers of similar rock type. Armed with such information we can qualitatively assess the advisability of various aquifer management propositions, drawing upon experiences from the most similar aquifers in a similar climatic setting elsewhere. In part to provide a first step towards this sort of analysis, Figure 3.5 summarizes the ranges of values of hydraulic conductivity which are commonly encountered in various rock types. It is immediately obvious that, although there is a general tendency for certain rock types to have higher hydraulic conductivities than others (for the reasons already discussed in Section 1.5.2), the full range of hydraulic conductivities exhibited by any one rock type can often vary over several orders of magnitude. This is why the identification of aquifers and aquitards (see Section 1.5.1) is such a subjective business. For instance, basalt lavas form excellent aquifers in many places, but have also been found to act as effective confining aquitards in others.

The reader will have noticed that the values of hydraulic conductivity cited in Figure 3.5 are expressed in units of meters per day. In practice, many different units of measurement have been used for hydraulic conductivity. The water industry in the USA continues to favor so-called "English Units," in which hydraulic conductivity is expressed in gallons per day per square foot (often abbreviated "gal/day/ft^2" or "gpd/ft^2"),[ii] while some authors in the USA prefer to use "feet per day" (e.g. Fetter 2001). In most of the rest of the world, units derived from the SI standard of meters and seconds are widely applied. In geotechnical engineering circles, the SI standard is strictly followed and hydraulic conductivity is expressed in units of meters per second (m/s). However, as there are very few rocks permeable enough to have hydraulic conductivities greater than 1 m/s, this practice condemns the user to perpetually thinking and calculating in large negative exponents. To make life easier, therefore, hydrogeologists working in the European water industry tend to express hydraulic conductivity in meters per day (m/day), as in Figure 3.5. This convention has the great advantage that nearly all aquifers of note tend to display hydraulic conductivities in the range from 1 to 1000 m/day, leading to easier mental arithmetic and to more

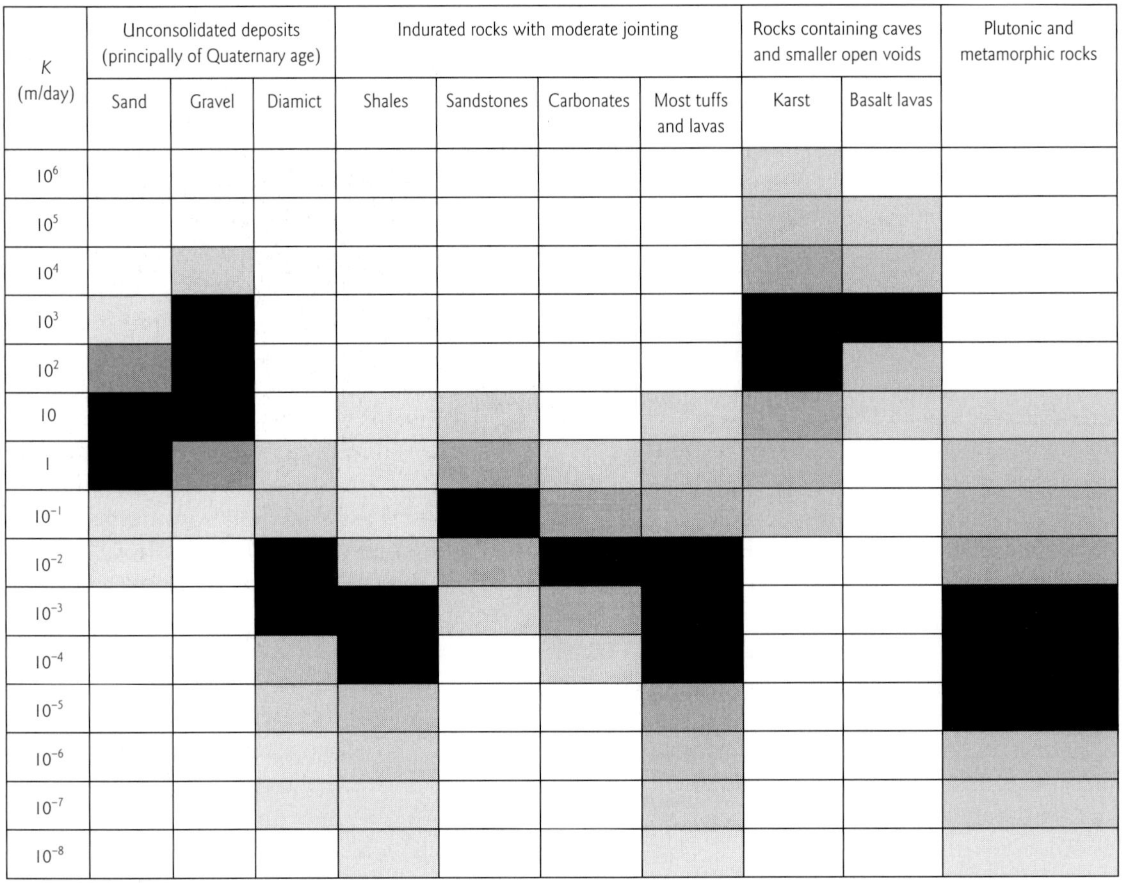

K (m/day)	Unconsolidated deposits (principally of Quaternary age)			Indurated rocks with moderate jointing				Rocks containing caves and smaller open voids		Plutonic and metamorphic rocks
	Sand	Gravel	Diamict	Shales	Sandstones	Carbonates	Most tuffs and lavas	Karst	Basalt lavas	
10^6										
10^5										
10^4										
10^3										
10^2										
10										
1										
10^{-1}										
10^{-2}										
10^{-3}										
10^{-4}										
10^{-5}										
10^{-6}										
10^{-7}										
10^{-8}										

Fig. 3.5 Ranges of hydraulic conductivities encountered in various rock types. The darker the shading, the more common are values in that range for the rocks indicated. "Diamict" refers to deposits in which large clasts are set within a finer-grained matrix. "Carbonates" includes limestones and dolostones. "Karst" refers to classic cave-bearing limestones, and also similarly weathered dolostones and evaporates (especially gypsum). The "Basalt lavas" referred to here are those which contain lava-tubes. The term "plutonic and metamorphic rocks" refers to granites, gneiss, schist, etc. The data upon which this figure was based were collated from datasets for all of North America, summarized by Back et al. (1988).

memorable comparative values for use in general discussions. Throughout this book hydraulic conductivity will be expressed in m/day. (Extensive listings of conversion factors to other units are provided in most advanced groundwater textbooks, such as those of Fetter (2001), Marsily (1986), and Freeze and Cherry (1979).)

Given the natural internal variability of rock masses, it is to be expected that hydraulic conductivity will vary spatially. Two intimidating-looking words are used to describe this variability: **anisotropy** and **heterogeneity**. If the hydraulic conductivity of an aquifer is **anisotropic**, then you would obtain a different value if you measured it in one direction than you would if you measured it in another. In most sedimentary and volcanic aquifers, by far the most important manifestation of anisotropy in K values relates to the stark contrast between "horizontal hydraulic conductivity" (K_h; i.e. K measured parallel to bedding), and "vertical hydraulic conductivity" (K_v; i.e. K measured perpendicular to bedding). Ratios of K_h/K_v very commonly exceed 10, and can often be much greater. While anisotropy can also be manifest within the horizontal plane of a flat-bedded aquifer, such patterns rarely persist over very large differences.

On the other hand, substantial changes in the magnitude of K are very common within and between the layers of a given aquifer. Variations of K from place to place in this manner are referred to as **heterogeneity**. In some cases, K will be found to vary by more than a factor of 10 within a single aquifer unit, and variations by smaller factors can be expected in all aquifers (Lu et al. 2002). Given that K varies over 13 orders of magnitude, however, internal variations by less than a factor of 10 are normally sufficiently modest to allow us to regard many of the less variable aquifers as being relatively **homogeneous**. Thorough analysis of the relative heterogeneity/homogeneity of an aquifer requires the application of sophisticated statistical methods based on probability theory and fractal geometry (e.g. Lu et al. 2002).

Major variations in K are commonly identified where we compare contiguous hydrostratigraphic units. While these are clearly examples of "vertical" heterogeneity, they are rarely discussed as such, for stark contrasts in K are one of the key criteria which are used in hydrostratigraphic studies (see Section 1.5.1) to distinguish different aquifers and aquitards from one another.

In the practical analysis of many real aquifers the following assumptions concerning anisotropy and heterogeneity are commonly made:

- It is typical to assume a strong vertical/horizontal anisotropy in K (which is principally used to justify using one-dimensional vertical flow representations of recharge, as described in Chapter 2).
- In the horizontal plane, most aquifers are assumed to be internally **isotropic**[iii] with respect to hydraulic conductivity and storage parameters, unless there is strong evidence to the contrary.
- Most aquifers are initially assumed to be **homogeneous**[iv] unless there are good grounds for quantifying heterogeneous distributions of K and/or storage parameters.

The **golden rule in practical analysis of groundwater flow** is: start simple and introduce complexities (such as heterogeneity and anisotropy) only where more simple explanations prove wholly inadequate.

3.2.3 Transmissivity

Although hydraulic conductivity is an extremely important characteristic of an aquifer, it will only be effective in contributing to the transmission of large quantities of water where it is developed in an aquifer of substantial saturated thickness.[v] For instance, if a bed of gravel has a very high K value but a saturated thickness of only 0.5 m, then it will transmit far less water than a sandy aquifer of much lower K but far greater saturated thickness. For instance, if the K of the thin gravel bed was 1500 m/day, and that of the sandy aquifer only 100 m/day, then supposing the later to be 25 m thick it would be expected to transmit more than three times as much groundwater[vi] than the far more permeable gravel bed. For many practical purposes therefore, it is precisely the combination of hydraulic conductivity and saturated thickness which we really need to know. This combination of hydraulic conductivity and saturated thickness is captured in the property **transmissivity**, which in strict terms is defined as the integration of the values of (horizontal) hydraulic conductivity between the base and top of the aquifer. In cases where hydraulic conductivity does not vary dramatically over depth, the process of integration amounts to nothing more than multiplying the mean hydraulic conductivity by the saturated thickness. This leads to the following, very common, definition of transmissivity (T):

$$T = K \cdot b \tag{3.2}$$

where b is the saturated thickness. With K in m/day and b in m, T will have units of m^2/day, which are the units used throughout this book. (Alternative units which are often found in the literature include "gal/day/ft", which remains in use in the US water industry, and ft^2/day.)

Given the dependence of T on saturated thickness, it is important to bear in mind that variations in water table elevation can significantly affect transmissivities in unconfined aquifers. In some cases, the effects are so marked that it is best to view T in such aquifers as a time-variant property rather than a constant value at a given point (which is the typical starting assumption). There are two particular hydrogeological settings in which water table fluctuations often lead to important changes in T over time:

- In thin (<10 m) sand and/or gravel aquifers, in which drawdowns induced by pumping can easily remove more than 20% of the saturated thickness, at least locally, within very short periods of time.
- In certain limestone aquifers with fracture permeability, in which fracture apertures are greatest in the shallower parts of the saturated zone (corresponding to the zone of natural water table fluctuation), so that lowering of the water table can dewater some of the most permeable fractures, leading to a sudden drop in the overall transmissivity of the aquifer.

Examples of these situations are presented by Rushton (2003).

Besides accounting for the importance of saturated thickness in making hydraulic conductivity effective, there is another very practical reason for expressing the water-transmitting properties of aquifers in terms of transmissivity. This is that most of the major methods of analysis of test-pumping data, which are our principal means of obtaining *in situ* estimates of the water-transmitting properties of aquifers (Section 3.4), directly yield values of transmissivity rather than hydraulic conductivity.

Finally, it is worth noting that the comments made in the preceding section in relation to the heterogeneity and anisotropy of K apply equally to T (e.g. Lu et al. 2002).

3.2.4 Recognizing and coping with turbulent flow

It is important to note that Darcy's Law (and therefore K and T values derived from its application) is firmly based on the assumption that groundwater flow is laminar (Section 3.1.2). In the vast majority of groundwater systems, this will indeed be the case. However, in certain rock types (see Section 1.5.2) caves or other large conduits may be present, and if hydraulic gradients are steepened in any way (e.g. by pumping, or simply by the approach of groundwater to a seepage face in a cliff-line) flow might well become turbulent. It is possible to approximately calculate the hydraulic conditions under which turbulence sets in (see, for instance, Fetter 2001, pp. 123–124). The key thing to note, however, is that the simple proportional relationship between flow rate (Q) and hydraulic gradient (i) given by Darcy's Law is rendered more sensitive under turbulent flow conditions, in which it is the *square* of the flow rate which is proportional to i (e.g. Forchheimer 1930). This has an extremely important practical consequence, in that the increase in flow rate needed to steepen the hydraulic gradient by a given amount will be far less under turbulent flow conditions than is the case where flow remains laminar. Because of this it is foolhardy to persist in using Darcy's Law alone where it is likely that a significant element of the local groundwater flow regime is turbulent (Dudgeon 1985). To do so can lead to greatly overpredicting:

- The amount of pumping needed to achieve a specified lowering of the water table.
- The area likely to be affected by water table decline as a result of pumping groundwater at a given rate.

Where proposed mine dewatering operations are assessed using predictions which ignore turbulent flow components, such gross overpredictions of impacts can lead to the prohibition of activities which would in fact never have caused widespread impacts (Dudgeon 1985). The economic penalties for failing to take turbulent flow seriously can be extremely serious. Given that turbulent flow effects have been successfully incorporated into several groundwater simulation packages in recent years (see, for instance, Dudgeon 1985;

Adams and Younger 2001a,b; Nuttall et al. 2002), there is no real excuse for continuing to ignore the existence of turbulent flow in certain groundwater flow settings.

3.2.5 Permeability with respect to saline waters and other fluids

As was explained in Section 3.2.2, the standard formulation of Darcy's Law (Equation 3.1) assumes that the fluid in question is fresh groundwater. Where the groundwater is actually saline (total dissolved solids $\gg$ 10,000 mg/L) and/or warm (>40°C), then its density and/or viscosity are likely to be sufficiently different from those of ordinary, fresh, sub-tepid groundwaters that the application of the simple version of Darcy's Law presented above (Equation 3.1) begins to become untenable. Similarly, if we wish to quantify the flow rates of liquids which are considerably more viscous than water (e.g. oils and other dense liquids), or to quantify gas flow rates, then Equation 3.1 cannot be used either. In such cases, it is necessary to re-cast Darcy's Law in a form which explicitly allows for the viscosity and density of the liquids/gases in question. At its simplest, the necessary re-casting can be achieved by substituting a parameter known as the **intrinsic permeability (κ)** in place of K in Equation 3.1. To obtain an appropriate value of κ (in units of m^2), it is necessary to multiply K (in m/s) by the kinematic viscosity of the liquid in question (in units of Pa/s), and then divide the product by the acceleration due to gravity (g, i.e. 9.81 m/s^2). As a final step, it would be usual practice to multiply the m^2 value by 0.987×10^{-15}, in order to obtain a final value in millidarcies (the unit of intrinsic permeability in widest use in the petroleum industry). Unless you become involved in some of the more complex fields of contemporary groundwater modeling,[vii] you are unlikely to need to work with intrinsic permeabilities very often. If you *do* get involved in such work, you will need to refer to specialist, advanced texts; Marsily (1986) is a good starting-point.

Nevertheless, given the abundance of intrinsic permeability data available for various rock types in the archives of the petroleum industry, it is sometimes handy to know how to convert an intrinsic permeability value into an equivalent hydraulic conductivity. If you have a value of intrinsic permeability in millidarcies, simply multiply this by 8×10^{-4} to obtain an equivalent hydraulic conductivity in m/day.

3.2.6 Limitations on Darcy's Law at low permeabilities

Just as Darcy's Law becomes invalid where flows are rapid and turbulent, it also tends to become invalid at very low permeabilities, where groundwater flow is extremely slow. This limitation on Darcy's Law has not been as widely studied as the departures at high velocity. However, it is clear that where pore-necks are narrower than about 10 μm, the movement of groundwater under the impulse of natural hydraulic gradients is often so slow that molecular diffusion will be more rapid. Under such circumstances, Darcy's Law will tend to become less useful as a descriptor of groundwater motion than Fick's Law of Diffusion.

3.3 Groundwater flow patterns

3.3.1 Groundwater flows from recharge areas to discharge zones

The lie of the land determines the physical framework within which groundwater flow systems develop. Yet it is precisely the configurations of hills, valleys, and many smaller-scale landforms which ultimately delineate the boundaries between which groundwater flow systems can develop (e.g. Coates 1990). Indeed it is one of the key "rules of thumb" of practical hydrogeology that **the disposition of the water table tends to mimic (in a subdued manner) the shape of the overlying ground surface**. In other words, the water table tends to be close to surface in valleys occupied by perennial streams, whereas it lies at higher absolute elevations (though usually further below ground level) beneath the adjoining hills.

It is as well to appreciate that the natural landscape is not an utterly passive determinant of groundwater flow patterns. Many landforms are themselves surficial expressions of processes of erosion (and, less commonly, sedimentary deposition) intimately associated with groundwater flow systems (e.g. Baker et al. 1990). The science of landforms is geomorphology. Thorough explorations of the interrelations between geomorphology and hydrogeology have already been published elsewhere (LaFleur 1984; Ford and Williams 1989; Higgins and Coates 1990; Brown 1995); here we will restrict ourselves to briefly appraising the roles of those geomorphological features which form the principal boundaries to most natural groundwater flow systems.

Recharge areas

As Chapter 2 makes clear, the literal starting points for the circulation of almost all groundwaters are those locations in which direct or indirect recharge occur, i.e. **recharge areas**. Classically, recharge areas are thought of as the hilltops in districts underlain by aquifers. While there is no denying that much recharge does occur beneath hills developed on the outcrops of aquifer materials, recharge will as readily occur on hill-flanks or even in valley floors, provided an infiltration pathway to the water table exists. Recalling the definition of indirect recharge (see Sections 2.2.2, 2.2.4, and 2.2.5), it is equally clear that channel beds often also function as recharge areas, especially in semi-arid and arid areas where the water table lies at some depth below the bed. In terms of natural groundwater head distributions, recharge areas are disarmingly simple to identify: they are those zones in an aquifer in which the head can be shown to be decreasing *away* from the ground surface. Conversely, those zones in which head tends to decrease *towards* the ground surface are generally natural **discharge zones**.

Discharge zones

Nature abounds in groundwater discharge zones (cf. Figure 1.2). Table 3.1 summarizes the partitioning of natural groundwater discharge between direct discharges to the sea (submarine groundwater discharge) and discharges which flow into rivers (including both springs and seepages through riverbeds) for six of the world's continents. Submarine groundwater discharge appears to predominate over discharge to rivers in the three continents which lie predominantly in the

Table 3.1 The partitioning of natural groundwater discharge pathways between the sea and rivers for six continents (derived from data presented by Herschy 1998). All figures are percentages of the estimated total groundwater discharge rate for the continent in question.

	Continent						
	Europe	Asia	Africa	North America	South America	Australia	Mean
% of total groundwater discharge entering rivers	62	58	23	56	44	42	47.5
% of total groundwater discharge flowing directly into the sea (submarine groundwater discharge)	38	42	77	44	56	58	52.5

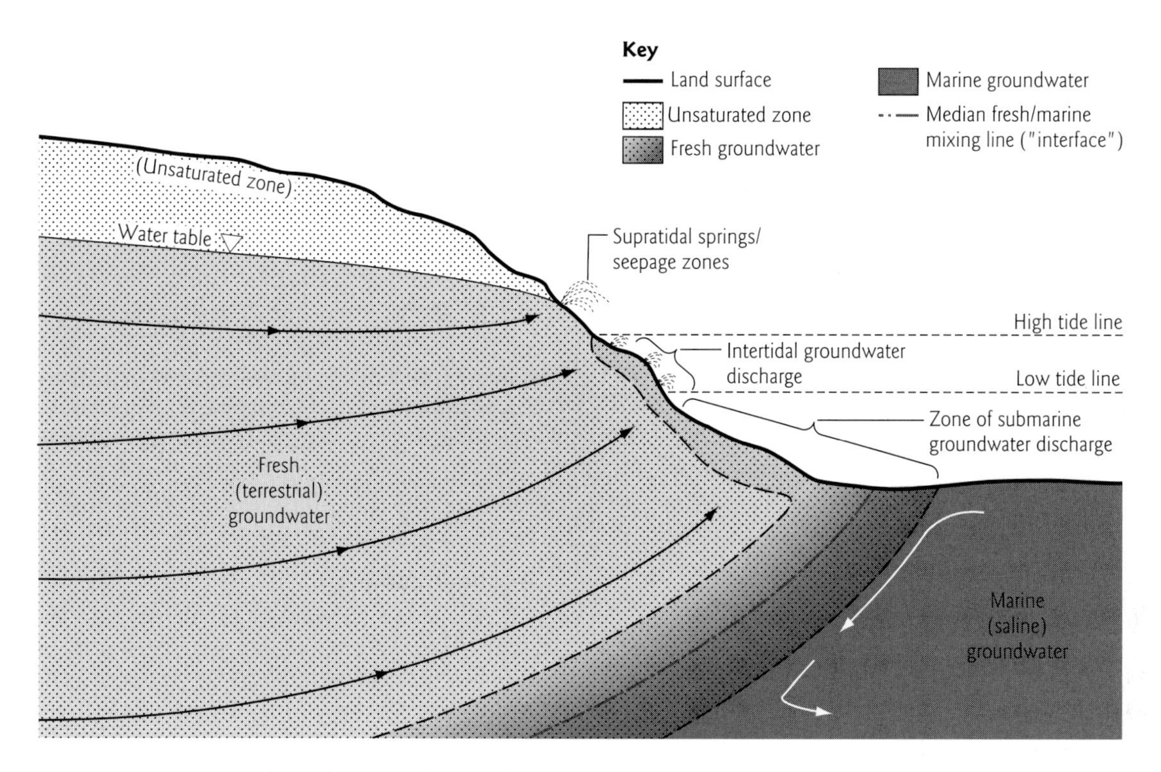

Fig. 3.6 Cross-section through a coastal groundwater discharge area, illustrating the typical supratidal, intratidal and submarine groundwater discharge zones. The extent of each zone depends very much on local geological conditions and tidal amplitude, though it is rare for the zone of submarine groundwater discharge to extend more than a few hundred meters from the low tide line.

southern hemisphere (i.e. Africa, South America, and Australia), whereas the opposite applies in the northern hemisphere (North America, Europe, and Asia). Overall, the global averages for both discharge routes are remarkably similar, indicating an approximately 50 : 50 split.

Coastal and submarine groundwater discharge zones

The dynamics of coastal groundwater discharge zones has been studied for many years, due to the sensitivity of many aquifers to invasion by sea water if wells near the coast are pumped too heavily (see Chapters 7, 8 and 11). In recent years, further interest in groundwater discharge below the low-tide line has arisen in the context of studies of the chemistry of the oceans (e.g. Burnett et al. 2003).

Figure 3.6 shows the typical modes of coastal groundwater discharge. Supratidal and intertidal groundwater discharges are often conspicuous, and no doubt account for a considerable quantity of coastal groundwater discharge. However, as Figure 3.6 shows, much coastal groundwater discharge actually occurs *below* the low-tide mark, in the form of groundwater rising up through the sea bed, usually within a few hundred meters from shore (see, for instance, Younger 1996). In some cases, where the contact between an aquifer and its overlying aquitard is exposed in the sea bed some kilometers from shore, it is possible for fresh groundwater discharge to occur far out to sea (e.g. Kooi and Groen 2001).

There are a number of geomorphological features of many coastlines which serve to greatly enhance their efficiency as "drains" to onshore aquifer systems:

- The embayed nature of many coastlines results in far longer aquifer/sea interfaces than one might suppose from the linear distance between two points along the coast. Hence, although Figure 3.6 appears to represent a section through a straight, linear fresh/saline groundwater interface, in plan view such interfaces will faithfully follow the shape of the coastline with all its twists and turns.
- Where an aquifer forms cliffs along the shoreline, coastal erosion processes can greatly enhance the permeability of the rocks in proximity to the cliff line, thus favoring convergence of groundwater on the coastline in preference to discharge to inland watercourses. Typical processes contributing to a localized increase in permeability include the formation of wave-cut caves, and the oversteepening of cliffs by basal erosion, leading to extensional deformation of the rock behind the cliff-line resulting in increased incidence of open fractures.
- The accumulation of coarse, clean sand deposits above rockhead in many coastal settings also provides a high-permeability medium at the coastal end of the aquifer, which further promotes convergence of groundwater on the coast.

Inland groundwater discharge zones: natural features

Springs, wetlands, ponds, lakes, streams, and rivers can all function as significant groundwater discharge zones. The nature and dynamics of these natural inland groundwater discharge features are discussed in detail in Chapter 5. In any one landscape it is common for one or more of each class of features to occur in close proximity to (and in hydrological continuity with) another. For instance, springs commonly give rise to the headwaters of streams, and also often flank the banks of streams and rivers over great distances. Other springs can often be seen to feed wetlands, ponds, and lakes, whence surface outflows usually carry water onwards to streams and rivers. Although spring discharges are conspicuous, in many catchments they actually account for a relatively small proportion of total inland groundwater discharge, with the bulk being accounted for by direct groundwater upflow through the bed sediments of surface water bodies (wetlands, ponds, lakes, streams, rivers).

The relative positions of such groundwater discharge features within the landscape exert a profound influence on patterns of groundwater flow (Box 3.2). Where a river is functioning as a groundwater discharge zone, for instance, groundwater flowing within the adjoining aquifer will converge towards both banks. In some cases, the patterns of surface-water features incised into aquifers will have been determined by external geological forces. For instance, the usual explanation for meanders incised deeply into bedrock is that the channel pattern must have been established by downcutting through previously overlying soft sediments, since removed. Similarly, erosion along lines of weakness such as faults can result in the incision of linear river reaches into aquifers.

In other cases, patterns of groundwater outflow can themselves strongly influence the patterns assumed by river systems (e.g. LaFleur 1984; Higgins and Coates 1990). Where groundwater outflow has greatly contributed to valley development, characteristic landforms can often be recognized, including:

- So-called "theatre-headed valleys," in which the uppermost reaches of valleys end in curved cliff-lines, formed by headward-retreat of the valleys in response to undercutting of these cliffs by outflowing groundwaters (e.g. Baker et al. 1990).
- "Light-bulb shaped valleys" (in plan view) containing low-density drainage systems, which are typically arranged in dendritic patterns of marked asymmetry (with few channels originating down-dip). In such valleys, the steep-sided near-channel part of the catchment (normally a distinct canyon) accounts for nearly all of the catchment area. (All of these features contrast markedly with those of valleys which have been principally carved by surface runoff.)

The miniature landforms which form below the high-water mark on beaches during periods of low tide (Box 3.2) often mimic characteristic "groundwater discharge landforms" recognized at much larger scales, not only on Earth (e.g. Higgins 1984; Baker et al. 1990) but also on Mars (e.g. Higgins 1984; Coleman 2003). Further

Box 3.2 Life's a beach and then you dye: observing groundwater discharge geomorphology and flow processes at the seaside.

An old friend of mine, who like me studied geology for his first degree, was once informing me of the sad news of the breakdown of his marriage. After expressing a suitable range of condolences and providing him with an appropriate helping of beer, I asked him what had gone wrong. "I blame geology" he said. "But how can geology be to blame?" I asked him. "Well," he told me "my wife loved beach holidays, and insisted we went to the coast at every opportunity. But she just couldn't get used to the embarrassment of me being the only person on the beach who always set up his deck-chair facing the cliffs, rather than the sea!"

So you have been warned. However, it *is* possible for the hydrogeologist to find informative entertainment on the seashore without running the risk of marital discord. What you have to do is find a sandy beach in an area with a decent tidal range (which unfortunately counts out most Mediterranean beaches). Next, arrange to arrive on the beach during a period of falling tide, and propose strolls along the foreshore at regular intervals. As you make your way across the wet sand, look carefully, and you will soon spot small dendritic drainage systems being etched into the sand by flowing water, which is washing the sand away grain by grain. Figure 3.7a is a typical example from my local beach. The first thing to note about these drainage networks is that they are entirely fed by the discharge of groundwater left behind in the sand by the retreating tide: there are no surface sources of water entering their catchments, to support the observed flows and concomitant erosion. Here we have prima facie examples of the development of (temporary) landforms by groundwater outflow. If you observe the particular valley patterns developing in the sand, you will soon be able to recognize landforms which resemble the "theatre-headed valleys" and "light bulb-shaped valleys" which typify landforms carved by discharging groundwater at much larger scales, both on Earth and on Mars (Higgins 1984; Coleman 2003).

A further instructive activity is to imagine the directions of flow of the discharging groundwaters *before* they reach the dendritic channels, i.e. while they are still within the sand. If you assume that groundwater enters the channels throughout their lengths, and does so predominantly by flowing in at right angles to the channel margins, you can easily add arrows indicating inferred groundwater flow directions to an observed pattern of channels (Figure 3.7b). The resulting patterns of deduced groundwater flow serve to illustrate the relative complexity of subsurface flow pathways in the immediate vicinity of channels which incise into an aquifer, a point that is examined in greater detail, in relation to full-scale aquifer systems, in Chapter 5.

If your partner has not already stormed off in irritation, and you have a toy spade to hand, you can take the analysis further by excavating "observation wells" in the sand around a naturally developed channel system to check the veracity of your deductions. Only the most determined beach hydrogeologist is likely to have the patience to accurately measure head differences between one "observation well" and another. However, if you have some cola or other intensely colored (harmless) liquid at your disposal, you could undertake a rudimentary "dye tracer" test by tipping some of that into one of your "observation wells" and looking carefully to see into which of the drainage channels it emerges. For many hydrogeologists this might well be the most fun they can have with their swimsuits on.

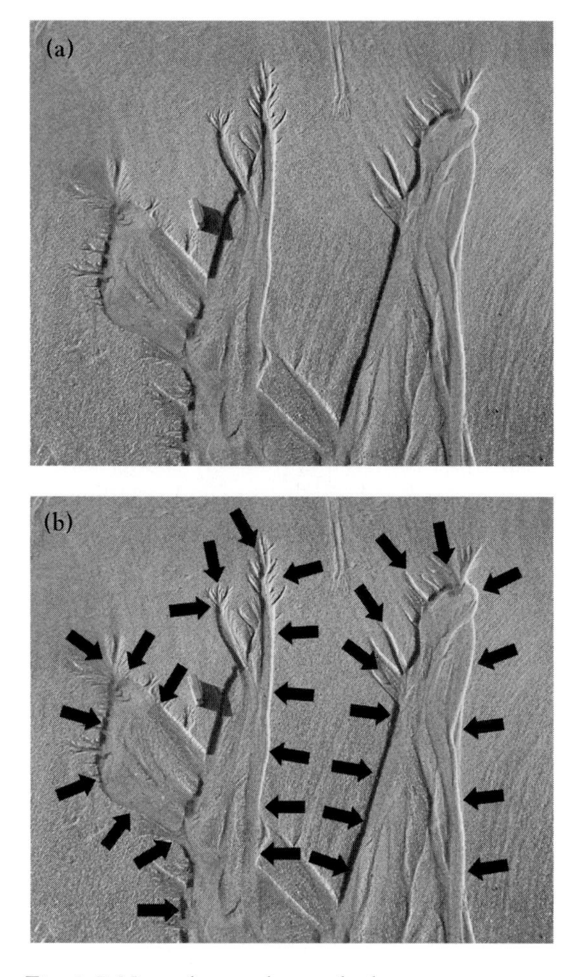

Fig. 3.7 Natural groundwater discharge microlandforms forming during low tide on South Shields Beach, northeastern England, on November 1, 2005. A folded Swiss army knife is shown for scale. In image (b), the raw image shown in (a) has been annotated with arrows indicating the inferred directions of groundwater flow towards the evolving erosional channels. For further discussion, see Box 3.2.

discussion of the modes of groundwater discharge into surface water bodies is reserved for Chapter 5.

Inland groundwater discharge zones: water wells

So far we have considered only *natural* discharge zones. However, in many aquifers natural groundwater discharge is already exceeded by artificial groundwater discharge to wells. Water wells come in a great range of shapes and sizes. For the most part they are approximately vertical excavations from the ground surface down to some position below the water table (or the base of the overlying aquitard where the shallowest aquifer at a given site is confined). The construction of wells has a very ancient pedigree. For instance, there are hand-dug wells still in use today in parts of Iran which were constructed more than 3000 years ago. Hand-digging of wells was the principal technique of well construction worldwide until the nineteenth century, when various types of drilling rig finally came into widespread use.

An interesting collation of early drilling techniques has been presented by Eberle and Persons (1978). The first drilling rigs operated by means of percussion, with a blade of some sort being repeatedly dropped into the hole to break up the strata into looser fragments which can then be removed using a long, narrow bucket known as a "bailer." Alternating periods of cutting and bailing of fragments from the hole eventually results in a deep hole suitable for completion as a well. While percussive drilling can be used to drill in consolidated materials, it is very slow in such applications, and is thus largely restricted nowadays to sinking wells in soft sediments. Further developments in drilling technology were stimulated by the phenomenal growth of the oil industry in the late nineteenth century. To meet the needs of the burgeoning oil industry in Oklahoma and Texas, where it was necessary to drill to depths of many hundreds of meters through consolidated sedimentary rocks, rotary drilling methods were developed. In this form of drilling, a specially shaped drill-bit is rotated in the bottom of the hole, with removal of fragments (known as "cuttings") by means of flushing the hole with a fluid. Because of the high pressures encountered in deep oil and gas reservoirs, the oil industry generally uses rather dense barite-based muds as flushing fluids during rotary drilling. However, in water-well drilling these muds have the substantial disadvantage of impairing the permeability of the aquifers penetrated by the well. A variety of other flushing fluids are therefore

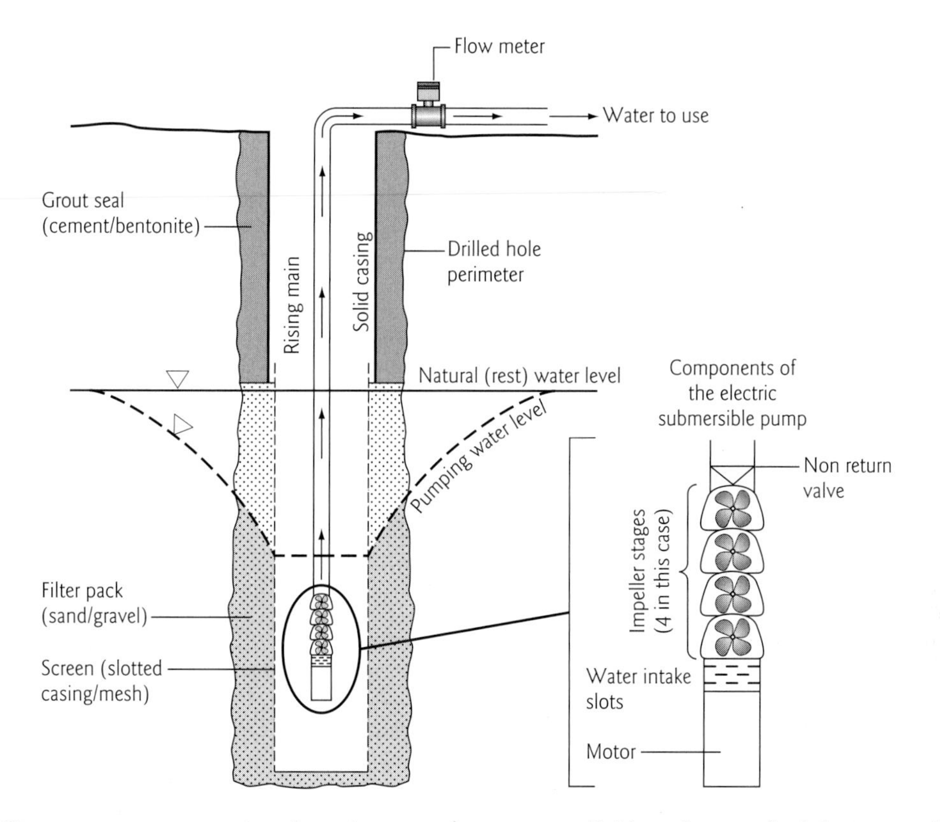

Fig. 3.8 Diagrammatic cross-section through a typical pumping well. Note the use of solid casing and a grout seal to prevent ingress of soil above the water table, and the use of a screen and filter pack to allow water to enter the well below the natural rest water level. If the well were to be equipped for long-term use, the rising main (the pipe taking the water from the pump to the surface) would be made to pass through a sanitary seal at ground level, to prevent accidental pollution of the well by inflow of polluted water from above.

used in water-well drilling, including water itself, aqueous suspensions of biodegradable polymers (which help prevent hole collapse, but are later easily removed from the well bore without permanently impairing aquifer permeability), and compressed air. The latter is now the most commonly-used flush in rotary water-well drilling. Comprehensive details of water-well drilling techniques are presented by numerous authors including Driscoll (1986) and Clark (1988).

Irrespective of the means by which a hole is sunk below the water table, if an adequate supply of water is to be obtained over a reasonable period of time it is necessary to carefully "complete" the well. **Well completion** involves installing pipework in the bored hole to achieve the following goals:

- To prevent the collapse of the hole.
- To prevent the excessive ingress of sediment particles.
- To maximize the ingress of water.

In general, these goals are achieved by installing a well lining with solid walls (known as **casing**) above the water table, and a lining with permeable, slotted walls (known as **screen**) below the water line (Figure 3.8). How casing and screen are constructed in practice can vary dramatically depending on the method of well construction used and the economic resources of the well owner. In wealthy countries, water supply wells are typically rotary-drilled and completed using standardized, commercially available casing and screen components (Driscoll 1986). These take the form of thread-ended lengths of precision-engineered pipework, either made of steel or (for less demanding applications) of some plastic-like synthetic material, such as high-density polyethylene (HDPE). By contrast, hand-dug water wells in poor countries are often completed by manual emplacement of relatively inexpensive materials (Watt and Wood 1979), such as brick-work (e.g. Wagner and Lanoix 1959), concrete rings cast on site, or a range of more "exotic" materials such as hand-woven bamboo screens (e.g. Pickford 1991, pp. 17–20).

Whatever materials are used for the casing and screen, the principles of well completion remain the same. The annulus between the casing and the borehole wall is generally tightly "grouted" (i.e. completely filled with lime-based cement), in order to prevent polluted drainage from migrating down to the aquifer via the well itself (Figure 3.8). To achieve satisfactory sealing of the casing against the wall rock, it is often worthwhile constructing the well in two stages. In the first stage, the well is drilled down to the water table (but no deeper), the casing is installed and grouted. After the grout has set (which typically takes 12 hours or so), the second stage commences with renewed drilling of the borehole (at a rather smaller diameter than before the casing was emplaced), down through the saturated zone to the final target depth. The screen is then lowered into the borehole, and packed around with gravel, which forms a permeable "filter pack" between the screen and the rock, permitting water to enter the hole, but preventing the ingress of excessive amounts of sediment.

Having thus "completed" the well, the remaining task before it can be used for routine pumping is termed **well development**. This involves agitating the water within the well so that any fine sediments within the screen and filter pack are dislodged and are thus removable by pumping. The necessary agitation can be achieved by rapid lowering and raising of plungers within the well, and/or by blasting the inside of the well with compressed air. A sustained current of compressed air within the hole can be used to "pump" the muddy water out of the well (a process known as **air-lifting**). These activities are repeated until such time that the water coming from the well runs clear, with no suspended sediment.

Having finalized well development, permanent pumping equipment can be lowered into the well (Figure 3.8), and water production can begin. Again, the choice of pump depends on the mode of well construction and the economic context. In typical water utility operations in developed areas, the most useful pump is likely to be an electric submersible pump (Figure 3.8), so-called because the electric motor which drives the impellers within the pump is itself submerged in the well, attached to the bottom of the pump unit. By contrast, in large-diameter hand-dug wells serving poor communities, hand-pumps are more likely to be used. (Indeed, expensive electric submersible pumps do not perform efficiently in hand-dug wells unless they are first fitted with an external "shroud," which ensures that water reaches the pump intakes only after flowing past the motor unit at high velocity, which is necessary for cooling of the motor.) Again, further details on appropriate pumping equipment are available in specialist books, with Driscoll (1986) describing pumps used in developed areas and Arlosoroff et al. (1987) and Reynolds (1991) describing hand-pumps used in developing countries.

While the vast majority of water wells in use around the world tend to conform to the vertical configuration just outlined, it is worth noting that a range of horizontal well designs do exist, and are used to good effect in certain circumstances. First in the lineage of horizontal wells are the ancient **Qanat** systems of the Persian and Arabic worlds (Wagner and Lanoix 1959; Todd 1980). These are long tunnels which are driven from piedmont areas into mountain-foot alluvial aquifers until they encounter the water table. As the Qanat presents a far more permeable pathway to flow than the aquifer itself, a certain amount of flow will decant into the tunnel and flow to the portal in the piedmont area, where it is typically used for irrigation. Water galleries which resemble Qanats in principle, but which have been constructed using mining methods more appropriate to hard-rock settings, are found throughout Spain (much of which was, of course, under Arabic rule for many centuries) and in the Canary Islands. In the latter location, more than 1000 km of galleries have been driven into the volcanic rocks of the major islands to intercept groundwater resources needed for public supply.

In some cases, horizontal pipes have been driven outwards from the base of conventional vertical wells to improve their yields. Notable examples of this approach include:

- "Ranney Collector Wells," numerous examples of which may be found in the alluvial aquifers of the central USA. In the rather thin bodies of saturated, unconsolidated sand and gravel exploited by these wells, the presence of lateral pipes can considerably increase the efficiency of water collection. A variation on the theme of Ranney Collector wells are the various types of "infiltration galleries" developed alongside rivers in developing countries (e.g. Wagner and Lanoix 1959).
- Limestone-well adit systems, large diameter (2 m or more) tunnels driven distances of several hundreds of meters from the sumps of shafts sunk into aquifer limestones. Widespread in the Chalk aquifer of northwestern Europe (e.g. Zhang and Lerner 2000), as well as in other limestone aquifers of the Old World, these adit systems typically date from the second half of the nineteenth century, when the contemporary cost of labor and the recent development of advanced mining techniques temporarily conspired to make their development economically viable.

We will repeatedly return to the topic of water wells throughout this book. The influence of pumping wells on groundwater flow patterns is described in Section 3.3.3. The manner in which measurements of water levels in pumping wells can be used to quantify transmissivity is outlined in Section 3.4. In Chapter 7 we will examine the use of water wells for public supply purposes, while the vulnerability of such wells to pollution is considered in Chapter 9. Finally, some of the wider implications of water well use are considered in the context of aquifer management strategies in Chapter 11.

3.3.2 Groundwater flow fields

Patterns of groundwater flow as a "field" phenomenon

We have already seen that the "driver" of groundwater flow is head (Section 3.1.2). As the sum of the elevation of the point of measurement and the water pressure in the rock pores at that point, head is clearly a physical entity which varies considerably from place to place. Although the simple version of Darcy's Law (Equation 3.1) which was introduced in Section 3.2.1 considers head variations only in one dimension, in real aquifers head varies in three dimensions. In this respect, head resembles many other physical entities found in other branches of physics, such as temperature and electromagnetic potentials. Most people I know have vivid memories of that early science class which every high school seems to run, in which the students get to play with bar magnets and iron filings. Do you recall what happens when you sprinkle iron filings around a rectangular bar magnet? As if by magic (but actually in response to magnetic potentials) the filings line up to form patterns of concentric rings around the bar, joining the positive pole of the magnet to the negative pole. Keep that image in mind and you have a good analog for the way in which groundwater flow occurs in two and three dimensions – as a field of parallel, curved trajectories transferring energy from one "pole" of the groundwater system (i.e. the recharge area) to the other pole (i.e. the discharge zone). Of course if an aquifer has more than one recharge area and/or more than one discharge zone, the details of the flow field can end up being rather more complex than the alignment of iron filings around a single bar magnet, but you get the idea.

So groundwater flow is one of those physical phenomena, like magnetism, electricity, and heat, which occurs in *fields*. Apart from helping us to visualize the manner in which head variations tend to occur within aquifers, the recognition that we can talk of groundwater "flow fields" proved to be crucial to the development of mathematical analysis tools for aquifers during the twentieth century. For instance, it was by analogy to long-established heat flow analysis methods that Charles V Theis (1935) first realized how analysis of water level behavior in the vicinity of pumping wells can be used to quantify transmissivity. In the late 1960s and early 1970s, the first tentative steps in the development of numerical modeling techniques for groundwater systems (e.g. Prickett and Lonnquist 1971) were based firmly on analogies to the pre-existing heat flow models described by Carslaw and Jaeger (1959). Indeed, during the 1970s numerous aquifer models were successfully developed by constructing large layouts of electrical capacitors and resistors and subjecting these to currents of various magnitudes (e.g. Rushton and Redshaw 1979). In using such "electrical analog models," groundwater flow was quantified by means of a direct comparison between Darcy's Law and Ohm's Law (which equates electrical current to the ratio of voltage to resistance). Although such electrical analog models have been almost wholly superseded by digital models now, it is important not to lose sight of the key fact that groundwater flows in a manner entirely analogous to heat transfer and electromagnetic conduction.

At root, the reason why groundwater movement occurs in flow fields comes back to the laminar nature of most subsurface flow. With no turbulent eddy currents and negligible momentum, groundwater tends to move in a very smooth manner, gently maintaining continuity between the recharge area in which it first seeped below the water table and the discharge zone, through which it will eventually leave the aquifer.

Contours of groundwater head, flow lines, flow nets, and flow tubes

We have already seen how groundwater head is measured in practice (Box 3.1). When we have collected a large number of simultaneous[viii] head measurements from different points in a given groundwater flow system, we are in a position to begin to delineate groundwater flow patterns. The first step is to contour the head values. This can either be done in plan view, thus producing a map of groundwater head variations (e.g. Figure 3.9a), or on a vertical cross-section through the aquifer, so that a profile of heads is produced (e.g. Figure 3.9b). Once a contour plot of head values has been produced, it is possible to proceed directly to determine groundwater flow directions within the field of view by adding **flow lines**. Flow lines are essentially long arrows representing the pathway which groundwater would be likely to follow through the part of the aquifer represented on the contour plot. The first flow line is added to a head contour map as follows:[ix]

- Pick any starting point on the highest value contour.
- Draw a line leaving this contour line at right angles and heading towards the next-highest value contour.
- Make sure the flow line crosses the next contour at right angles.
- Repeat the last two steps until the line reaches the lowest valve contour.

All subsequent flow lines can be added in the same manner; this is precisely how the flow lines shown on Figure 3.9a were constructed. The lateral spacing between adjoining flow lines is entirely a matter of user preference. Where a very detailed knowledge of flow patterns is needed, close spacing of flow lines is warranted. Where the aim of the exercise is to obtain rather general estimates of groundwater flow rates, more widely spaced flow lines may well suffice.

An assemblage of head contours and flow lines together constitute a **flow net**.[x] Analysis of flow nets is one of the principal means of quantifying groundwater flow. While we shall explore in detail how this is done in Chapter 10, a simplified introduction to flow net analysis is given here to pave the way for illustrative use of flow nets in Chapters 4 through 9. The area enclosed between two adjacent flow lines is termed a **flow tube**. All of the water within a flow tube will discharge through the plane corresponding to its lowest value contour line. It is possible to manipulate Darcy's Law to allow calculation of the rate of groundwater discharge from the end of a flow tube. The logic is as follows. First, recall

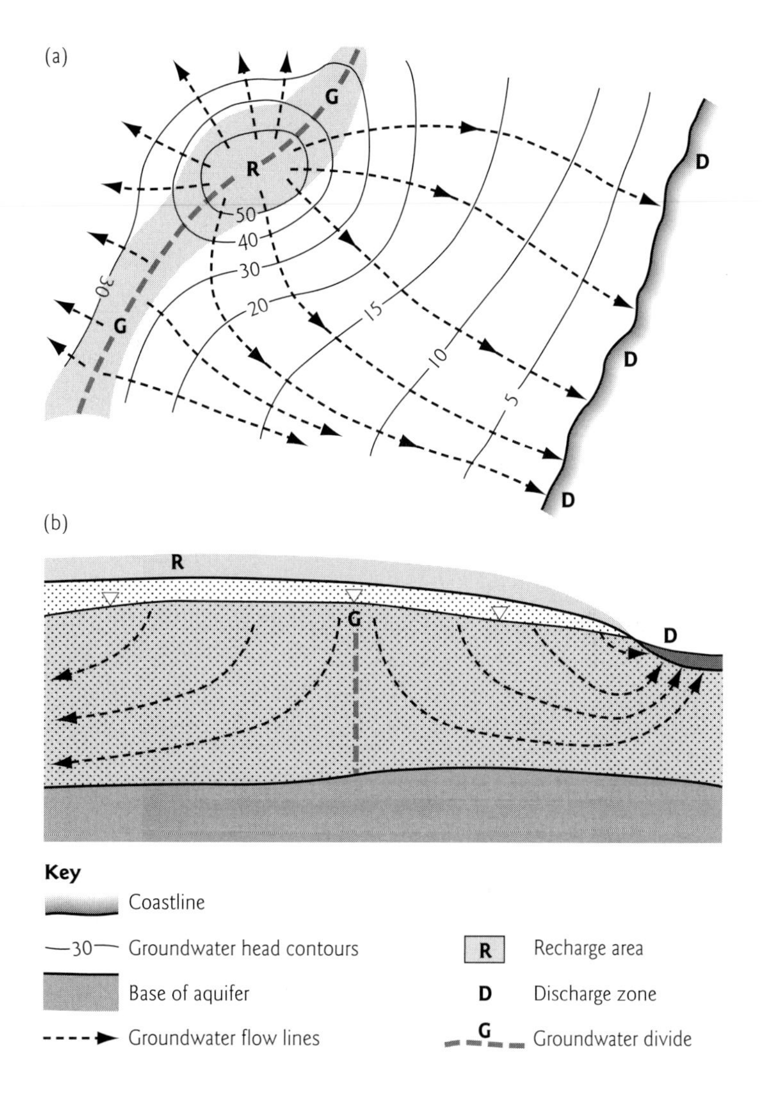

Fig. 3.9 Flow nets representing patterns of groundwater movement between recharge areas and discharge zones. (a) Flow net in plan view, with a groundwater divide separating westward and eastward flow systems. The eastward flow system terminates in a linear discharge zone corresponding to the coastline (cf. Figure 3.5). (b) Flow net in cross-section (profile) view, revealing that although flow is near-horizontal in much of the flow system, significant vertical components of flow do occur in the vicinity of recharge areas and discharge zones. Note also the difference in extent of identified recharge areas between (a) and (b). In case (b) the aquifer is everywhere open to the atmosphere and the recharge zone extends across the entire outcrop. The restriction of the recharge area to the uppermost part of the flow system in (a) likely reflects the presence of less permeable soil cover overlying the aquifer elsewhere.

the basic form of Darcy's Law ($Q = K \cdot I \cdot A$; Equation 3.1). Now, the use of K in Darcy's Law is often inconvenient when we are dealing with real aquifers, because most large-scale methods of aquifer characterization yield values of transmissivity (T) rather than K (see Section 3.4). Bearing in mind the simplification that $T = K \cdot b$, (Equation 3.2), and that b denotes the saturated thickness of the aquifer, then for a flow tube of width w we can express the area A in Darcy's Law as: $A = w \cdot b$. With $Q = K \cdot i \cdot w \cdot b$, and $T = K \cdot b$, then we can simplify Darcy's Law to read:

$$Q = T \cdot i \cdot w \tag{3.3}$$

Equipped with Equation 3.3 and a map showing groundwater contours and flow lines, we are in a position to begin quantifying groundwater flow. Indeed, because flow tubes split the aquifer up into subsections, we are equipped to detect spatial variations in the rate of groundwater flow at different points within the aquifer. We shall further explore this, and other aspects of flow net analysis, in Chapter 10.

Before closing this introduction to groundwater flow patterns, consider again the flow line patterns shown in the head profile in Figure 3.9b. It is evident that, except in the recharge area and the discharge zone at the right, the flow lines tend to become more or less horizontal. Construction of head profiles similar to Figure 3.9b for many aquifers all over the world has revealed that horizontal flow predominates in regional-scale aquifers. Except in close proximity to recharge areas and/or discharge zones, it is therefore often reasonable to assume that most flow is indeed horizontal. This assumption also turns out to be highly convenient when it comes to quantifying groundwater flow using a wide range of analytical techniques (see, for instance, Section 3.4 and Chapter 10). The importance of this assumption was recognized approximately simultaneously, early in the twentieth century, by two well-known European hydraulic engineers. In their honor, the assumption that regional groundwater flow is predominantly horizontal in orientation has long been named the **Dupuit–Forchheimer Assumption**.

3.3.3 The effects of pumping wells on groundwater flow patterns

When a well is pumped, the water level within the well is obviously going to drop. This in turn means that we have just created a new, local minimum in the elevation of the water table. Given that groundwater flows towards points of minimum head, flow can be expected to converge on a pumping well. In order to understand the wider effects of pumping wells on groundwater flow patterns, it is worth thinking through what happens when a pump begins to run in a well which was previously at rest. When a well has not been pumped for a long time, the water level in the well will correspond almost exactly to the head in the aquifer immediately outside of the well. As soon as we switch a pump on, we will begin to remove water from the well. In the very early stages of pumping, it's as if the well doesn't realize it's sitting in an aquifer, and for the time being we might as well think of the well as a tall bucket with vertical sides. At this stage, provided the pump is running at a constant rate, the water level will also drop at a constant rate (and rapidly at that). If we were to pump at 100 L/s, for instance, in a bucket of 200 mm diameter, then for every hundred liters of water removed from the well, the water level would drop by 0.159 m; in other words the rate of water level drop will be 0.159 m/s. When, during the first few minutes of pumping a well, we observe the water level dropping like this, in direct proportion to the elapsed time since the start of pumping, we can be confident that we are simply removing water from storage within the well. This type of behavior is known as the **well-bore storage effect**.

However, the further the water level in the well drops, the greater the difference becomes between the head in the aquifer just outside the well and the water level inside. With a hydraulic gradient towards the well now established, water will begin to flow into the well. In accordance with Darcy's Law, the greater the head difference, the greater the rate of inflow will be. This has a very important consequence: as pumping continues beyond the initial stages,[xi] the increasing inflow of groundwater results in a decreasing rate of water level decline within the well. Figure 3.10a illustrates how these changing circumstances appear if we draw a graph of water level versus time for a pumping well. Eventually, the rate of inflow from the aquifer will exactly match the rate of pumping, and the water level within the pumping well will stabilize. In some cases this might happen quite quickly – for instance, where a very modest pumping rate is applied to a well penetrating a prolific aquifer. In the majority of cases, it may take months or even years of pumping before the water level in the pumping well really stabilizes, if it ever does. However, given that the rate of water level decline becomes so slow after pumping has been sustained for a long period of time, the water level will often seem to be stable even though monitoring over a period of months might well reveal it to be still creeping downwards.

In discussing water level changes caused by pumping, the term **drawdown** is normally used to describe the observed decline in water level. Drawdown is the difference between the initial water level in a given well and the observed water level at any specific time during a period of pumping. Drawdown can be measured in the

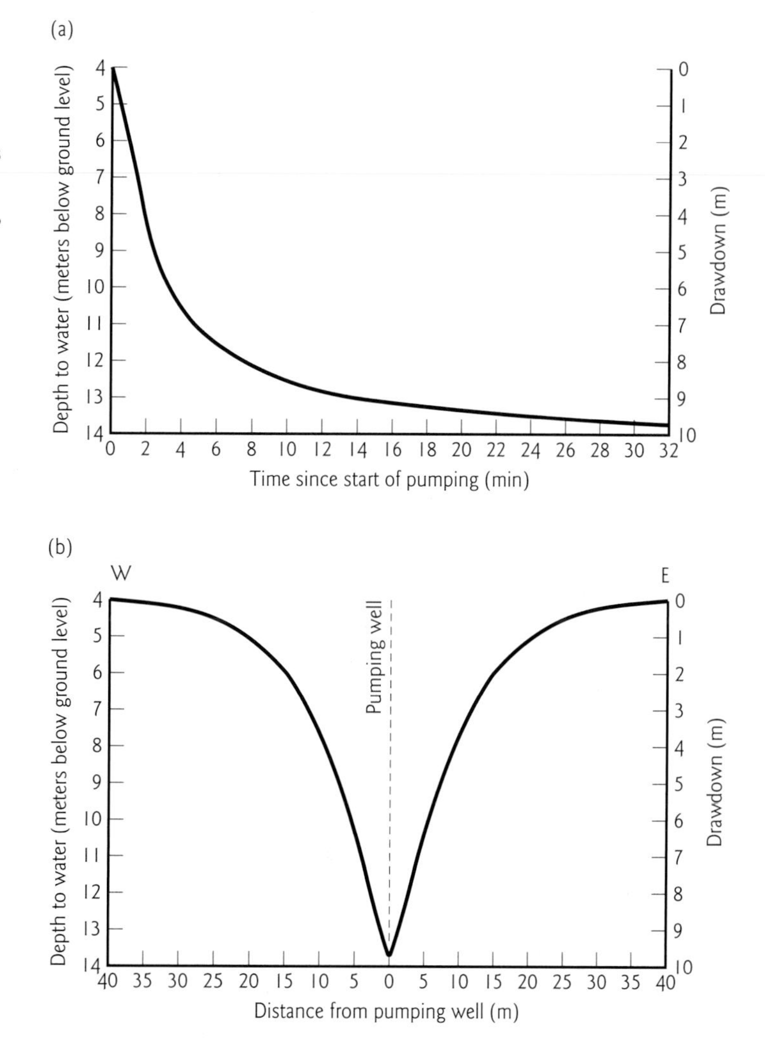

Fig. 3.10 Typical response of aquifers to pumping. (a) A plot of drawdown versus time ("time–drawdown curve"), measured in a pumping well. The early part of the drawdown behavior (first 2 minutes) shows a rather linear response, attributable to depletion of well-bore storage. Subsequently, a curved time–drawdown response is seen, with drawdown developing far more slowly the longer pumping persists. (b) A plot of drawdown versus distance east and west of a pumping well. The shape of this curve is determined by interpolation from point-measurements in a number of observation wells (not shown) at various distances from the pumping well. Note that the distance–drawdown relationship is virtually symmetrical around the well. As we would presumably find the same for a line of observation wells oriented north–south (i.e. at 90 degrees to the trend of the line represented here), it is easy to imagine that the 3D shape of the drawdown void is conical, hence the term "cone of depression."

pumping well itself and/or in any wells penetrating the same aquifer, in which water levels show a response to the abstraction of water by the pumping well.

Having considered how pumping affects water levels over time in any one well, let us now consider the effects of pumping on the surrounding aquifer. If the aquifer is very permeable and the rate of pumping rather modest, it may be that there will be no detectable drawdown in nearby wells. In most cases, however, observation wells will reveal that (at any one time during pumping) drawdown is greatest close to the pumping well and becomes progressively smaller the further away from the well it is measured. Figure 3.10b illustrates this point. Looking again at Figure 3.10 as a whole, it is immediately apparent that the graph of drawdown versus time in any one well (Figure 3.10a) closely resembles the profile of drawdown versus distance at any one time (Figure 3.10b): in both cases, the graph is steepest near the origin. We have already seen why the time–drawdown curves adopts this form; but why is the distance–drawdown response not a simple straight line? The answer lies in the radial nature of flow to a well.

Figure 3.10b is a cross-section through an aquifer, with the line of section having been cunningly selected to pass right through the pumping well. Another line of section perpendicular to this would be expected to yield a very similar plot. This is because flow converges radially on the pumping well from all directions. With water approaching the well from all directions, the closer the water gets to the well, the smaller will be the cross-sectional area through which it must squeeze. For instance, if the pumping well has a diameter of 200 mm, and pierces 50 m of the saturated zone, then the total surface area of the well in contact with the aquifer equals 31.42 m^2 ($= \pi \cdot 0.2 \cdot 50$). At a distance of 10 m from the well, the cross-sectional area (A) of aquifer through which flow is converging will equal 1570 m^2 ($= \pi \cdot 10 \cdot 50$), at 20 m it will equal 3142 m^2, and so on. Now recall Darcy's Law: given that A decreases with proximity to the pumping well, while the total amount of water converging on the well remains the same, then if Darcy's Law continues to hold true,[xii] the hydraulic gradient must become ever-steeper, the closer the water gets to the well.

It is therefore inherent in radial groundwater flow that the hydraulic gradient steepens with proximity to a pumping well, resulting in the characteristic drawdown profile shown on Figure 3.10b. If one imagines the water table adopting the same profile in all radial directions from the well, then in three dimensions the drawdown patterns around a well must result in a depression or "hollow" in the water table, which takes the form of a steeply tapering cone. Such a **cone of depression** typically develops around each pumping well in an aquifer, resulting in significant local perturbations in the water table. Figure 3.11 illustrates how a cone of depression shows up in a groundwater head contour map. In Figure 3.11a, an entirely circular pattern is shown, which is what one would expect if the water table were absolutely flat and the aquifer completely homogeneous. The outer limit of the cone of depression is at an equal distance from the well in all directions, a distance customarily referred to as the **radius of influence** of the well, which notionally delineates the entire **borehole catchment** or **capture zone**, i.e. that portion of the aquifer which feeds water to the well in question. In most real cases, however, the water table is sloping (albeit gently) and more water feeds into the cone of depression on its up-gradient side than on its down-gradient side (Figure 3.11b). The resultant asymmetry of cones of depression in real aquifers needs to be taken into account in the planning of protection zones intended to safeguard the quality of pumped groundwaters (see Section 11.4).

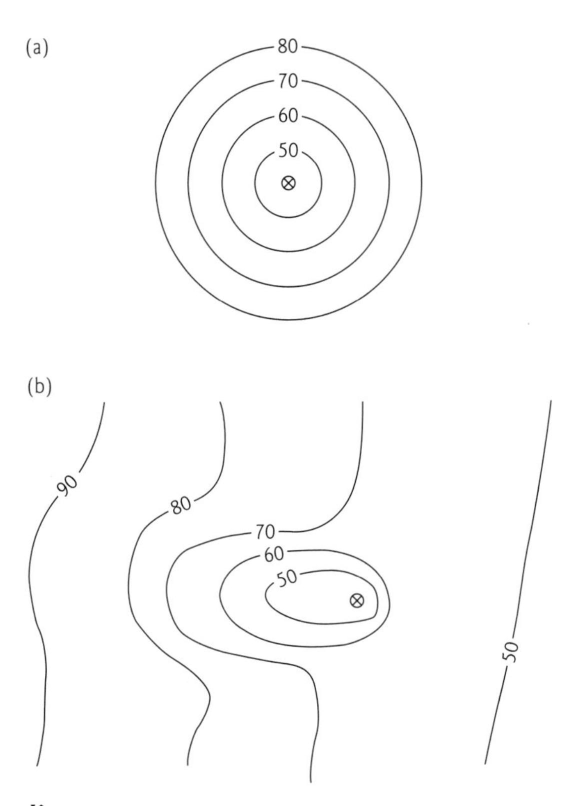

Key

 Pumping well, with drawdown to 45 m above sea level

Fig. 3.11 Cones of depression on groundwater head contour maps. (Contour labels are in meters above sea level.) (a) An idealized case, with an utterly flat water table and completely symmetrical distance–drawdown response (cf. Figure 3.10b). (b) A more realistic case, with a natural sloping water table in which the cone of depression is distorted into a more elliptical shape, due to the preferential supply of water from the higher-head area to the left of the pumping well.

3.4 Quantifying the hydraulic properties of aquifers

To make it possible for us to apply Darcy's Law to calculate real groundwater flow rates, we need to be able to quantify three things:

1 The hydraulic gradient.
2 The cross-sectional area through which flow occurs.
3 The hydraulic conductivity and/or transmissivity.

We have already seen how quantification of 1 and 2 is approached in practice, using field measurements of head (Box 3.1) to facilitate production of head contour maps, from which hydraulic gradients can be easily calculated. But what of hydraulic conductivity and transmissivity?

The most obvious starting-point for direct measurement of hydraulic conductivity is Darcy's laboratory apparatus (Figure 3.4). However, in real-world applications laboratory measurements of hydraulic conductivity using such apparatus is beset with three difficulties:

1 The fact that such tests can only be carried out on *disturbed* samples of soil or rock, which will usually have changed significantly in their porosity and permeability characteristics following excavation.
2 Laboratory columns typically test very small samples, which are unlikely to represent the true variability of the material in the field.
3 Much groundwater flow occurs via fractures (see Section 1.2 and Figure 1.3c), which are very rarely adequately represented in laboratory samples of aquifer material.

While there *are* some situations in which laboratory column tests can yield reliable values of K, for the most part they are not to be recommended in practice. Where such laboratory tests are to be undertaken, it is important to note that the constant-head setup used in Figure 3.4 only yields accurate results for materials of rather high hydraulic conductivity. For materials such as silts and muddy sands, which can be expected to yield K values less than about 0.01 m/day, an alternative approach is more reliable, in which the total head of water imposed across the column is allowed to decline naturally over time. Details of these so-called **falling head tests**, together with protocols for laboratory permeametry using **constant-head tests** (as in Figure 3.4), are given by Freeze and Cherry (1979) and many other authors.

Even less accurate than laboratory column tests are estimates of K derived from particle size distribution (PSD) analyses of sediments using empirical estimation techniques. The basic principle of such techniques is that the distribution of pore-neck sizes in a given sediment is likely to closely reflect the range of particle sizes of its constituent grains. An extensive literature exists on these estimation techniques (see Cronican and Gribb 2004), with most hydrogeologists favoring the calculation of K as a function of the "tenth percentile grain diameter" (d_{10}), i.e. the grain diameter which exceeds that of only 10% of the entire volume of the sediment. For instance, it is widely assumed that K can be calculated by a simple equation of the form $K = a \cdot d_{10}^{b}$. In the original formulation of Hazen (1892), the coefficient a was set equal to 1 and the exponent b equal to 2. More recent re-evaluations have revealed that a and b can take a range of values depending on the geological setting (see Uma et al. 1989 and Shepherd 1989), with a averaging 1.72 and b ranging mainly between 0.05 and 1.18. In the realm of construction dewatering design, estimates of K based on PSD data are often the only values obtainable from pre-existing site investigation data, but they must always be used with great caution. Other empirical relationships between rock properties and K have been developed, such as a formula for estimating K values for indurated sandstones by calculation from observed porosities determined by microscopic examination of thin sections of rock (Younger 1992). A similar approach has since been applied to soils (Lebron et al. 1999). In some types of investigation (e.g. detailed analyses of multiphase flow in aquifers contaminated with hydrocarbons), pore-level data of this type may be useful. It is also sometimes interesting to calculate what proportion of the field-measured K of an aquifer can be ascribed to intergranular flow, as opposed to flow through fractures (e.g. Price et al. 1982). Even where the *absolute* values of K determined in laboratory studies are dubious, they can at least reveal the *relative* permeabilities of a range of soils, i.e. identifying which are more permeable than others. However, in the vast majority of water resources investigations, laboratory measurements or estimates of K are unlikely to be of widespread practical use.

Wherever possible, it will always be best to determine K and/or T from data obtained by field testing of wells. The method of choice is **test pumping**, in which a borehole is pumped at a known rate and the resultant drawdown is measured. Essentially, test pumping is an exercise in quantifying the shape and rate of development of the cone of depression, from which it is possible to calculate transmissivity. There are two basic patterns of test pumping in common use: step-drawdown testing and constant-rate testing. **Step-drawdown tests** are essentially an engineering tool used for assessing the performance efficiency of pumping wells. Section 7.3.1 briefly describes their implementation and interpretation. For purposes of determining accurate, representative values of transmissivity, the technique of choice is **constant-rate test pumping**. As the name of this method suggests, it involves pumping the well at a single constant rate, and monitoring the resultant drawdowns. It is possible to perform constant-rate tests by measuring drawdowns only in the pumped well, but far greater insights will be gained by monitoring drawdowns in one or more observation wells located within the radius of influence of the pumping well. A wide range of analytical techniques has been developed to facilitate the interpretation of constant-rate pumping tests (see Kruseman and de Ridder 1991). The foundations for virtually all current methods for interpreting constant-rate pumping tests were laid by Theis (1935), who developed a method to calculate T and S by interpretation of time–drawdown data for constant-rate pumping of a well fully penetrating a horizontal, confined aquifer, which is assumed to have homogeneous and isotropic transmissivity, and to be of infinite areal extent. At first glance, these ideal conditions seem unlikely to apply in many real aquifers. However, they are often "true enough" in practice, particularly where the cone of depression is of modest extent. This is because a small cone of depression will not have traversed many zones of different transmissivity during its development, and neither will the radius of influence have yet come into contact with the outer boundaries of the aquifer. Even where boundaries *are* encountered by the expanding cone of depression, the deviation of real-time-drawdown behavior from the anticipated "ideal" response predicted by the Theis (1935) method provides powerful evidence for the existence and hydrological functioning of **recharge boundaries** (i.e. rivers/lakes connected to the aquifer) or **barrier boundaries** (e.g. impermeable faults, outcrop zones, etc). Furthermore, various adaptations of the basic Theis approach have been developed to allow various types of "nonideal" conditions to be taken into account (Kruseman and de Ridder 1991), including boundary effects and the depletion of saturated thickness due to drawdown in a thin unconfined aquifer. For the most part, however, it is impressive how often the basic Theis (1935) formulation proves adequate for analysing constant-rate test pumping data in a wide range of field settings.

Although the full intricacies of test-pumping interpretation are beyond the scope of this book, it is worthwhile mentioning one of the most powerful adaptations of Theis' (1935) basic method, namely the **Jacob Method** (Cooper and Jacob 1946). At its simplest, it is possible to use the Jacob Method as described in Box 3.3 to quantify transmissivity. A further advantage of the Jacob method is that plots of time versus drawdown on semi-logarithmic paper (so-called **Jacob plots**; Figure 3.12a) can provide powerful visual evidence for the presence of aquifer boundaries. A barrier boundary (e.g. a fault or some other feature that brings the aquifer into contact with low-permeability rock) causes drawdown to steepen beyond that which we would expect from extrapolation of the Jacob straight line (Figure 3.12b).

Box 3.3 Calculating *T* and *S* using the Jacob Method – a simplified approach.

Within a few years of Theis' (1935) breakthrough, two American researchers (Cooper and Jacob 1946) realized that if we take a piece of semilogarithmic paper and plot data from a constant-rate pumping test such that drawdown is on the arithmetically spaced y-axis and time is on the logarithmically spaced x-axis, then for all but the earliest datapoints (say, those gathered in the first 10 minutes of the test, which will be prone to the well bore storage effect), the data plot should form a straight line (Figure 3.12a). The gradient of this straight line can be expressed by determining the change in drawdown over any one log-cycle of time (e.g. between 0.1 and 1 days, or between 1 and 10 days). This measure of gradient is usually denoted by the symbol Δs. Transmissivity can then be easily calculated, as follows: Multiply the pumping rate of the well by 0.183, then divide the result by Δs. For instance, if the well is pumped at a rate of 1750 m^3/day, and we observe that the change in drawdown between 0.1 days and 1 day is 2.90 m, then we can calculate as follows:

$$0.183 \cdot 1750 = 320.25$$

$$T = 320.25/2.90 = 110 \text{ m}^2\text{/day}$$

Provided the time–drawdown data were obtained from a monitoring well (as opposed to the pumping well itself), we can go on to calculate storativity (S) as follows. First, extrapolate the straight line portion of the data-plot backwards until it cuts the zero drawdown level (see Figure 3.12a). Read out the corresponding value of time at this point of intersection (marked as t_0 on Figure 3.12a, in which case the value is seen to be about 0.025 days). Next, multiply this value by 2.25 T, i.e. $0.025 \cdot 2.25 \cdot 110 = 6.1875$. Finally, divide this value by the square of the distance from the monitoring well to the pumping well. In this case the monitoring well is 20 m from the pumping well, so we get:

$$S = 6.1875/(20)^2 = 6.1785/400 = 0.015.$$

Conversely, a recharge boundary (such as inflow from a river or a lake) leads to a lessening of the drawdown compared with that which the extrapolation of the Jacob straight line would predict (Figure 3.12c). Clearly where boundary effects become evident during the later stages of a pumping test, T should be calculated only using the straight-line portion of the data set, prior to the onset of any deviation from linearity. It is worth noting that records of water level recovery after the cessation of pumping can be used to calculate T in a closely analogous manner (for details see Kruseman and de Ridder 1991).

Where it is not possible to conduct a pumping test, either because no pump is available or because the available wells are too narrow to take a pump (which will often be the case with monitoring wells and piezometers), a range of single-well tests are available which yield point estimates of K. In such tests, a known volume of water is either suddenly added to or suddenly removed from the borehole, and the subsequent recovery of water levels back to their original elevation is monitored. Sudden removal of water is usually accomplished using a long, narrow bucket-like container known as a bailer, and the

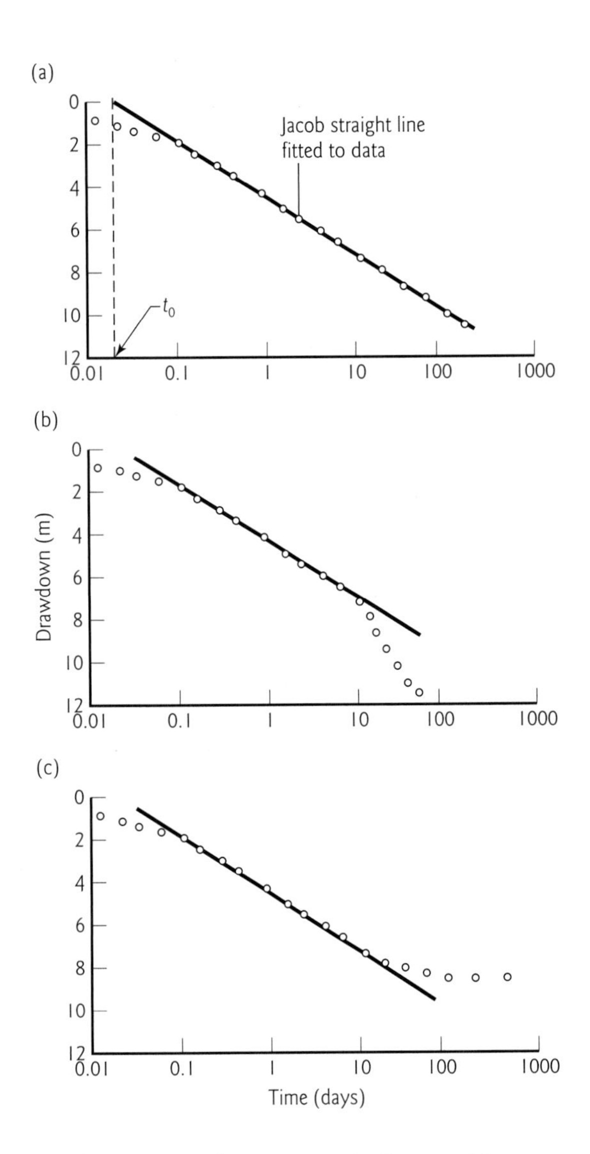

Fig. 3.12 (*Left*) "Jacob plots": one of the main test-pumping analysis tools in widespread use, comprising a plot of drawdown (on an ordinary arithmetic graphical axis) versus time (on a logarithmic axis). (a) Simple case, with a linear time–drawdown response after the initial depletion of well-bore storage. As the straight line relationship persists for the entire duration of the pumping data set, we can conclude that the aquifer is more extensive than the maximum extent of the cone of depression, even after 200 days of pumping. (See Box 3.3 for an explanation of the value t_0.) (b) Barrier boundary case, where a marked steepening of the time–drawdown plot occurs (in this example after 10 days), indicating that the cone of depression has encountered an effectively impermeable boundary at one of the outer edges of the aquifer. (c) Recharge boundary case, in which a shallowing of the time–drawdown plot occurs, indicating that further expansion of the cone of depression has been halted due to an encounter with an abundant source of further recharge, such as a lake or river in hydraulic contact with the aquifer.

resultant test is known as a **bail test**. Alternatively, where a known volume (i.e. a "slug") of water is suddenly added to the well (by tipping in from above), the monitoring of the decline of water levels back to their original state is known as a **slug test**. In fact, thanks to Archimedes Principle, it is possible to conduct slug tests using a solid plunger in place of the slug of water. Because the amounts of water added or removed in slug/bail tests are modest, only a small zone of the surrounding aquifer is tested during such operations. For this reason, the values of K obtained are not usually as representative as those obtained by means of test pumping.

The preceding discussion has concerned only the determination of T or K. The other key aquifer parameter is the storativity (or specific yield in unconfined aquifers). It is possible to directly determine S values using test pumping analysis methods. This can only be done where time–drawdown data have been obtained from an observation well; analysis of data from a pumping well will yield highly misleading S values due to the combined influence of well bore storage effects and deviations from laminar flow conditions near and within the pumping well. If T has been calculated using the Jacob method, S can be calculated as explained in Box 3.3. Where test pumping data are *not* available, storativity is best estimated from geological information (see Younger 1993).

Endnotes

i Alternative terms synonymous with "head" include "hydraulic head," "groundwater head," "groundwater potential," and "potential".

ii Ironically, since the US gallon = 0.832 English gallons, these units are arguably not especially "English."

iii Isotropic is the opposite of anisotropic; e.g. if K_h is said to be isotropic, this means that K_h values at any one point would be the same irrespective of the direction of groundwater flow.

iv "Homogeneous" means that there are no significant spatial variations in the parameter concerned.

v In confined aquifers "saturated thickness" simply equals the aquifer thickness. In unconfined aquifers it equals the water table elevation minus the elevation of the aquifer base; as such it varies over time due to water table fluctuations.

vi i.e. per unit width of aquifer, and assuming both are subjected to equal hydraulic gradients.

vii Complex groundwater modeling projects include, for instance, simulating the interactions between water, gas, and oil in a contaminated aquifer (which is an example of a **multi-phase flow** problem).

viii Given the slow rates of flow in most aquifers, head measurements made at a number of different points in the aquifer on a single day can safely be regarded as "simultaneous" for purposes of analysis.

ix Note that the rules as given here are strictly applicable only where the aquifer is isotropic (see Section 3.2.2 and endnote iii); in anisotropic aquifers it is possible for flow lines to cross equipotentials at oblique angles (see Freeze and Cherry 1979).

x If the aquifer is homogeneous and isotropic and the contour interval is constant, then by setting the spacing of flow line starting points equal to the spacing of contours, we will end up with a flow net made up of squares (or at least "near-squares," with curvilinear edges).

xi The "initial stages," during which well bore storage effects are likely to be important, seldom last more than 10 minutes in most real-world test-pumping situations.

xii This assumes of course that K is also more or less constant within the area of aquifer influenced by the well.

4

Natural Groundwater Quality

Talaes sunt aquae quam terrae perfluunt
(Waters become like the ground through which they have flowed).
(Pliny the Elder, 74 CE)

Key questions

- What does a water analysis mean?
- What are the main natural chemical constituents of groundwaters?
- How can groundwater chemistry best be displayed, classified, and interpreted using simple graphical techniques?
- What happens to the chemistry of rain water as it soaks through the subsurface to become groundwater?
- How does groundwater quality change along flow paths within aquifers?

4.1 How to read a water analysis

4.1.1 Why bother reading water analyses?

Water analyses can seem intimidating the first time you look at one: unfamiliar names, chemical symbols whose meanings you can't quite recall, abbreviations denoting strange units. It's enough to put many people off, and there are indeed many hydrogeologists who studiously avoid dealing with water quality if they can possibly get away with it. However, to ignore water quality is to throw away some of the most compelling evidence for past and present movement of specific groundwaters. It's like watching a movie in black and white instead of color. So what do you need to know in order to read a typical analysis of a groundwater, such as that shown in Figure 4.1?

4.1.2 Types of parameters

The various parameters listed on a typical water analysis fall into a number of categories. Normally listed first on an analysis report sheet are the

ID number EX/	Sample date	pH units	Conductivity µS/cm @ 25°C	Suspended solids	Total alkalinity as $CaCO_3$	Total hardness as $CaCO_3$	Calcium hardness as $CaCO_3$	Magnesium hardness as $CaCO_3$	Chloride as Cl (total)	Total sulfur as SO_4 (total)	Calcium as Ca (total)	Calcium as Ca (dissolved)	Magnesium as Mg (total)	Magnesium as Mg (dissolved)	Sodium as Na (total)	Potassium as K (total)	Nickel as Ni (total)
Units:			µS/cm	mg/L	mg/L	mg/L	mg/L	mg/L	mg/L	mg/L	mg/L	mg/L	mg/L	mg/L	mg/L	mg/L	mg/L
Method codes:		WSML3	WSML2	WSLM10	WSLM12	Calc	Calc	Calc	KONENS	ICPWATER	ICPWATER	ICPWATER	ICPWATER	ICPWATER	ICPWATER	ICPWATER	ICPWATER
Method reporting limits:			100	5	2				1	0.06	0.01	0.01	0.03	0.03	0.01	0.01	0.02
UKAS accredited:		Yes	Yes	Yes	Yes	Yes	Yes	Yes	Yes	Yes	Yes	Yes	Yes	Yes	Yes	Yes	Yes
0424710	Nov 2, 04	7.1	8900	6	1230	667	348	307	2220	714	141	139	76.3	73.0	1800	37.3	<0.02
0424711	Nov 2, 04	7.6	9080	<5	1440	676	342	310	2290	768	141	137	78.6	73.7	1920	39.9	<0.02

Client name	On site	Water sample analysis
Contact		
Date printed		26-Nov-04
Report number		EXR/045462
Table number		1

Fig. 4.1 An example of typical groundwater analyses (in this case for two samples from the same aquifer) produced by a commercial laboratory. The second row of the table (Method codes) identifies which standard operating procedures (SOPs) have been used for each measurement (identified by various acronyms). The detection limits for each parameter using the identified procedure are specified in the third row (Method reporting limits), and the row labeled "UKAS accredited" specifies whether the reported measurement is in accordance with an independent analytical accreditation program (the UK national program in this example).

physicochemical parameters (Section 4.2.1) such as temperature, *p*H, specific electrical conductance, and other physical properties of the water which determine and/or reflect its overall chemical behavior. Some so-called **collective parameters** (Section 4.2.5) are also often listed along with the physicochemical properties, most notably: alkalinity, acidity, hardness, organic carbon (total and/or dissolved), and total dissolved solids (TDS). These parameters are measured directly in the laboratory, but are regarded as "collective" because the values they take depend on the concentrations of more than one dissolved substance. Next come the various **chemical constituents** which have been measured by the analyst. Although the normal presumption is that the elements in question were originally present in the dissolved state, it is common to find that an analysis will report both "total" and "dissolved" concentrations. In practice, the distinction between the two is that "total" concentrations are measured without prior filtration of the sample, whereas "dissolved" concentrations are measured on subsamples of the water *after* they have been filtered (normally so that they pass through a filter paper with 0.45 μm diameter pores). As we shall see (Section 4.4.2), considerable uncertainty surrounds the actual definition of the dissolved state, and in some cases a substantial proportion of the reportedly "dissolved" fraction of a given element may actually be present in the form of minute suspended particles, known as **colloids**. In natural waters, almost all dissolved constituents carry an electrical charge; such charged constituents are known as **ions**. (There are very few nonionic dissolved constituents in most groundwaters; the most prominent example is silicon, which is usually reported as the uncharged molecule SiO_2.) It is usual to categorize ions according to whether they carry a positive or negative charge: **cations** are positively charged (mainly dissolved metals), while **anions** are negatively charged (mainly comprising nonmetals and their compounds). On analytical report sheets it is common practice to segregate the cations from the anions. Within each category, a distinction is often also made between **major** and **minor** species;[i] those species present at concentrations below 0.01 mg/L are sometimes referred to as **trace elements** (or more specifically as **trace metals** in the case of cations).

Many groundwater analyses will go no further than listing the types of parameters already mentioned. However, in some cases dissolved gases will also have been measured. Most widely measured and very commonly reported is **dissolved oxygen (DO)**, though other gases such as methane (CH_4), carbon dioxide (CO_2), radon (Rn), and a range of volatile organic compounds may be listed, especially if the sample was analyzed as part of an investigation of contamination.

Finally, as we shall see in see Section 4.2.7, certain types of investigation involve analysis for atomic-scale variants of specific elements known as isotopes. The purpose of the investigation in question determines which (if any) isotopic constituents are measured and reported.

4.1.3 Units of measurement and how to convert between them

Table 4.1 lists the principal units of measurement used on routine groundwater quality analysis reports. Concentrations of "dissolved" species are usually expressed in **milligrams per liter (mg/L)** (or **micrograms per liter (μg/L)** in the case of minor and trace constituents). To make much use of reported values, it is necessary to be able to convert from one set of units to another. A very thorough explanation of the various units used to report groundwater analyses, and the methods for converting values between them, is provided by Hounslow (1995), and only the key essentials will be explained here.

The main thing to remember when making any conversions of concentration units is that one liter of water weighs one kilogram. Bearing this in mind, unit conversions are very straightforward in many cases. For instance, as long as a groundwater contains less than about 10,000 mg/L of total dissolved solids, then mg/L concentrations will be equal in value to **parts-per-million (ppm)** concentrations. (The same goes for the equivalence between μg/L and **ppb (parts per billion)**;

Table 4.1). (The difference between the two types of units are that mg/L and μg/L express concentrations on the basis of "mass per unit volume of liquid," whereas ppm and ppb take a "mass per unit mass of liquid" approach.)

If a hydrogeologist is never called upon to do more than assess whether a given water quality analysis complies with limit values laid down by law (see Chapter 7), then he or she might never need to move beyond simple conversions of this type. However, whenever more detailed investigations of water chemistry are warranted, more extensive unit conversions are almost always needed (Box 4.1). The most fundamental transformation is to convert mg/L concentrations into an equivalent **molar concentration**. This is easily done: to obtain a solute concentration in **millimoles per liter (mmol/L)**, it is necessary only to divide the mg/L concentration by the relative atomic mass of the element[ii] in question (Box 4.1). This simple conversion is extremely powerful, as it opens the door to calculating the quantities of a given element which take part in the various chemical reactions occurring in groundwater systems. The reason for this is that chemical reactions occur by interactions between individual atoms, rather than between, say, spoonfuls of elements. Given that the mass of atoms vary greatly from one element to another, it is not sensible to try to directly calculate the reaction of a gram of one substance with a gram of another. Rather, we have to calculate in terms of *moles*, for reasons explained in Box 4.1.

Given the low concentrations of many dissolved substances in groundwaters, it will often be convenient to make calculations of reacting quantities of solids and solutes directly in terms of millimoles per liter. However, where necessary, mmol/L concentrations are readily converted into **moles per liter** (mol/L) simply by dividing them by 1000.

It should be noted that converting mg/L concentrations directly into mmol/L values is in itself a crude approximation. This is because analytical measurements tell us the total amount of a given substance present, not how that substance is actually present in solution. The true mode of occurrence of different elements in solution is determined by an array of complex interactions between dissolved species with different charges. For real species, more accurate estimates of their true mmol/L concentrations (which are known as their **activities**) are best obtained by using thermodynamic models to simulate the interactions between charged species in solution, a process known as speciation modeling (see Section 10.5.3).

One further unit conversion calculation is very common in practice: this is the conversion of concentrations to **milliequivalents per liter (meq/L)**. This conversion is applicable only to *charged species*, i.e. ions, as it is essentially a measure of the number of "moles of charge" available for participation in a range of electrochemical reactions. Conversion to meq/L is achieved simply by the multiplying the mmol/L concentrations by the **valence** (i.e. the charge) of the ion. For instance, if we have 25.6 mmol/L of Ca^{2+}, then given that the valence is 2, we can easily calculate that we have $2 \times 25.6 = 51.2$ meq/L of Ca^{2+}. (Converting directly from mg/L, we must multiply by the valence and divide by the relative atomic/molecular mass.) Fortunately, most major cations and anions do not vary in valence, so that constant conversion factors can be established for many dissolved species (Table 4.2). Exceptions arise in relation to certain metals (especially iron, manganese, copper, and chromium) and with certain anion compounds (notably those of sulfur and nitrogen) which can change their valence depending on the presence or absence of oxygen or other substances which readily exchange electrons with others. Apart from their utility in various geochemical calculations, meq/L values are invariably used in preparing graphical representations of groundwater quality (Section 4.3.1) and in assessing the reliability of laboratory analyses (Section 4.1.4).

Table 4.1 shows that acidity, alkalinity, and hardness (as defined in Section 4.2.5) are conventionally reported in units of "mg/L as $CaCO_3$ equivalent." Given that none of these three parameters necessarily correlates with dissolution or

Table 4.1 Principal units and typical concentration ranges of common constituents of natural groundwaters.

Constituent or property	Units	Symbol	Typical range of values in most groundwaters	Extreme values (and where encountered)
Temperature	Degrees Celsius (= degrees centigrade)	°C	10–25°C	300°C (under confined conditions in active volcanic areas)
Conductivity	microsiemens per centimeter*	μS/cm	15–3000 μS/cm	> 100,000 μS/cm (in hypersaline arid basins)
*p*H	(no units customarily reported)		6.5–8.5	–3.5 (negative *p*H due to evaporation of water following pyrite oxidation) > 12 (confined settings in ultrabasic rocks/serpentine)
Eh	millivolts†	mV	+100 to +300 mV	–200 mV (in the presence of decaying organic matter) +500 mV (in the presence of powerful oxidants such as Fe^{3+})
Dissolved oxygen (DO)	Percent saturation *or* milligrams per liter‡	$\%_{sat}$ *or* mg/L	0 to 25$\%_{sat}$ 0 to 3 mg/L	100% saturation 12 mg/L (in shallow, permeable aquifers)
Hardness§	milligrams per liter as calcium carbonate equivalent *or* milliequivalents per liter	mg/L as $CaCO_3$ *or* meq/L	10–500 mg/L as $CaCO_3$ 0.2–10 meq/L	3000 mg/L as $CaCO_3$ (or 60 meq/L) (in ground containing gypsum/anhydrite)
Alkalinity‖	milligrams per liter as calcium carbonate equivalent *or* milliequivalents per liter	mg/L as $CaCO_3$ *or* meq/L	10–500 mg/L as $CaCO_3$ 0.2–10 meq/L	1000 mg/L as $CaCO_3$ (or 20 meq/L) (sulfate-rich groundwaters in deeply confined carbonate aquifers)
Acidity¶	milligrams per liter as calcium carbonate equivalent *or* milliequivalents per liter	mg/L as $CaCO_3$ *or* meq/L	0–150 mg/L as $CaCO_3$ 0–3 meq/L	5000 mg/L as $CaCO_3$ (or 50 meq/L) (in acidic mine drainage, epecially where evaporation is also an important process)
Calcium (Ca^{2+})	milligrams per liter	mg/L	10–500 mg/L	2500 mg/L (in formerly acidic waters since neutralized by dissolution of calcite)

Magnesium (Mg^{2+})	milligrams per liter	mg/L	5–400 mg/L	10,000 mg/L (in the vicinity of hypersaline evaporative lakes in arid areas)
Sodium (Na^+)	milligrams per liter	mg/L	10–1000 mg/L	150,000 mg/L (in deep-seated brines found in many oilfields and other ancient sedimentary basins)
Potassium (K^+)	milligrams per liter	mg/L	1–50 mg/L	150,000 mg/L (groundwaters in contact with deposits of sylvite (KCl))
Bicarbonate (HCO_3^-)[‖]	milligrams per liter	mg/L	0–400 mg/L	1000 mg/L (deeply confined waters in carbonate aquifers)
Chloride (Cl^-)	milligrams per liter	mg/L	10–1000 mg/L	250,000 mg/L (in deep-seated brines found in many oilfields and other ancient sedimentary basins)
Sulphate (SO_4^{2-})	milligrams per liter	mg/L	10–500 mg/L	3000 mg/L (in acidic mine drainage and in gypsum karst drainage)
Nitrate (NO_3^-)	milligrams per liter *or* milligrams per liter as N equivalent	mg/L NO_3-N mg/L	10–60 mg/L 2–15 mg/l NO_3-N	150 mg/L (40 mg/L NO_3-N) (in heavily-polluted recharge in arable farming areas)
Fluoride (F)	milligrams per liter	mg/L	0.1–3 mg/L	20 mg/L (in groundwaters in some areas of recent alkalic volcanism)
Silica (SiO_2)	milligrams per liter	mg/L	6–12 mg/L	150 mg/L (in some hydrothermal spring waters)

[*] millisiemens per centimeter (mS/cm) are used for more saline groundwaters (1 mS/cm = 1000 μS/cm).

[†] Eh values can be negative as well as positive; values sometimes quoted in volts (V) (1 volt = 1000 mV).

[‡] mg/L are closely equivalent to ppm (parts per million, which equals mg per kg of water) – the distinction between the two only becomes important in more saline groundwaters where the density of a liter of water begins to significantly exceed 1 kg/L Note that 1 ppm = 1000 ppb (parts per billion) and also that 1 ppb ≈ 1 μg/L (valid for the same conditions as the mg/L = ppm equivalence).

[§] Older units such as "degrees of hardness" are rarely used these d°ays (1 degree of hardness ≈ 189 mg/L as $CaCO_3$).

[‖] It is common to calculate bicarbonate (mg/L HCO_3^-) from alkalinity (mg/L as $CaCO_3$) by the expression $HCO_3^- = 1.2$ (alkalinity) – note, however, that this relationship is only *approximate*, and speciation modeling is a better approach for many groundwaters.

[¶] See Section 4.2.5 for an explanation of how acidity can be calculated from *p*H and dissolved metals concentrations.

Box 4.1 Don't get yourself into a hole: why moles matter!

In a balanced chemical reaction each of the species taking part will be expressed in units of moles. For instance, consider the following simple, but very important, reaction:

$$CaCO_{3(s)} + H^{+}_{(aq)} \leftrightarrow Ca^{2+}_{(aq)} + HCO^{-}_{3(aq)}$$

This equation describes the dissolution of solid (denoted by the symbol $_{(s)}$) calcite ($CaCO_3$) by reaction with protons (H^+) in aqueous solution (denoted by the symbol $_{(aq)}$) to form aqueous calcium (Ca^{2+}) and bicarbonate (HCO_3^-) ions. More specifically it shows that *one mole of calcite* reacts with *one mole of protons* to liberate *one mole each* of calcium and bicarbonate to solution.

It was a celebrated Italian pioneer of chemistry, Amedeo Avogadro (1776–1856), who first explained how we can easily calculate the number of atoms present in a given mass of a specific chemical element. Key to making these calculations is a property known as relative atomic mass (RAM), values for which have been accurately determined for all known chemical elements, and which are listed in the Periodic Table (an up-to-date version of the Periodic Table is available on-line at www.webelements.com). RAM has no units, as it is a measurement of the mass of an atom of a given element relative to hydrogen, which is assigned a value of 1. In practice, all we need to remember is that one mole of any element is equal to the same number of grams as its RAM. Take for instance calcium, which has a RAM of 40. This means that one mole of calcium will weigh 40 g. For substances containing more than one element, such as calcite, the equivalent to RAM is "relative molecular mass" (RMM), which equals the sum of the RAMs of the constituent elements. Returning to the dissolution of calcite, therefore, given the RAMs of Ca (40), C (12), O (16), and H (1), then we can calculate the RMM of calcite as 40 + 12 + 3(16)) = 100. Thus in terms of *masses* of reacting substances, then by examining the equation at the top of this box, we can calculate that 100 g of calcite (= 1 mole) would react with only 1 g of hydrogen (= 1 mole) to yield 40 g of dissolved Ca^{2+} (= 1 mole) and 61 g of bicarbonate (= 1 mole). Clearly if we had tried to calculate the consequences of the above equation under the misapprehension that the reacting quantities were grams rather than moles, we would have come to utterly wrong conclusions. Moles matter.

If we take another important reaction, such as the oxidation of pyrite (FeS_2) by oxygen (O_2) in the presence of water:

$$FeS_{2(s)} + 3.5O_{2(g)} + H_2O \rightarrow Fe^{2+}_{(aq)} + 2SO^{2-}_{4(aq)} + 2H^{+}_{(aq)}$$

then we can see that in this case 1 mole of pyrite reacts with 3.5 moles of oxygen molecules to yield 1 mole of dissolved ferrous iron (Fe^{2+}), but *2 moles* each of sulfate (SO_4^{2-}) and protons (H^+). What if we started with 500 g of pyrite? Given the RMM of pyrite (= 120), then 500 g equals 4.17 moles. If 3.5 times as much O_2 (RMM = 32) must react with this amount of pyrite to drive the reaction to the right as shown, then this requires 14.58 moles of O_2, which would equate to $14.58 \times 32 = 467$ g of O_2. The result would be the liberation of $2 \times 4.17 = 8.34$ moles each of SO_4^{2-} (equating to 800.6 g) and H^+ (equating to 8.3 g).

Calculations of this kind are part of a branch of chemistry known as **stoichiometry**, and they are the starting point for virtually all hydrogeochemical interpretations of water analyses. The two examples given demonstrate the importance of converting mg/L concentrations to equivalent molar quantities before proceeding to interpret the origins and potential reactivity of real groundwaters.

Table 4.2 Conversion factors from mg/L to meq/L and mmol/L for major cations and anions and selected other species commonly found in groundwaters. (Adapted from Hem 1985.)

Element and reported species	Conversion factors, mg/L . . .*	
	to meq/L	to mmol/L
Aluminum (Al^{3+})	0.11119	0.03715
Ammonium (NH_4^+)	0.05544	0.05544
Arsenic (As)	–	0.01334
Barium (Ba^{2+})	0.01456	0.00728
Bicarbonate (HCO_3^-)	0.01639	0.01639
Boron (B)	–	0.09250
Bromide (Br^-)	0.01252	0.01252
Cadmium (Cd^{2+})	0.01779	0.00890
Calcium (Ca^{2+})	0.04990	0.02495
Carbonate (CO_3^{2-})	0.03333	0.01666
Chloride (Cl^-)	0.02821	0.02821
Copper (Cu^{2+})	0.03147	0.01574
Fluoride (F^-)	0.05264	0.05264
Hydrogen (H^+)	0.99216	0.99216
Hydroxide (OH^-)	0.05880	0.05880
Iodide (F^-)	0.00788	0.00788
Iron (ferrous) (Fe^{2+})	0.03581	0.01791
Iron (ferric) (Fe^{3+})	0.05372	0.01791
Lead (Pb^{2+})	0.00965	0.00483
Lithium (Li^+)	0.14407	0.14407
Magnesium (Mg^{2+})	0.08229	0.04114
Manganese (Mn^{2+})	0.03640	0.01820
Mercury (Hg)	–	0.00499
Molybdenum(Mo)	–	0.01042
Nickel (Ni)	–	0.01704
Nitrate (NO_3^-)	0.01613	0.01613
Nitrite (NO_2^-)	0.02174	0.02174
Phosphate (PO_4^{3-})	0.03159	0.01053
Phosphate (HPO_4^{2-})	0.02084	0.01042
Phosphate ($H_2PO_4^-$)	0.01031	0.01031
Potassium (K^+)	0.02558	0.02558
Selenium (Se)	–	0.01266
Silica (S_iO_2)	–	0.01664
Sodium (Na^+)	0.04350	0.04350
Strontium (Sr^{2+})	0.02283	0.01141
Sulfate (SO_4^{2-})	0.02082	0.01041
Sulfide (S^{2-})	0.06238	0.03119
Uranium (U)	–	0.00420
Zinc (Zn^{2+})	0.03059	0.01530

* Multiply mg/L value by the factor indicated to make the conversion to desired units.

precipitation of $CaCO_3$, it is arguably more reasonable to express them in meq/L. The two sets of units are readily interconvertible:

Concentration in meq/L
= (Concentration in mg/L as $CaCO_3$)/50

Finally, since most groundwaters have a *p*H close to 7 (Section 4.2.1), under which condition the alkalinity of the water is utterly dominated by the bicarbonate content (Section 4.2.5), it is common practice to calculate the bicarbonate concentration (mg/L HCO_3^-) simply by multiplying the alkalinity (in mg/L as $CaCO_3$) by a factor of 1.2. (For some *caveats* on this point, see Section 4.2.5.)

4.1.4 Analytical methods and quality control measures

The importance of understanding field and laboratory procedures

To truly appreciate what is shown on a groundwater analysis such as Figure 4.1, it is important that the end-user have some appreciation of the manner in which the reported values have been obtained. For instance, I have often been able to detect analytical errors in data emanating from various laboratories simply because I am aware that dilution of samples with deionized water is common practice, in order to bring sample concentrations within the preferential limits of analytical devices. Failure to multiply the machine outputs to account for prior dilution is a very common source of erroneous reporting from laboratories. This is not the only source of error, however. Even if you have no intention of carving out a career as a laboratory scientist, it is nevertheless vital that you familiarize yourself with analysis techniques and equipment. A detailed explanation of best practice in groundwater sampling and analysis is beyond the scope of this book; the interested reader is referred to Fetter (1999) for a description of North American practices and Environment Agency (2003) for details of current European recommendations. However, a few pointers are given below to alert you to some of the questions you ought to be asking when scrutinizing analytical results.

Analytical methods

The process of water analysis actually begins in the field at the time of sample collection. This is because a number of physicochemical characteristics of groundwaters change rapidly during transit of samples from the field to the laboratory and during subsequent storage. The minimum suite of parameters which should *always* be measured on site are *p*H, temperature, and specific electrical conductance (commonly referred to simply as "conductivity" – Section 4.2.1). Fortunately all three of these are easily measured on site using robust electronic meters. If the investigator has a particular interest in oxidation-reduction processes (see discussion of Eh in Section 4.2.1), it is also advisable to measure redox potential (Eh) on site, as well as dissolved oxygen. Again, both of these can be measured using electronic meters, though the accuracy of Eh measurements is often dubious, except where the water in question is rich in dissolved iron (see Younger et al. 2002a). On site measurement is also to be recommended for alkalinity (Section 4.2.5), though in this case accurate determinations demand field titration. While robust, portable titrators are now widely available, their successful use requires considerable skill on the part of the operator.

When collecting samples for later laboratory analysis, it is common practice to fill at least two clean PVC bottles for each sampling point. One of the two bottles will contain a few drops of concentrated nitric acid, which will dissolve in the water, lowering the *p*H and preventing precipitation of cations during sample transit and subsequent storage. The other bottle is typically left without any preservative. If there are good reasons for undertaking filtration on site, it is also desirable that both unfiltered and filtered samples be collected in pairs of bottles as described, so that any changes in quality induced by filtration can be detected by later laboratory analysis.

Back in the laboratory, analysis of the samples typically proceeds as follows. For the determination of metals, the most commonly used techniques involve one or other form of "spectrometry", i.e. the measurement of the spectral ranges of light emitted or absorbed by metals which have been restored to their uncharged, elemental forms by exposure to high amounts of energy. Most metals determinations prior to about 1990 were made using **atomic absorption spectrophotometry (AAS)**, in which the energy source is a flame of burning acetylene. For each metal of interest, a beam of light at a specific frequency (unique to each metal) is aligned to pass through the flame, so that its intensity can be measured using a light meter (i.e. a "photometer") located on the other side of the flame. After a spray of sample water has been released into the flame, the reduction in intensity of the light beam is recorded, and converted into an equivalent concentration of absorbing metal atoms. In most modern laboratories a technique known as **inductively coupled plasma (ICP)** is now the preferred option. In this case the sample is sprayed into a stream of argon gas, which passes through externally heated quartz tubes, heating up to form a "plasma" at temperatures of around 7000°C, in which the various metals can be separated from one another when they are subsequently exposed to powerful electromagnetic currents (provided by induction coils). Abundances of metals are then determined using optical devices or mass spectrometers. ICP machines have major advantages over AAS in that they can: (i) analyze for a large number of different elements at once; (ii) achieve very low detection limits for most metals; and (iii) accommodate large numbers of samples per hour.

For the analysis of anions the most common analytical technique at present is **ion chromatography** (IC). This separates the various ions according to their relative affinity for a static adsorbent material lining the walls of a long tube (known as a "column"). IC is fast and easy to use for major anions, and offers low detection limits for most compounds of interest in groundwater studies. A range of other techniques exist for anions that are not readily analyzed by IC (Clesceri et al. 1998).

Little or no analysis of organic compounds tends to be carried out for natural, unpolluted groundwaters, with the exception of occasional analysis of total and dissolved organic and inorganic carbon. However, in contaminated environments, the bulk of the overall analytical burden may well relate to the determination of concentrations of a range of synthetic organic compounds, principally by means of gas chromatography–mass spectrometry (GC-MS). Details of these and all other analytical methods commonly applied to groundwater samples can be found in the exhaustive compilation of Clesceri et al. (1998).

Quality assurance and quality control issues

Most scientists glaze over at the mention of the twin spectres of QA and QC. Between them they conjure up images of mountains of paperwork and hours of poring over dusty ledgers and spreadsheets. No one goes into science out of enthusiasm for such tasks. Yet assuring that the raw data with which we work are of sufficient quality to make them useful for our wider purposes is of fundamental importance to the meaning and credibility of everything else we do. Quality is a bit like good health – you don't realize how much you depend on it until it's gone.

Much of the drudgery of QA and QC can be alleviated by paying close attention to the initial design of a laboratory quality assurance plan. If procedures are well defined at the outset, reliable techniques will always be used and appropriate checks on analytical accuracy will be built in to the daily routine. One of the most useful steps to take in defining a laboratory QA plan is to specify the **standard operating procedures (SOPs)**. Fortunately, much of the workload in this regard can be avoided by simply following established, internationally agreed "standard methods." One of the most widely used compendiums of standard methods applicable to the analysis of groundwaters is has been published by the American Public Health Association (Clesceri et al. 1998). This comprehensive volume has gone through 20 revisions and thus benefits from a very long history of checking and updating in the light of experience. It includes techniques for the determination of all relevant parameters using a range of analytical equipment, allowing for the fact that not all laboratories are equipped to equal standards.

Besides establishing and adhering to reliable SOPs, a number of other quality control measures can be usefully included in analytical routines including:

- Routine analysis of replicate samples.
- The deliberate addition of "spikes" of known amounts of specific chemical elements to aliquots of real samples, in order to assess the precision of measurement techniques.
- The frequent analysis of externally provided "certified reference materials" (which in the case of groundwater analysis will be waters with chemical compositions close to those of the waters of interest); these are usually readily obtainable from the central laboratories of national governments.
- Occasional analysis of "blanks", i.e. distilled deionized water which has been subjected to the same range of field and laboratory preparation steps as the real samples; these can help identify where contamination is entering into the analytical process and resulting in false positive measurements.

Once the analysis is complete, a final test of quality is provided by invocation of the **principle of electroneutrality**, which states that a water cannot carry a net electrical charge (positive or negative), but must always be electrically neutral. Given that most dissolved species carry a charge, electroneutrality demands that the sum of equivalents of positively charged species matches the sum of equivalents of negatively charged species. It is possible to take advantage of this principle to check the credibility of a water analysis. This is done by calculating the **cation-anion balance** (CAB) of the water,[iii] which is defined as:

$$\text{CAB (\%)} = 100 \cdot \frac{\text{(sum of cation concentrations)} - \text{(sum of anion concentrations)}}{\text{Sum of cation + anion concentrations}}$$

where all concentrations are expressed in meq/L. If a CAB value is less than 5%, then the analysis can be regarded as sufficiently accurate for all uses. If a CAB lies in the range 5–15%, then the analysis should be used with caution, while those analyses with CAB values greater than 15% cannot really be regarded as being sufficiently reliable to justify using them for serious scientific purposes.

4.2 Chemical characteristics of natural groundwaters: origins and significance

4.2.1 Physicochemical characteristics

Temperature

It is very easy to measure temperature to within ±0.1°C using electronic sensors, which are so robust that they rarely require calibration and can thus be left unattended for extended periods of time. Knowledge of groundwater temperatures is essential for the correct interpretation of solution chemistry, especially for assessing the tendency for minerals to dissolve in, or precipitate from, a given groundwater.

What controls groundwater temperatures? Logically, one might expect the natural temperature of a given groundwater to reflect the temperatures of incoming recharge waters. In broad terms, this is indeed found to be the case, at least in very shallow permeable aquifers in which the shallowest groundwaters are typically warmer than the waters deeper in the saturated zone during the summer, but cooler than them in winter. However, there is a general tendency for groundwaters between about 5 m and 150 m below ground surface to closely approximate the local mean annual air temperature. A combination of factors contributes to the tendency of the subsurface environment to "average" the temperatures of incoming waters in this way:

- The high specific heat capacity of water,[iv] which results in a significant "lag" between an abrupt change in air temperature and the consequent change in the temperature of recharge waters.
- The insulation against short-term extremes of temperature offered by the uppermost layers of soil.
- The upward transmission of warmth from the deep subsurface.

Beyond this shallow zone of seasonal influence, groundwater temperature generally increases steadily with depth, so that deep-seated groundwaters tend to be significantly warmer than those lying close to the water table. The rise in temperature with increasing depth (which is termed the **geothermal gradient**) averages approximately 2.0–2.5°C per 100 m depth. In some geological settings, such as active volcanic areas and areas undergoing active fault movement, the geothermal gradient will be far greater. Less extreme (but still elevated) geothermal gradients are also commonly associated with the presence of granites or other rocks naturally rich in unstable radionuclides, decay of which releases heat that is then conducted towards the Earth's surface (though fortunately the associated ionizing radiation is generally *not* emitted in tandem).

In places where all recharge occurs by direct infiltration (see Section 2.2.2) groundwater temperatures increase smoothly with depth. More erratic temperature–depth profiles are typical of areas in which much of the recharge is indirect (cf. Figure 2.3), because preferential flow paths can rapidly introduce cool recharge waters to depths at which surrounding, directly sourced recharge has already attained warmer temperatures. Similarly, if a deep-seated thermal groundwater has access to a fast-flow pathway (such as a fault plane), it is possible for localized peaks of high temperature to occur at certain depths. Continuous measurement of temperature down a deep borehole (which is readily achieved using a digital thermometer suspended on a cable) can thus help identify zones of significant inflow of waters with different thermal (and hydrological) histories, providing powerful evidence of groundwater flow pathways which could never be deduced from head measurements alone. This is particularly true where caves are present within aquifers, as these can rapidly deliver recent recharge deep within the saturated zone. It has thus been found that, by monitoring how the temperature of limestone spring waters varies over time, the extent of preferential flow within the aquifer can be reliably deduced (e.g. Birk et al. 2004).

Conductivity

Although strictly termed "specific electrical conductance," in practice the term "conductivity" is very widely used. The ability of a given water to conduct electricity is directly proportional to the amount of dissolved, charged species (ions) which it contains. It is very easy to measure the conductivity of a water sample, using robust and inexpensive hand-held electronic meters. Conductivity values are normally expressed in the units of microsiemens per centimeter (μS/cm), or else for more saline waters, in millisiemens per centimeter (mS/cm). (1 mS/cm = 1000 μS/cm). The relationship between ionic content and conductivity is direct: if one takes the sum of meq/L cation concentrations in a water and multiplies this by 100, the resulting number should approximate very closely (i.e. ± 10%) the conductivity of that water expressed in μS/cm. (Obviously, because of the principle of electroneutrality, the same calculation can be done equally well using the sum of meq/L anion concentrations.) As we shall see in Section 4.2.5, it is also possible to directly estimate the total dissolved solids (TDS) content of a groundwater from its conductivity.

Groundwaters exhibit a very wide range of conductivity values. For instance, shallow groundwaters found in the soil zone of mountain basins in humid areas typically have conductivities in the range 10–50 μS/cm, reflecting the paucity of solutes in rainwater. Groundwaters in major aquifers in lowland temperate regions usually display conductivities in the range 150–1000 μS/cm, reflecting an increase in the total solute content due to dissolution of common minerals. Where highly soluble minerals such as gypsum ($CaSO_4 \cdot 2H_2O$) or halite (NaCl) are present, far higher solute contents quickly develop, resulting in conductivities of thousands to tens of thousands of millisiemens per centimeter. Similarly high conductivities can also develop in hot countries due simply to direct evaporation from the water table, in areas where it lies less than about 2 meters below ground level. Where marine waters have invaded aquifers, conductivities as high as 55 mS/cm are common, reflecting the high solute content of the sea. Ancient groundwaters found at great depth in certain sedimentary basins have been found to have conductivities as high as 350 mS/cm.

Conductivity is easy to measure on site, and because it provides good clues about the presence of distinctive bodies of groundwater it is very widely measured during routine hydrogeological fieldwork. This in turn allows more focused sampling of the different types of groundwater in a study area.

*p*H

*p*H is the most common measure of the acidity/alkalinity balance in a solution. It is a measure of the availability in solution of hydrogen ions (H^+), also known as "**protons**"; this is why *p*H is sometimes referred to as an indicator of the "proton acidity" of a groundwater (see Younger et al. 2002a). In formal terms, *p*H is defined as the negative logarithm (to base 10) of the hydrogen ion activity (in moles/liter). Values commonly fall in the range between 0 and 14,[v] normally reported without units. As previously noted, accurate estimates of activity are most rigorously obtained using thermodynamic modeling. In most cases of practical concern, however, only negligible errors will be introduced by assuming that the concentration and activity of H^+ are equivalent. The key process which governs the proton balance, and therefore *p*H, is the dissociation of the water molecule H_2O:

$$H_2O \leftrightarrow H^+ + OH^-$$

In other words, water molecules can split to release both protons (H^+) and hydroxide ions (OH^-) to solution. When the concentrations of protons and hydroxide ions are equal, the logarithm of the molar H^+ concentration yields a value of 7. In other words, a *p*H value of 7 denotes a water which is neither acidic nor alkaline, and can thus be described as having **neutral** *p*H. In practice few waters have a *p*H of precisely 7, and we therefore refer to waters with a *p*H in the range between 6.5 and 8.5 as being **circum-neutral**.

In practice, groundwaters with a *p*H below 6.5 can be termed **acidic**. In acidic waters, little OH^- is present in solution, whereas H^+ is abundant. This typically occurs where the meq/L concentration of major anions is not balanced by an equivalent quantity of major cations. Dissociation of water occurs, releasing protons to solution to maintain electroneutrality. Most rainwaters, and therefore many groundwaters, are at least slightly acidic. Even in the absence of industrial pollution,[vi] rainwater *p*H commonly falls in the range 5–6. This is due to the tendency for atmospheric carbon dioxide to dissolve in rainwater to form carbonic acid. Usually, reaction of acidic rainwater with minerals and organic matter present in soil raises *p*H to around 7. However, where acidic rainwater infiltrates without encountering any reactive minerals (for instance in a soil composed mainly of quartz sand), then the resultant groundwater can be expected to have a *p*H of 6 or less. Far lower *p*H values are commonly found in two principal hydrogeological environments. First, where recently recharged groundwaters have interacted with peats and/or other acidic soils rich in organic matter but virtually devoid of mineral matter, *p*H can range as low as 3.3 (e.g. Banas and Gos 2004). Even lower *p*H values, down to around 2 (e.g. Banks et al. 1997), are associated with dissolution of the oxidation products of the common iron disulfide (FeS_2) minerals, pyrite and marcasite (Box 4.1). To put some of these acidic *p*H values in perspective, lemon juice has a *p*H around 2.5, vinegar around 2.8, cola and orange juice about 3, tomato juice around 4, and black coffee 5.

Waters with a *p*H above 8.5 are considered **alkaline**. In alkaline waters, the dissolved concentrations of OH^- greatly exceed those of H^+. This normally occurs where the meq/L concentration of major cations in solution greatly exceeds that of the major anions. In order for the overall electroneutrality of the water to be maintained, water molecules will dissociate to release sufficient hydroxide ions to balance the positive charge exerted by the major cations. In natural groundwaters, alkaline *p*H values most commonly lie in the range 8.5–9.0. This is a common range in many confined limestone aquifers, for instance. Few natural hydrogeological settings yield strongly alkaline waters. Two exceptions are: deep-seated groundwaters, long isolated from the atmosphere, which have equilibrated with a category of igneous rocks ("ultrabasic rocks") which contain abundant calcium-rich silicate minerals (e.g. Barnes et al. 1978); and groundwaters emerging from lime-rich coaly mud beds previously subject to spontaneous combustion, forming natural quick lime (Khoury et al. 1985). These waters typically have *p*H values as high as 12.5, and an overall solution chemistry dominated by Ca^{2+} as the major cation and OH^- as the major anion. Again, some perspective on the *p*H values typical of these alkaline groundwaters is afforded by comparison with some everyday substances, for instance milk (*p*H ~ 7), sea water (8), baking soda (9), and domestic bleach (11).

Three aspects of *p*H value interpretation deserve special emphasis.

1 It is important always to remember that *p*H is logarithmic in nature, so that a one-unit change in *p*H corresponds to a tenfold change in proton concentration. In view of this, an apparently "modest" *p*H change needs to be evaluated with a due sense of proportion.
2 The balance of dissolved cations and anions determines the degree to which water molecules must dissociate in order to maintain electroneutrality. Because strongly ionizing cations or anions can affect the H^+–OH^- balance, *p*H is rightly regarded as being only one component of the *total acidity* of a given water. While ambient *p*H is certainly the most useful single index of the acidity–alkalinity balance, the small atomic mass of H^+ in comparison with the dissolved metals means that it tends to make a rather modest contribution to the total dissolved mass of acidity-generating ions. In other words "*p*H" is *not* the same parameter as "acidity". We shall explore this crucial distinction a little further in Section 4.2.5.

3 The rates of many geochemical reactions are strongly *p*H-dependent. For instance, dissolution of carbonates and silicates occurs much more rapidly at low *p*H than at high *p*H. Conversely, adhesion of most cations to mineral surfaces occurs much more rapidly at circum-neutral to moderately alkaline *p*H than it does at low *p*H. *p*H is thus rightly considered to be a "master variable" in many geochemical environments: to know *p*H is to be able to predict many aspects of solution chemistry.

Eh

Eh is a measure of the status of "**redox**" reactions in a given water, and as such Eh is sometimes also known as "**redox potential**." The term "redox" is actually a contraction of the two words "reduction" and "oxidation," which respectively refer to the gain and loss of electrons by reacting ions. A typical example of a redox reaction (the oxidation of pyrite by atmospheric oxygen) has already been presented in Box 4.1. We can label this reaction as shown below to clearly identify which substances are losing electrons ("**electron donors**") and which are receiving electrons ("**electron acceptors**"). Alternative, classical terminology is given in the final line below the reaction, in which the substances which are doing the oxidizing (**oxidants**) are distinguished from those which are being oxidized (**reductants**).

$$FeS_{2(s)} + 3.5O_{2(g)} + H_2O \rightarrow Fe^{2+}_{(aq)} + 2SO^{2-}_{4(aq)} + 2H^{+}_{(aq)}$$

electron acceptor (reductant)	electron donor (oxidant)	electron donor (oxidant)

Being essentially a measure of the status of electron distribution between potentially interacting ions, Eh is an electrical potential, and it is therefore usually measured in millivolts.

If you find the concept of Eh a little difficult to grasp at first, you are in good company: most hydrogeologists find Eh difficult to deal with, not least because it is extremely difficult to measure accurately in the field (see Section 4.1.4 "Analytical methods"). Indeed, so unreliable are many Eh measurements of groundwaters that the values obtained cannot be interpreted in strictly quantitative terms. The most one can normally say is that well-oxygenated waters, in which most cations are in their most highly charged forms (e.g. with iron present as Fe^{3+} rather than as Fe^{2+}), tend to display high values of Eh ($\gg$100 mV). On the other hand, in waters utterly devoid of dissolved oxygen, in which cations are in their least-charged form, Eh tends to be low (<100 mV), or even negative.

In the same manner that *p*H indicates the activity of H^+ ions in solution, it is also possible to define an oxidation-reduction potential which directly represents the theoretical electron transfer potential of a solution. Using this approach, a parameter known as ***pe*** has been defined. Unfortunately, as free electrons do not occur as such in solution, *pe* cannot be measured directly. However, it can be calculated from Eh and other parameters. Readers desiring to know more about *pe*, Eh, and other ways of quantifying redox phenomena should consult Schüring et al. (2000) and Christensen et al. (2000).

4.2.2 Solutes versus colloids

Generally, it is easy to understand the difference in state between solids, which are merely suspended in water, and substances which are truly dissolved. Think about what happens if you sprinkle table salt into a glass of clear water. At first, you can plainly see the individual particles of salt: stir the water and you will see the particles of salt swirling around. However, within a few seconds the particles begin to diminish in size and number. Before half a minute has passed, the water will appear completely clear once more, with no suspended particles of salt visible at all. The salt has dissolved in the water. That is to say, the previously solid salt (NaCl) has broken down ("dissociated") into individual charged atoms (ions) of sodium (Na^+) and chloride (Cl^-), all of which are too small to be visible to the human eye (even with the aid of powerful microscopes). The ions are thoroughly mixed in amongst the water molecules with which they are now interacting electrostatically. Nice and simple. The dissolved state is obvious in the case of sodium chloride table salt. When a substance is truly dissolved in this manner, it is considered to be a **solute**.

But what about more complex substances? What if the substance in question breaks down not to individual ions, but to large molecules, which are many times larger than water molecules? This is what happens when we add milk to tea, for instance: the milk molecules do not truly dissolve, rather they form a dense suspension of particles which are much large than water molecules and which make the tea cloudy (reflecting light and thus making the tea appear paler). If you add sour milk to tea, the size of the resulting suspended particles is even greater, so that very large curds will float on the liquid surface. In groundwaters, some compounds behave rather like fresh milk in a cup of tea. This is true, for instance, of the very large carbon-bearing molecules (usually referred to as **"humic and fulvic substances"**)[vii] which occur in groundwaters with a high content of natural organic matter. Other complex substances present in groundwater interact with other ions to create very large molecules, which again may actually occur in suspension rather than in true solution. Such molecules (or clusters of adhering molecules) are referred to as **colloids**. While many colloids are present in natural groundwaters, they have been most widely studied in groundwaters polluted by synthetic organic compounds, such as pesticides and solvents (see Chapter 8). Pathogenic organisms, such as viruses, typically also behave as colloids rather than as solutes. Failure to recognize colloids can lead to oversight of important processes by which apparently uncharged fractions of certain substances can be transported through groundwater systems (see Stumm and Morgan 1996, p. 819).

In strictly physical terms, the distinction between colloids and solutes is best made in terms of the presence or absence of **chemical potentials**, which are measures of the energy contents of given dissolved substances, as functions of temperature, pressure, and composition. Using this convention, Stumm and Morgan (1996, p. 819) define a solute as a "species for which a chemical potential can be defined." By contrast, a colloid is a small particle (<100 μm) (but still larger than a solute) which is devoid of any definable chemical potential. As direct measurement of the chemical potentials of individual solutes is exacting, to say the least, operational definitions of colloids are often based on particle diameter alone. The selection of an appropriate cut-off diameter between solutes and colloids is arguable. Common sampling and analysis techniques involve filtering water through 0.45 μm diameter pores. Although it is generally held that this is too large a diameter to provide a true distinction between solutes and colloids, the very small particles which will pass through a 0.45 μm filter do not readily settle from suspension, and therefore behave in a manner very similar to true solutes (see Stumm and Morgan 1996, p. 825). More rigorous differentiation between colloids and solutes is rarely attempted outside of research projects. In the majority of cases it will be sufficient to bear in mind that part of the chemical load of a groundwater may well be colloidal in nature, and that this might affect the interpretation which we place upon analytical results.

4.2.3 Major cations

The cations which are present in the greatest concentrations (almost always greater than 1 mg/L) in most groundwaters are calcium (Ca^{2+}), magnesium (Mg^{2+}), sodium (Na^+), and potassium (K^+). The concentration ranges of these cations in most freshwater aquifers are summarized in Table 4.1. Modest concentrations of all four elements are introduced to aquifers in rainwater, although evaporative concentration during the recharge process seldom raises any of them above about 20 mg/L. The dissolution of minerals present in the soil and bedrock are the major natural sources of all four.

Calcium and magnesium are predominantly sourced from dissolution of carbonate minerals, especially calcite ($CaCO_3$, which can also contain significant quantities of Mg) and dolomite ($CaMg(CO_3)_2$), both of which are abundant in limestone terrains. Calcite is also a common cementing phase in many sandstones. In some sedimentary sequences, beds comprising the highly soluble minerals gypsum ($CaSO_4 \cdot 2H_2O$) and anhydrite ($CaSO_4$) can act as important sources of dissolved Ca. In some cases, the dissolution of gypsum over geological time has led to the development of extensive cave systems (Klimchouk et al. 1996) of equal magnitude to the more common cave systems found in limestones.

Many silicate minerals are also important sources for Ca^{2+} and Mg^{2+} in groundwaters. For instance, in the so-called ultrabasic and basic igneous rocks[viii] (including basalt lavas), dissolved Ca and Mg are derived from the weathering of anorthitic plagioclase ($CaAl_2Si_2O_8$), diopsidic pyroxene ($CaMgSi_2O_6$), and forsteritic olivine (Mg_2SiO_4). In more acidic igneous rocks[viii] such as rhyolite tuffs and granites, and in the many sedimentary rocks derived from them, common sources for Ca and Mg include hornblende ($Ca_2Mg_4Al_2Si_7O_{22}(OH)_2$) and biotite mica ($K(Mg,Fe)_3(Si_3Al)O_{10}(OH)_2$). In contrast to the carbonate minerals which tend to dissolve completely in water without depositing any new solid phases (so-called **congruent dissolution**), these silicate minerals are all subject to **incongruent dissolution**, in which the release of Ca^{2+}, Mg^{2+}, and SiO_2 to solution is accompanied by simultaneous precipitation of clay minerals. The formation of clay minerals effectively traps nearly all of the aluminum (and much of the SiO_2) in solid form.

Silicate weathering is also a common source for dissolved Na^+ and K^+. Most plagioclase feldspars contain at least some Na^+, and in many acidic igneous rocks[viii] and associated sediments, the Na-rich plagioclase predominates, with a composition close to that of pure albite ($NaAlSi_3O_8$). The same types of rock are similarly enriched in potassium feldspars ($KAlSi_3O_8$) of various varieties, including sanidine (common in rhyolite tuffs and lavas), orthoclase (which forms the conspicuous pink mega-crystals in many granites), and microcline (common in many coarse-grained granites and hydrothermal veins). Because Na^+ and K^+ are both so soluble, neither of them form carbonate minerals. However, they are abundantly present in the minerals halite (NaCl) and sylvite (KCl), which are both common constituents of ancient "evaporite" (i.e. salt lake) deposits, formed under hyperarid conditions. When such minerals gain access to modern groundwaters, they tend to dissolve so vigorously that they yield many thousands of mg/L of Na^+ and/or K^+ to solution.

Readers seeking further details on the sources and sinks for major cations in natural waters are recommended to consult the comprehensive account of Hem (1985).

4.2.4 Major anions

The anions which are present in the greatest concentrations (all > 1 mg/L) in most groundwaters are bicarbonate (HCO_3^-), sulfate (SO_4^{2-}) and chloride (Cl^-). The ranges of concentrations of these anions in most freshwater aquifers are summarized in Table 4.1. Although modest concentrations of all three anions are introduced to aquifers in rainwater, even after evaporative concentration during the recharge process their rainwater-derived concentrations seldom exceed about 20 mg/L.

Bicarbonate dissolved in groundwaters is derived from two principal natural sources:

- Biogenic: CO_2 is released into the soil atmosphere, and thus into waters draining through the soil, both directly from plant roots and (more importantly) by the microbial degradation of soil organic matter. At circum-neutral *p*H, CO_2 dissolves in water to form bicarbonate as follows: $CO_{2(d)} + OH^-_{(aq)} \leftrightarrow HCO^-_{3(aq)}$.
- Mineral: the dissolution of the same carbonate minerals which release Ca^{2+} and Mg^{2+} to solution also yield abundant dissolved HCO_3^- (see the first reaction listed in Box 4.1).

Sulfate dissolved in groundwaters has two principal natural sources:

- Weathering of sulfide minerals, most commonly pyrite (see the second reaction in Box 4.1).
- Weathering of gypsum and/or anhydrite, as already mentioned in relation to Ca^{2+} release.

Besides being a common *source* of dissolved SO_4^{2-}, gypsum also serves to impose an upper limit on sulfate concentrations in most groundwaters, due to its maximum solubility limit of around 2500 mg/L, which is frequently reached in waters which receive their SO_4^{2-} from pyrite oxidation (see Younger et al. 2002a).

Chloride is one of the least reactive solutes found in groundwater systems, and as such it has very few natural mineral sources. Clearly where the evaporite minerals halite and sylvite are encountered by flowing groundwaters (see preceding section), very high concentrations of Cl^- can result. Beyond these evaporite minerals, few others dissolve to release Cl^-. One exception is sodalite ($Na_4(Si_3Al_3)O_{12}Cl$), a mineral found in some (rather rare) alkaline igneous rocks. Given the sparsity of mineral sources for Cl^-, it can often be a very effective index of the degree of evaporation a given groundwater must have undergone after first arriving at the soil surface as rainwater (see the final two paragraphs of Section 2.2.3). In many hydrogeological settings, concentrations of Cl^- much greater than can be accounted for by evaporative concentration of rainwater can be taken to indicate that the groundwater in question actually represents a mixture of different water sources. Sea water is, of course, very rich in chloride (averaging 18,980 mg/L), so that Cl^- concentrations can be a sensitive indicator of the intrusion of marine groundwaters into terrestrial aquifers (see Chapters 7 and 8).

Many ancient groundwaters trapped at depth in sedimentary aquifers are notably rich in Cl^-. In some cases, these ancient waters can be shown to be trapped sea waters which entered the aquifer during periods of the Quaternary when the relative positions of the land and sea were rather different than at present (e.g. Elliot et al. 2001). In other cases, the high Cl^- content of a deep-seated groundwater may reflect: (i) an ancient history of evaporation in the near-surface environment; (ii) dissolution of evaporite rocks at depth; or (iii) enrichment of solute concentrations due to natural **membrane filtration** (Freeze and Cherry 1979, pp. 292–295). This is thought to occur at very great depths in some sedimentary basins, where the natural head gradient forces groundwater to flow through a mudstone bed which has pores so small that they even prevent the migration of solutes. The result is that a hyperconcentrated brine accumulates on the up-gradient side of the mudstone bed.

4.2.5 Some other important components

Silica

A common natural inorganic component of nearly all groundwaters, which is very often present at concentrations in the range 1–20 mg/L, and yet cannot be classed as a major cation or anion, is **silica (SiO_2)**. Solid silica is an extremely common component of many rocks, as the mineral quartz, grains of which form the bulk of most sandstones and unconsolidated sand deposits the world over. Despite its ubiquitous occurrence, however, quartz is highly insoluble and thus contributes virtually no SiO_2 to solution. Rather, the source of most dissolved silica is the incongruent dissolution of silicate minerals, as discussed in Section 4.2.3. This is such a prolific source of dissolved SiO_2 that it is present at saturation concentrations in the vast majority of groundwaters. Haines and Lloyd (1985) provide further insights into the controls on SiO_2 occurrence in groundwaters.

Alkalinity

The **alkalinity** of a groundwater is one of its most important characteristics, as it represents the ability of the water to resist acidification. As was noted in Section 4.1.4 ("Analytical methods"), alkalinity measurement involves a titration of the raw groundwater with a strong acid (normally sulfuric) until the *p*H has been lowered to 4.5. By analytical definition, therefore, a water with a *p*H less than 4.5 is regarded as having zero alkalinity. In the *p*H range 4.5–8.0, the alkalinity of most groundwaters is dominated by the bicarbonate content. The dominance of bicarbonate lessens as *p*H rises further, and once *p*H exceeds 8.5, CO_3^{2-} becomes the dominant form of alkalinity. At very high *p*H (>11), OH^- dominates the total alkalinity. Sometimes other dissolved components contribute significantly to the alkalinity; for instance, dissolved hydrogen sulfide (HS^-) is common in deep groundwaters totally devoid of oxygen. For the most part, however, alkalinity can be regarded as a surrogate measure of bicarbonate concentration (see the end of Section 4.1.3 for an explanation of how to calculate HCO_3^- concentrations from reported alkalinities).

Acidity

There is more to **acidity** than meets the eye. Most newcomers to hydrogeochemistry presume that acidity is fully accounted for by quoting the *p*H value. However, as was mentioned in Section 4.2.1, this is an incorrect presumption. The reason is that there are many metals which tend to resist any rise in *p*H in their host water by either releasing protons by hydrolysis (e.g. Fe^{3+}, Al^{3+}) and/or by consuming hydroxide ions from solution as they precipitate solid phases at circum-neutral *p*H (e.g. Fe^{2+}, Fe^{3+}, Al^{3+}, Mn^{2+}, Zn^{2+}, Cu^{2+}, Cd^{2+}). Given that the measurement of acidity involves titrating water with a strong alkali (usually NaOH) until it reaches a high-*p*H end-point (usually 8.5), any dissolved substance which reacts to resist the rise in *p*H must be regarded as a component of the total acidity. This understanding is crucially important in the context of the management of groundwater in mining areas, in which acidity generated by pyrite oxidation is a common problem (e.g. Younger et al. 2002a). It was within that field of activity that the following method of calculating the total acidity of a water from knowledge of its dissolved components was developed:

$$\text{Total acidity (meq/L)} = 1000(10^{-pH}) + \{Fe^{2+}\} + \{Fe^{3+}\} + \{Mn^{2+}\} + \{Zn^{2+}\} + \{Al^{3+}\} + \{Cu^{2+}\}$$

where each of the values given in curved parentheses (i.e. "{ }") is the concentration of the corresponding ion in meq/L. Cations listed in this formula that are not present at high concentrations (which is often the case for Cu^{2+}, for instance) can simply be omitted from the calculation. On the other hand, if other metals (e.g. Cd^{2+}, Ni^{2+}, Cr^{6+}) are present in significant concentrations, they may be easily added to the formula. To obtain a total acidity value in mg/L as $CaCO_3$ equivalent, multiply the result obtained using the above formula by 50.

Hardness

The **hardness** of a water is a reflection of the concentrations of Ca^{2+} and Mg^{2+} in solution. It can be easily calculated from the dissolved concentrations of these two cations as follows:

$$\text{Total hardness (mg/L as } CaCO_3) = 2.5[Ca^{2+}] + 4.1[Mg^{2+}]$$

In this formula the square brackets denote the concentrations of each cation in mg/L. An abundance of Ca^{2+} and Mg^{2+} in a water makes it highly prone to precipitate white or creamy yellow "lime scale" ($CaCO_3$) when it is heated. This tendency greatly limits the utility of hard waters (see Chapter 7). The total hardness of a water is often subdivided into two components, namely the **temporary hardness** (which can be removed by boiling, during which the Ca^{2+} and Mg^{2+} combine with the bicarbonate alkalinity in the water to precipitate carbonate scale) and the **permanent hardness** (which remains even after boiling). Given that temporary hardness depends on the availability of alkalinity, we can summarize the basis for distinguishing between temporary and permanent hardness as follows:

- If there is no alkalinity, all hardness is permanent hardness.
- If alkalinity exceeds total hardness, all of the hardness is temporary (i.e. there is zero permanent hardness).
- If total hardness exceeds alkalinity, then temporary hardness equals alkalinity and permanent hardness equals the difference between total hardness and alkalinity.

Total dissolved solids and salinity

The **total dissolved solids (TDS)** content of a water is the most common measure of its overall degree of **mineralization**, i.e. its content of dissolved mineral matter. It is also the best measure of the **salinity** of a groundwater. Although TDS is best determined by evaporation (Section 4.1.4 "Analytical methods"), it is also possible to estimate it by summing the concentrations of the individual dissolved components of the water (provided that the analysis as a whole has a cation–anion balance less than 5%). More commonly, TDS is estimated by multiplying the conductivity of the water by some factor in the range 0.55–0.75, with the median value of 0.65 being a common choice. If this method is to be used extensively in a given groundwater study, a correlation analysis of observed TDS and conductivity values should be used to obtain a reliable local estimate of this conversion factor. Having obtained a TDS value, it is possible to classify the water as follows (Freeze and Cherry 1979):

Fresh water: TDS < 1000 mg/L.
Brackish water: 1000 mg/L < TDS < 10,000 mg/L.
Saline water: 10,000 mg/L < TDS < 100,000 mg/L.
Hyper-saline water (or "brine"): TDS > 100,000 mg/L.

It should be noted that sea water generally has a TDS of around 35,000 mg/L, and that water becomes too salty to drink when the TDS exceeds about 2500 mg/L. High TDS greatly limits the utility of a water for various purposes (see Chapter 7).

Minor ions

Having identified the major cations and anions present in most waters, it is important to note that there are a number of other ions which are usually present at concentrations in the range 0.01–1 mg/L, but which are occasionally present at far higher concentrations, such that they can locally be regarded as major ions. Substances in this category include ferrous iron (Fe^{2+}), manganese (Mn^{2+}), nitrate (NO_3^-), ammonium (NH_4^+), hydrogen sulfide (HS^-), fluoride (F^-), and boron (B^{3+}). As we shall see in Chapter 7, every one of these ions is problematical in relation to water use.

Trace ions

Virtually the entire periodic table can potentially occur in groundwaters at low concentrations (typically < 0.01 mg/L). Where the ions in question are toxic to humans (e.g. Cd^{2+}, Hg^{2+}) and/or to wildlife (e.g. Al^{3+}, Zn^{2+}, Cu^{2+}) their importance in practical terms (see Chapter 7) may far outweigh their contribution to the TDS of the water.

4.2.6 Natural organic compounds

Organic compounds are defined as "covalent" compounds of carbon, i.e. compounds in which electrons are shared between adjoining carbon atoms. In nature, such compounds are typically synthesized biologically, hence the name "organic." Most unpolluted groundwaters contain little in the way of organic compounds. It is usual to quantify the overall concentration of organic compounds in a water by measuring the **total organic carbon (TOC)** and the **dissolved organic carbon (DOC)** contents (Table 4.1). The TOC is a sum of the DOC plus any colloidal organic compounds. The TOC of most groundwaters rarely exceeds 5 mg/L. Most of the TOC is accounted for by **humic and fulvic substances (HFS)**, which are large and complex molecules derived from degradation of plant material in the soil zone.[vii] Characterization of the detailed chemical structures of HFS compounds is a very demanding research task, which is unlikely ever to become sufficiently straightforward and inexpensive for it to be undertaken routinely during the laboratory analysis of water samples. HFS tend to carry a net negative charge in solution, which makes them highly prone to associate with dissolved cations to form organo-metallic complexes.

In a few natural groundwater settings, e.g. peat lands and other terrains with soils very rich in organic matter, HFS can form an important element of the overall TDS. Groundwater fed by recharge through peat and similar soils can often pick up a high TOC (>10 mg/L). The presence of large quantities of dissolved HFS is often indicated by anomalous cation–anion balances, where these are calculated using the conventional major cation and anion concentrations only. If the negative charge of the HFS is neglected, many peat groundwaters yield cation–anion balances which are excessively positive. Accounting for the contribution of the HFS to electroneutrality requires the application of a specialist geochemical modeling program known as WHAM (**W**indermere **H**umic **A**queous **M**odel), which not only speciates waters rich in TOC, but also predicts the likely partitioning of metals between the dissolved state and organometallic complexes (Tipping 1994; Lofts and Tipping 2000).

In general, the field of organic hydrogeochemistry is dominated by the scientific study of pollutant organic compounds derived from human activities. Some of the principal categories of organic pollutants are discussed in Chapters 9 and 11. For a robust introduction to the terminology of pollutant organics, the reader should consult Hounslow (1995); a thorough grounding in the key scientific issues relating to these pollutants is provided by Fetter (1999).

4.2.7 Useful isotopes

To understand what an "**isotope**" is, it is necessary to recall some of the basics of atomic structure. An atom consists of a **nucleus**, in which protons and neutrons reside, and **orbitals**, which are occupied by electrons. When an atom bears no electrical charge (i.e. it is in the **nonionized** form), the total number of protons in the nucleus will exactly match the total number of electrons in the various orbitals. However, as we have already seen, electrons can be added to or removed from atoms during chemical (redox) reactions, resulting in **ionization**, i.e. giving the atom a net charge (positive in the case of cations, negative in the case of anions). While the electrons can enter or leave the outer orbitals relatively freely, the same does not apply to the protons in the nucleus. Irrespective of the chemical transformations any atom or ion of a particular element has undergone, its nucleus will always contain exactly the same number of protons as other atoms or ions of the same element. It is thus precisely the number of protons which really defines the identity of each element in the periodic table. However, the number of neutrons in the nucleus can vary (within narrow bounds) from one atom to another without affecting the identity of the element. Thus the nucleus of a carbon atom normally contains 6 protons and 6 neutrons, the combined weights of which result in carbon having an average relative atomic mass (RAM) close to 12. However, a small proportion of all carbon atoms present in the universe have an extra neutron in their nucleus, so that the RAM = 6 protons + 7 neutrons = 13. A carbon atom with a RAM of 13 is said to be an isotope of carbon. This particular isotope is generally referred to as "carbon-13," normally written in abbreviated form as ^{13}C. Both carbon-12 (^{12}C) and ^{13}C are physically stable, and will not decay under normal environmental conditions: they are classic examples of **stable isotopes**.

Not all isotopes are stable. For instance, where the nucleus of a carbon atom contains 8 neutrons plus the usual 6 protons (giving a RAM of 14), the resulting isotope is carbon-14 (^{14}C), which is famously unstable. The nucleus of each ^{14}C atom will eventually undergo a fundamental change: one of the neutrons will partially breakdown, becoming a proton in the process, releasing a packet of energy equivalent to an electron (known as a beta-particle). The result is a nucleus with 7 neutrons and 7 protons. As the number of protons has changed from 6 to 7, the atom has also now changed identity: it is no longer a carbon atom, but a nitrogen atom (^{14}N, which is stable). Those atoms of a particular element which will break down in this manner are known as **radioactive isotopes**, and the process whereby their nuclei are transformed is known as **radioactive decay**. Not all unstable nuclei undergo radioactive decay at the same instant; some will do so relatively soon after formation, others a long time later. Overall, however, the process of decay "averages out" to a predictable rate. In the case of ^{14}C, for instance, if we start counting atoms at a particular point in time, then after 5720 years only half of the original number of ^{14}C atoms will remain – the rest will have already decayed to form ^{14}N. The number of atoms will halve again after a further 5720 years. We can thus say that ^{14}C has a **half-life** of 5720 years. All other radioactive isotopes have their own characteristic half-lives, ranging from 50,000 million years in the case of potassium-40 down to only 3.82 days in the case of radon-222.

In hydrogeological studies, lots of useful information can be gained by studying **environmental isotopes**, i.e. those isotopes which are present naturally in the environment, as opposed to those introduced deliberately for experimental purposes (Hiscock 2005). For instance, using our knowledge of the half-life ^{14}C, it is possible (though by no means easy) to estimate how many centuries have passed since a given water first entered the subsurface (for several well-documented examples, see Lloyd and Heathcote (1985) and Hiscock (2005)). For several decades after the start of large-scale atmospheric testing of atomic weapons in 1953, fairly accurate dating of shallow groundwaters and waters in the unsaturated zone was possible by means of studying tritium (^{3}H), a radioactive isotope of hydrogen which was released in great quantities by the atomic bomb explosions. Large peaks in tritium concentrations were observed in waters which had entered aquifers after 1953 by infiltration of rainwater (Freeze and Cherry 1979, pp. 136–137). However, given that tritium has a relatively brief half-life (12.3 years) and given also that there has been no further atmospheric testing of atomic weapons since 1962, tritium is now ceasing to be a useful groundwater tracer.[ix] A number of other "radiometric" groundwater dating techniques exist (e.g. Mazor 1991; Elliott et al. 1999). However, because of the costs of the necessary analysis and the very high level of skill needed to interpret the results, radiometric dating of groundwaters is restricted mainly to research projects.

Studies of stable isotopes can shed much light on the details of both water movement patterns and geochemical processes (Lloyd and Heathcote 1985; Mazor 1991; Hiscock 2005). The principal stable isotopes which are studied in order to infer physical processes affecting groundwater flow patterns are oxygen-18 (^{18}O) and deuterium (^{2}H, or simply "D"), as both of these isotopes are to be found in the water molecule itself. As water evaporates, the heavier molecules tend to get left behind, so that water which has been subject to much evaporation can be expected to contain higher proportions of ^{18}O and ^{2}H than fresh rainwater. Armed with this information, it is sometimes possible to identify "old" and "new" components of groundwater in environments where vigorous mixing can occur, such as in upland stream runoff source areas (Buttle 1994).

Another important use of isotope studies is in identifying the influence of microbial processes on hydrogeochemical reactions. Just as the heavier isotopes tend to get "left behind" during evaporation, the heavier isotopes of nutrients (especially such as ^{13}C, ^{34}S, and ^{15}N) tend to pass through many metabolic chains more slowly than their lighter counterparts (^{12}C, ^{32}S and ^{14}N). Shifts in the relative proportions of heavy and light isotopes can provide evidence of the extent and rates of microbial intervention in geochemical cycles.

4.2.8 Dissolved gases

Direct measurements of the quantities of gases dissolved in groundwater tend to be made only as part of specialized investigations, such as:

- Determining the **dissolved oxygen (DO)** content of groundwater which is to be pumped into a surface water course to support flows during the dry season; if the water is too low in oxygen, it will be detrimental to aquatic life.
- Assessments of the nature of gas seen bubbling out of groundwater intended to be used for public water supply; for instance, it is important to be able to distinguish between effervescence due to degassing of carbon dioxide (CO_2) and potential explosion risks associated with the (far less common) evolution of methane (CH_4).
- Evaluations of the potential for release of **volatile organic compounds (VOCs)** from groundwaters polluted with organic chemicals (e.g. Fetter 1999).
- Studies of gaseous hydrogen (H_2) release from deeply anoxic groundwaters undergoing *in situ* remediation (see Christensen et al. 2000).

- Investigations of dissolved noble gases, both in relation to the health risks posed by radon (Rn), and in relation to the use of relative abundances of various noble gases as indicators of past climatic conditions during the recharge of an ancient groundwater (Elliott et al. 1999).

While measurement of dissolved oxygen is easily made in the field using electronic meters, for most other dissolved gases it will be necessary to carefully extract the gas from the water and then submit the samples for laboratory analysis.

4.3 Displaying and classifying groundwater quality

4.3.1 Converting raw data

Many commonly used techniques for graphically displaying groundwater qualities require that the concentrations of constituent ions be expressed in meq/L. We have already seen (Section 4.1.3) how to calculate meq/L from raw data expressed in mg/L. For ease of reference, Table 4.2 gives conversion factors for the principal cations and anions used on most groundwater composition plots.

Several plotting techniques further require that the meq/L concentrations be converted into percentage values. Most commonly, the requirement is for the meq/L quantity of each cation to be expressed as a percentage of the sum of the meq/L concentrations of all major cations, and for the meq/L quantity of each anion to be expressed as a percentage of the sum of the meq/L concentrations of all major anions.

4.3.2 Representing groundwater quality on maps

The simplest way of representing groundwater quality information on a map is to contour the concentrations of a particular substance of interest. The problem with this approach is that it is difficult to represent more than two or three substances on each map without the view becoming extremely cluttered. Where spatial variations in overall major ion chemistry are to be investigated, it is better to use one of the following plotting techniques, all of which require the data to be in meq/L.

The simplest mapping symbols are pie-charts (e.g. Figure 4.2a), arranged so that cations and anions plot in the two separate semicircles (Hem 1985). (Given concentrations in meq/L, if the analysis has a decent cation–anion balance, then the two semicircles should be equal in size.) The diameter of each pie-chart can be scaled in proportion to the TDS (or the sum of meq/L concentrations if preferred). Alternatively, bar-charts can be used (Figure 4.2b), with separate columns representing the cations and anions (Lloyd and Heathcote 1985). Another less obvious plotting symbol is the "Stiff diagram" (Figure 4.3), so-called in commemoration of its originator (Stiff 1951). A Stiff diagram is an elongate polygon, the precise shape of which is determined by "joining the dots" corresponding to the meq/L concentrations of each major ion on a template (Figure 4.3a). The template need only be shown in the legend of a map, with the individual symbols (Figure 4.3b) plotted on the map without clutter.

4.3.3 Groundwater classification diagrams

The interpretation of groundwater flow systems is greatly aided by use of a range of graphical diagrams which allow us to easily spot similarities and differences between separate water analyses. The variations under study may either relate to repeated samples of the same borehole over time, or samples collected from various points in a groundwater flow system at roughly the same time.

The most widely used diagram of this type is the Piper diagram (Figure 4.4a) (Piper 1944). To plot an analysis on a Piper diagram, the cations and anions are first plotted separately in the triangles at bottom left and right respectively, and then lines are drawn upwards from the plotting positions within both triangles (parallel to the outer edges of the upper diamond) until they meet within the upper diamond (Figure 4.4a).

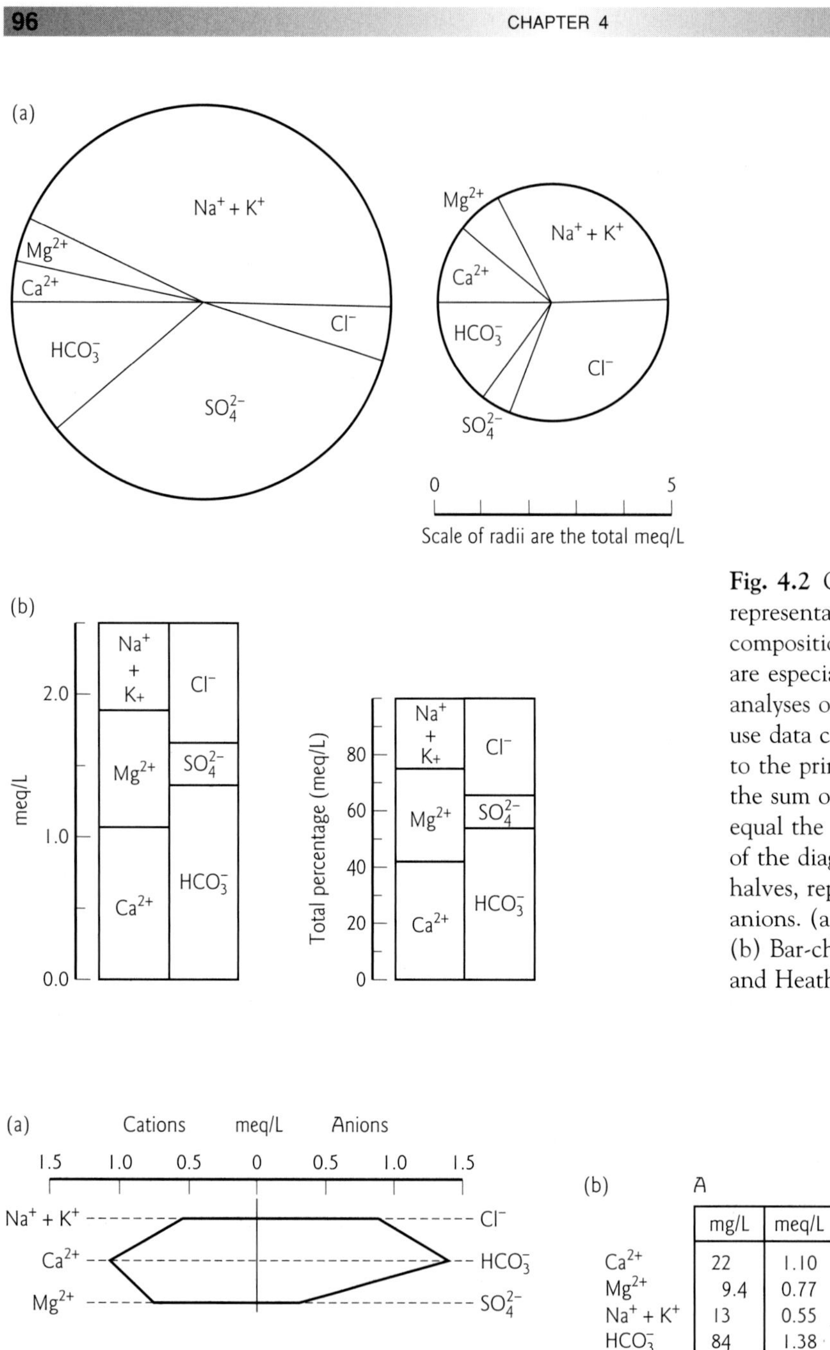

Fig. 4.2 Conventional graphical representations of the chemical compositions of natural waters, which are especially suited for plotting analyses on maps. Note that the plots use data converted into meq/L. Due to the principle of electroneutrality, the sum of cations in meq/L must equal the sum of anions. Hence both of the diagrams have equally sized halves, representing the cations and anions. (a) Conventional pie-chart. (b) Bar-chart. (Adapted after Lloyd and Heathcote 1985.)

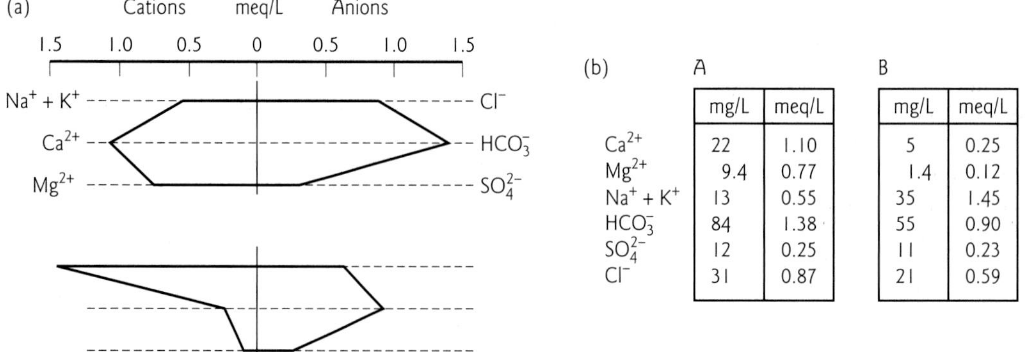

	A mg/L	A meq/L	B mg/L	B meq/L
Ca^{2+}	22	1.10	5	0.25
Mg^{2+}	9.4	0.77	1.4	0.12
$Na^+ + K^+$	13	0.55	35	1.45
HCO_3^-	84	1.38	55	0.90
SO_4^{2-}	12	0.25	11	0.23
Cl^-	31	0.87	21	0.59

Fig. 4.3 Stiff diagrams, a form of graphical representation especially useful for plotting hydrochemical analyses on maps. (a) The template shows the axes on which individual Stiff diagrams are drawn. In this case, analysis A is plotted on the template. Below, the second Stiff diagram without axis labels (as typically used on a map) displays analysis B. (b) Analyses used on the stiff diagrams in (a). (Adapted after Lloyd and Heathcote 1985.)

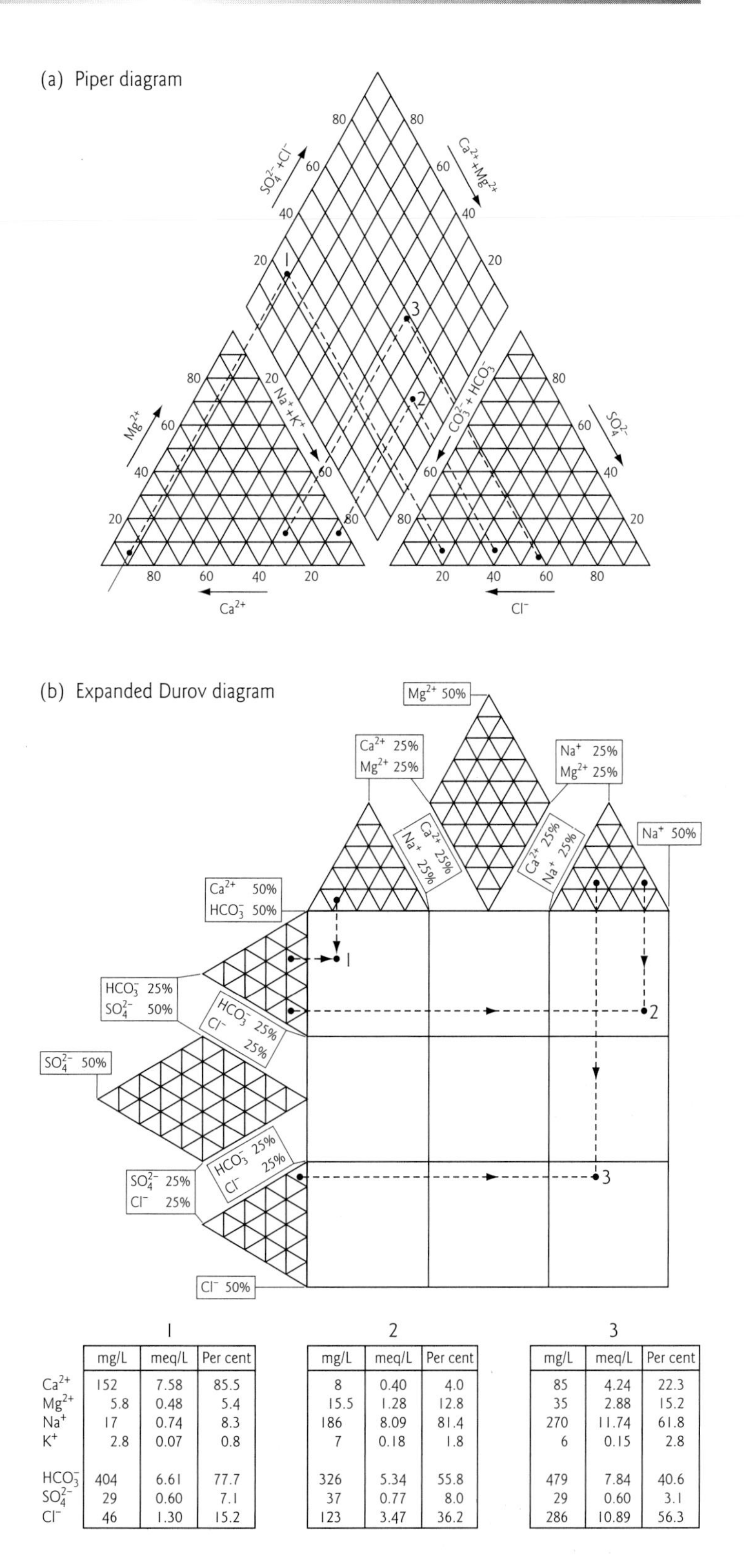

	1			2			3		
	mg/L	meq/L	Per cent	mg/L	meq/L	Per cent	mg/L	meq/L	Per cent
Ca^{2+}	152	7.58	85.5	8	0.40	4.0	85	4.24	22.3
Mg^{2+}	5.8	0.48	5.4	15.5	1.28	12.8	35	2.88	15.2
Na^+	17	0.74	8.3	186	8.09	81.4	270	11.74	61.8
K^+	2.8	0.07	0.8	7	0.18	1.8	6	0.15	2.8
HCO_3^-	404	6.61	77.7	326	5.34	55.8	479	7.84	40.6
SO_4^{2-}	29	0.60	7.1	37	0.77	8.0	29	0.60	3.1
Cl^-	46	1.30	15.2	123	3.47	36.2	286	10.89	56.3

Fig. 4.4 The two classic trilinear hydrochemical plotting diagrams widely used in hydrogeology. Analyses 1, 2, and 3 are plotted on both diagrams for comparison. (a) A Piper diagram. Cation and anion concentrations are separately plotted in the lower left and lower right triangles respectively, and then lines are projected parallel to the outer edge of the upper diamond until they meet within the diamond, to define the overall plotting positions for each of the samples (after Piper 1944). (b) Expanded Durov diagram (the slight difference in plotting procedures compared with Piper diagrams is explained in the text; after Lloyd and Heathcote 1985).

Similar in form to the Piper diagram, but yielding clearer final plots which are especially well suited to deducing the processes contributing to the chemical evolution of groundwaters in a given flow system, is the **Expanded Durov Diagram** (Figure 4.4b) (Lloyd and Heathcote 1985). Apart from the fact that the central plotting area is a square rather than a diamond, the principal difference between Piper and Expanded Durov diagrams is that the percentages of individual ions used in preparing Durov plots are calculated as *percentages of total ions* (i.e. anions plus cations), rather than percentages of the separate totals of cations and anions as in the Piper plot.

For certain types of groundwater, Piper and Durov diagrams do not offer sufficient discriminatory power. This is so, for instance, in the case of saline waters (Hounslow 1995). In order to differentiate between alternative sources of saline water which may be invading a freshwater aquifer, a **Hounslow diagram** is used (Figure 4.5a). If an individual groundwater sample is plotted on a Hounslow diagram, it is possible to infer the

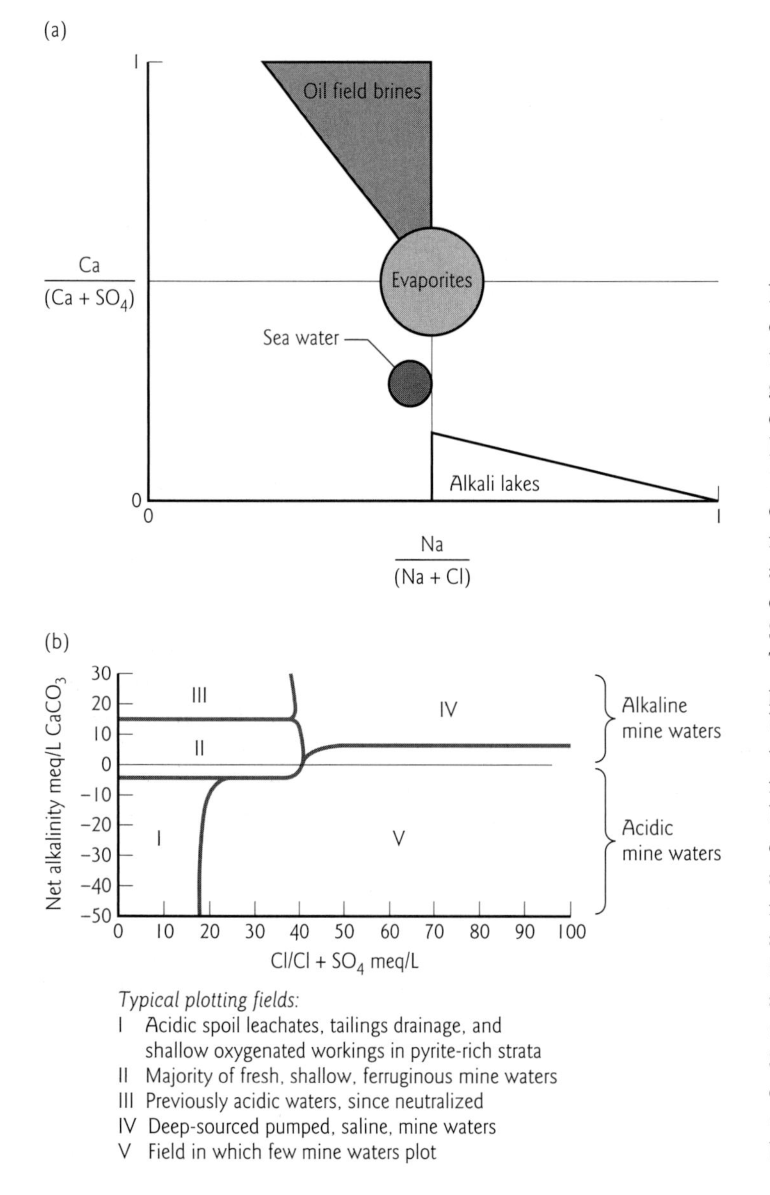

Fig. 4.5 Appropriate geochemical classification diagrams for certain types of highly mineralized groundwater. (a) A Hounslow diagram (also known as a "Brine Differentiation diagram"; Hounslow 1995) for deducing the likely origins of saline groundwaters. The ionic ratios plotted on the two axes are calculated from the molar concentrations (mol/L) of Ca, Na, SO_4, and Cl. (b) A Younger diagram. This diagram allows meaningful plotting of groundwaters affected by pyrite oxidation (especially mine waters, but potentially also drainage from acid-sulfate soils), and has been used to track the geochemical changes occurring during natural attenuation of acidity in mine water remediation studies. The "net alkalinity" parameter on the y-axis is calculated by subtracting the total acidity from the total alkalinity (both expressed in meq/L) (see Section 4.2.5 and Table 4.1 for details on these two parameters) (modified after the form proposed by Rees et al. 2002).

most likely source of any saline water which is affecting its composition by its proximity to one or more of the fields of saline water types plotted on the diagram. A further example of a specialized groundwater classification diagram comes from the field of minewater management. In many mining contexts, the acidity–alkalinity balance of a groundwater is of greater practical importance than the balance of conventional major and minor cations. Mixtures of saline and acidic waters are also of considerable importance in many cases. The diagram shown in Figure 4.5b (which was dubbed a **Younger diagram** by Rees et al. (2002)) captures these two factors, allowing easy differentiation between various types of groundwater found in many mining areas.

A range of other groundwater plotting techniques exist, further details of which may be found in the works of Hem (1985), Lloyd and Heathcote (1985), and Hounslow (1995).

4.3.4 Hydrochemical facies

Having mastered the presentation and comparison of hydrochemical data, it is possible to systematically distinguish one type of water from another. The handiest way to do this is to assign waters to **hydrochemical facies** which can be defined as zones within a groundwater system which display distinctive combinations of cation and anion concentrations. These "zones" are normally geographical, but can also be defined temporally where the quality of groundwater at a particular point changes over time. For the most part, hydrochemical facies can be identified simply by identifying the predominant cation(s) and anion(s) in a given water on the basis of "percent of total meq/L" of each category. Looking at the table in Figure 4.4, water 1 can be said to be of "Ca-HCO_3 facies," water 2 is "Na-HCO_3 facies," and water 3 is "Na-Cl-(HCO_3) facies." Two important points about facies definition are illustrated by the last example. First, if we attempted to classify facies using mg/L rather than meq/L, we would wrongly conclude that HCO_3^- is the dominant anion, whereas, because of the difference in atomic mass between the two, Cl^- actually dominates the anion complement. Second, if we have more than one anion present in high concentrations, we can list more than one in the facies definition. In this case HCO_3 is listed in parentheses, implying that it is a significant component of the anion complement even though Cl dominates.

Clearly, if the relevant data have already been plotted on a Piper or Expanded Durov diagram, it is feasible to define hydrochemical facies by inspecting the plot. For the case of an Expanded Durov diagram, for instance, the nine numbered fields in the main square of the diagram (Figure 4.6) can be related to hydrochemical facies as shown in Table 4.3.

4.3.5 Eh–*p*H diagrams

We have already noted the profound influence which *p*H and redox conditions exert on the mobility of a range of inorganic solutes. It is often helpful to directly examine the relationship between Eh, *p*H, and the behavior of specific ions or compounds in a given water. The best way to do this is to construct an Eh–*p*H diagram, such as the example given in Figure 4.7. The interpretation of Eh–*p*H diagrams is relatively straightforward, as explained in Box 4.2.

It is of course possible to construct Eh–*p*H diagrams to represent the behavior of a very wide range of elements (or groups of elements). A substantial compilation of such diagrams is given by Brookins (1988). However, as published examples usually relate to behavior at 25°C and to specific molar quantities of elements which are unlikely to precisely match those of any particular groundwater which you might be studying, the interpretation of published Eh–*p*H diagrams needs to be approached with considerable caution, especially given the major uncertainties which beset most measured values of Eh (see Section 4.2.1). Furthermore, Eh–*p*H diagrams display *equilibrium* conditions and do not tell us how quickly the mode of occurrence of a given solute will change following a change in Eh and/or *p*H. Notwithstanding these *caveats*, the recent advent of computer programs which can rapidly derive and

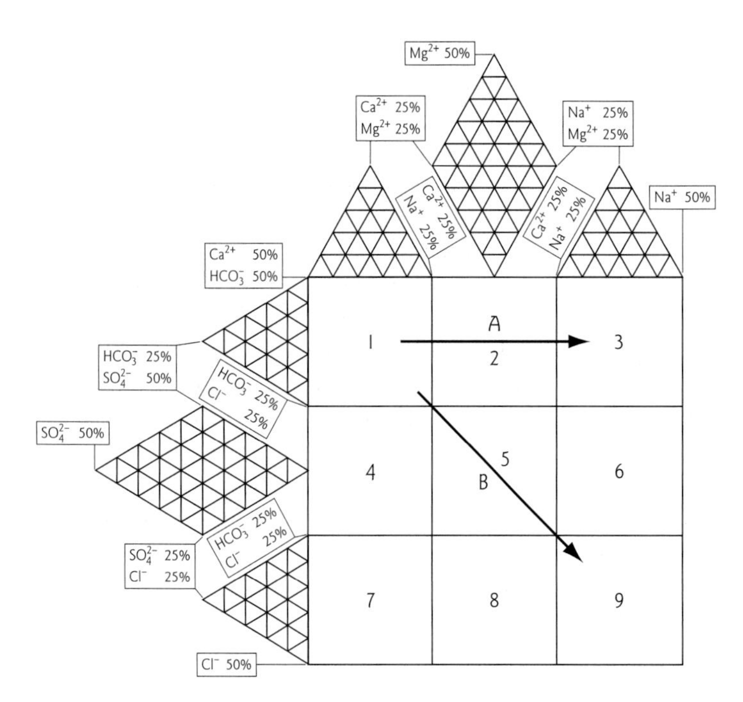

Fig. 4.6 Labeled plotting fields on an Expanded Durov diagram (cf. Figure 4.4b), which can be used to identify different hydrochemical facies in groundwater systems. The nine numbered fields in the central plotting square are identified and explained in Table 4.3. If plotted samples along a groundwater flow line (as defined on a flow net) show a clear trend from one facies to another, then it is possible to infer the predominant processes affecting the evolution of groundwater chemistry. Two examples of trend lines are shown here: Trend A is typical of ion-exchange during groundwater flow (as explained in Section 4.4.2), while Trend B is typical of mixing between fresh and saline groundwaters (see Section 4.4.3).

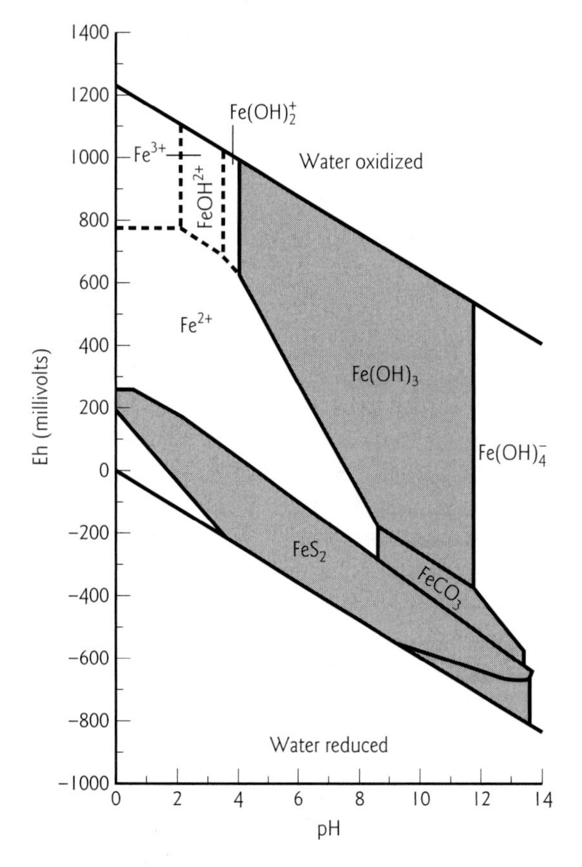

Fig. 4.7 (*Left*) An example of an Eh-*p*H diagram (modified after Hem 1985), in this case indicating the predominant modes of occurrence of species of iron (Fe), as either solid mineral phases (shaded fields) or dissolved aqueous species (unshaded fields), for various *p*H and Eh conditions. The lines dividing the shaded and nonshaded fields indicate equilibrium conditions between the solid phases and aqueous species. This diagram has been drawn assuming a temperature of 25°C, 1 atmosphere of pressure, and the following dissolved constituents: 0.1 mg/L Fe, 100 mg/L S, and 100 mg/L C (inorganic).

display Eh–*p*H diagrams for almost any user-specified conditions is gradually increasing the practical utility of such diagrams.

4.4 The evolution of natural groundwater quality

4.4.1 From rainwater to groundwater

We are used to thinking of rainwater as pure and unsullied (or at least we were in the days

Table 4.3 Summary of hydrochemical facies associated with the nine numbered fields on the Expanded Durov diagram shown in Figure 4.6.

Field number	Predominant facies*	Typical occurrences of these facies in real aquifers
1	Ca-HCO_3	Shallow, fresh groundwaters in recharge areas in a wide range of aquifer types†
2	Mg-HCO_3	Typical of the "leading-edge" of sea water intrusion into shallow unconfined aquifer‡
	Mg-Ca-HCO_3	Shallow, fresh groundwaters in aquifers composed (at least in part) of dolomite†
3	Na-HCO_3	Shallower portions of regional confined aquifers; waters deduced to have been affected by ion exchange (see Section 4.4.2)†
4	Ca-SO_4	Gypsum-bearing sedimentary aquifers,† and groundwaters affected by oxidation of pyrite and other sulfide minerals§
5	No clear facies	Normally the result of mixing of two or more different facies
6	Na-SO_4	Rare; can result from mixing of ancient Na-rich groundwaters with pyrite oxidation waters‖; also formed by intensive evaporation of waters which have previously lost their Ca and HCO_3 to calcite precipitation¶
7	Ca-Cl	Associated with the invasion of shallow, fresh aquifers by sea water‡; less commonly, can form by dissolution of the very rare evaporite mineral tachyhydrite ($CaCl_2$)
8	Mg-(Na)-(Ca)-Cl	Mixing of fresh and saline waters‡; possible influence also of reverse ion exchange†
9	Na-Cl	Influence of sea water, ancient saline groundwaters, or dissolution of halite (NaCl)††

* It may be necessary to refer to the raw data to determine the precise facies in some cases.
† See Lloyd and Heathcote (1985) for further discussion.
‡ See Appelo and Postma (1993), especially their Chapter 5.
§ See Nuttall and Younger (1999).
‖ See Hattingh et al. (2002).
¶ See Dudgeon (2005).
†† Discrimination between these various alternatives requires further analysis, such as plotting a Hounslow diagram (Figure 4.5a).

before the growing awareness of atmospheric contamination). While it is true that solute concentrations in rainwaters are significantly lower than in groundwaters, from a hydrochemical perspective we cannot simply ignore the chemical composition of rain and snow. Table 4.4 summarizes mean values and ranges of concentrations for major constituents of rainwaters from a wide range of monitoring sites on three continents. Most of the solute load of rainwaters is derived from the release of aerosols from bursting bubbles at the surface of the sea (or inland water body) from which they derived their moisture. It is thus not surprising that systematic variations

Box 4.2 Interpretation of Eh–*p*H diagrams.

All Eh–*p*H diagrams share some common features. Besides the universal convention of plotting *p*H on the *x*-axis and Eh on the *y*-axis, all Eh–*p*H diagrams which are drawn to cover the full Eh range from −1000 mV to +1400 mV will feature two parallel diagonal lines enclosing all of the plotted fields showing the occurrence of solutes and solids. As shown in Figure 4.7, above the top diagonal line (which runs from Eh = +1200 mV at *p*H = 0, across to Eh = +400 mV at *p*H = 14) water is so oxidized that it becomes unstable (it will begin to spontaneously form hydrogen peroxide H_2O_2), and discussion of "aqueous species" loses meaning. Below the bottom diagonal line (which runs from Eh = 0 mV at *p*H = 0, across to Eh = −820 mV at *p*H = 14), water again becomes unstable, this time breaking down to release hydrogen gas (H_2). It is therefore usual only to plot aqueous species between these two diagonal lines.

Turning to our specific example, Figure 4.7 illustrates the behavior of iron in groundwaters for the following specific conditions: temperature = 25°C; total Fe ≈ 100 μg/L; total sulfur ≈ 100 mg/L as SO_4^{2-}; total dissolved carbon dioxide ≈ 60 mg/L as HCO_3^-. The Eh and *p*H conditions over which it is possible for iron to be dissolved as free cations (i.e. Fe^{2+} and Fe^{3+}) are seen to occupy the left hand side of the diagram. This shows that free Fe^{3+} (**ferric iron**) is only soluble at very low *p*H (<2) and under highly oxidizing conditions (Eh > 780 mV). However, some Fe^{3+} can remain in solution at higher *p*H (≤4) in the form of hydroxide complexes (i.e. $FeOH^{2+}$ and $Fe(OH)_2^+$). The reduced form of free iron, Fe^{2+} (**ferrous iron**), is soluble over a much wider *p*H range than ferric iron (≤8.5), but only at relatively low Eh (<780 mV at *p*H 0, and <400 mV at *p*H 6). The shaded fields on the diagram show the solid mineral phases which limit the solubility of iron under various Eh–*p*H conditions. Across a very wide range of Eh conditions (−100 mV to +1000 mV), where *p*H is circum-neutral or greater, the principal limitation on the presence of dissolved Fe is the precipitation of ferric hydroxide ($Fe(OH)_3$), which occurs in a range of forms from amorphous orange sludge ("ochre") to black, crystalline goethite. In a narrow window at low Eh and high *p*H (>9), the predominance of CO_3^{2-} as the principal anion means that the precipitation of $FeCO_3$ (the mineral siderite) becomes the principal solid "sink" for Fe. However, under most low-Eh conditions, and across a very wide *p*H range, the precipitation of iron disulfide (FeS_2), either in amorphous form or as the minerals pyrite and marcasite, exerts a powerful limitation on the availability of dissolved Fe^{2+}.

Table 4.4 Rain water chemistry: a summary* (all values in mg/L, save for *p*H).

	*p*H	TDS	Ca	Mg	Na	K	Cl	HCO_3	SO_4	NH_4	NO_3	SiO_2
Mean	5.3	13	0.75	0.25	1.2	0.30	1.8	2.3	1.9	0.37	0.6	0.38
Range	4.1–5.6	4.8–35	0.0–1.42	0.0–0.5	0.26–2.46	0.14–0.37	0.2–30	1.9–3.0	0.4–3.7	0.0–0.48	0.1–2.0	0.0–0.9

* Derived from compilations of analytical results presented by Freeze and Cherry (1979) and Appelo and Postma (1993).

in rainwater chemistry can be correlated with distance from the coast. For instance, chloride concentrations are always much higher near the coast (reaching 30 mg/L in some cases) than they are many kilometers inland (with 1–2 mg/L being common at distances in excess of 200 km inland; Appelo and Postma 1993).

Rain water does not represent the only input of solutes to the unsaturated zone. Settlement of wind-borne dust particles, and the adsorption of atmospheric gases to mineral and plant surfaces, both constitute important sources of chemical loading to the soil surface. Collectively, these processes are known as **dry deposition**. It is far harder to measure chemical loadings derived from dry deposition than those arriving in rainfall. However, where accurate studies have been made, the results usually indicate that loadings due to dry deposition are of similar magnitude to those derived from rainfall. Indeed for specific contaminants, such as the heavy metals Pb, Cd, and Zn, dry deposition may be a more important source than rainwater inputs (see Appelo and Postma 1993).

Solutes arriving in rainfall, or dry-deposited species which dissolve in rainfall when it hits the surface, provide the starting point for the evolution of groundwater chemistry. As we have already noted, during the recharge generation process (see Chapter 2) evaporation significantly increases the concentrations of all solutes in the water. If we briefly contemplate Table 4.4 once more, it is easy to estimate the resultant concentrations from the loss of, say, 60%, 70%, or 80% of the water to evaporation. In doing such calculations, it is important to remember that *p*H is the negative log of hydrogen ion concentrations, so that evaporative concentration can be expected to significantly *lower p*H. On this basis, then, we can expect most fresh recharge waters to be relatively acidic (*p*H of 5 or less) with TDS contents of the order of 50 mg/L. These are precisely the sorts of values which we find in shallow groundwater systems in peat lands, for instance (e.g. Soulsby et al. 1998; Banas and Gos 2004), or where groundwater accumulates within pure quartz sand dunes solely by direct recharge (Appelo and Postma 1993).

However, in most hydrogeological settings, water does not infiltrate very far into the unsaturated zone before it encounters significant quantities of geochemically reactive solids, which proceed to dissolve in the water and alter its chemistry.[x] In many cases, the *p*H will be promptly raised by dissolution of calcite or (where calcite is lacking) by incongruent dissolution of silicate minerals (Section 4.2.3). Biogenic production of carbon dioxide is also an important process in the soil zone (Section 4.2.4), which can add further bicarbonate alkalinity to neutralize recharge waters. The concomitant microbial activity can strip much of the original dissolved oxygen from the waters as well. Of course in some soils further acidification of infiltrating water can occur, for instance if pyrite is present in the soil profile (as occurs in **acid-sulfate soils** which flank the coastlines of many mid-latitude countries; see Section 9.3.5). However, in most hydrogeological settings recharge arrives at the water table pre-neutralized and relatively low in dissolved oxygen.

4.4.2 Rock–water interactions in the saturated zone

Given the generally low velocities typical of flow in the saturated zone, there is normally ample time for even relatively slow geochemical reactions to substantially alter groundwater chemistry. For instance, dissolution of calcite under laboratory conditions normally requires no more than 24 hours to approach equilibrium. In a limestone aquifer, groundwater will typically be in contact with calcite for many years or even centuries, so that equilibrium is the norm rather than the exception. Only where very rapid, preferential flow pathways exist below the water table (for instance, where there are caves) are we ever likely to find groundwaters at depth which have not equilibrated with the carbonate minerals in the overlying sequence. The dissolution of silicate minerals generally occurs millions of times more slowly than carbonate minerals (see Younger et al. 2002a, p. 104). However, where groundwater residence times are of the order of decades or centuries, even silicate dissolution reactions can reach equilibrium. As the dissolution of both carbonates and silicates neutralizes acidity, there are very few pervasively acidic groundwaters (though note the significant exceptions mentioned in Section 4.2.1).

The flip-side of the coin is that *precipitation* of a number of minerals can impose an upper limit on the dissolved concentrations of certain ions. For instance, the upper limit of dissolved SO_4 is often imposed at around 2500 mg/L by precipitation of gypsum ($CaSO_4 \cdot 2H_2O$). Similarly, in the absence of sulfate, the upper limit of calcium solubility is frequently imposed at around 250 mg/L by precipitation of calcite.

Besides mineral dissolution reactions, another class of geochemical reactions can significantly alter groundwater compositions during flow through both the unsaturated and saturated zones. These are the so-called **surface processes** by which ions either adhere to, or are released from, the surfaces of aquifer solids. Adhesion of a solute to a solid surface is termed **adsorption**. The more familiar term "absorption" strictly refers to the passage of an ion *inside* a solid material (i.e. into microscale pores within the solid material). Because "adsorption" and "absorption" are difficult to distinguish in practice, at least at the scale of an aquifer as opposed to in a test-tube, the all-embracing term **sorption** is often used to refer to any process other than mineral precipitation which results in an ion becoming immobilized in close association with solid material. The opposite process, whereby previously sorbed ions are released back into solution, is referred to as **desorption**. In some cases, sorption of one ion can be matched by simultaneous desorption of another. Where this occurs, the overall process is referred to as **ion exchange**. Numerous ion exchange reactions are known in nature (see Appleo and Postma 1993; Langmuir 1997). The classic example of this phenomenon, which is widely documented from major aquifers worldwide, involves Ca^{2+} and Na^+. At depth in many ancient sedimentary aquifers (especially in confined zones), adsorbed Na^+ is present in abundance on mineral surfaces, presumably reflecting an earlier period in geological time when the aquifer sediments were bathed in sea water. As fresh groundwater, rich in dissolved Ca^{2+}, penetrates the depths of such an aquifer, a preferential sorption process occurs. Given the 2+ charge on the calcium ion and the fact that it has an atomic mass nearly twice that of Na^+ with its single charge, the mineral surfaces exert a greater electrostatic attraction for Ca^{2+} ions than for Na^+. Consequently, Ca^{2+} ions are sorbed onto the mineral surfaces while Na^+ ions are released into solution. The net result is that an incoming groundwater of Ca-HCO_3 facies is transformed into Na-HCO_3 facies, a process sometimes referred to as "natural softening" (since it removes the principal cause of hardness, Ca^{2+}, from solution).

The propensity of aquifer materials to participate in sorption/ion exchange reactions varies from one substance to another. A measure of this propensity, known as the **cation-exchange capacity (CEC)**, is used widely in soil science investigations. The highest CEC values are associated with sedimentary organic matter, especially in the soil zone, where there is much fresh humus (i.e. the relatively stable organic matter left behind after aerobic microbial decomposition of plant and animal refuse). Next in importance are the oxides and hydroxides of iron and manganese. Finally, clay minerals are also powerful sorbents, with CEC varying systematically from one class of clay minerals to another. While other minerals, such as silicates and carbonates, do participate in surface processes, they are far weaker sorbents than the substances listed above. For more detailed information on sorption and desorption reactions and the interpretation of CEC values, the reader is referred to Langmuir (1997).

It is often possible to identify the occurrence of ion exchange processes within an aquifer by plotting a series of analyses on a Piper diagram or an Expanded Durov diagram. For instance, Trend "A" marked on Figure 4.6 shows an example of a series of waters in a sandstone aquifer which range in quality from Ca-HCO_3 facies in the recharge area to Na-HCO_3 facies in the confined portions of the aquifer, where water quality has been modified by Ca-Na ion exchange, presumably involving clay minerals within the sandstone.

4.4.3 Mixing of fresh and saline groundwaters

We have already noted that groundwater beneath the sea bed tends to have a salinity similar to that of the overlying ocean, and that where the hydraulic head conditions are appropriate, marine groundwaters can readily invade terrestrial aquifers. This is a significant management issue for coastal aquifers (see Chapter 7). However, saline groundwaters are not restricted to coastal environments: they are present at depth in many thick sedimentary aquifers. The occurrence of deep saline groundwaters has been studied extensively in Europe, where thick bedrock aquifers are exploited for public water supply far more extensively than in North America. It is generally found that below depths of a few hundred meters, most groundwater is at least brackish, if not thoroughly saline. Comprehensive discussions of fresh–saline water interactions in a range of aquifer settings are provided by Lloyd and Heathcote (1985) and Appelo and Postma (1993). A number of geochemical diagnostic tools can be used to identify:

- The origins of different sources of saline water, using both major-ion chemistry (see Figure 4.5a and Hounslow 1995) and the presence of trace concentrations of "halide" ions, particularly bromide and iodide (e.g. Elliot et al. 2001).
- Incipient indications of saline intrusion, long before the TDS itself rises noticeably (Appelo and Postma 1993).

Plotting a series of groundwater analyses on a Piper diagram, or more usefully on an Expanded Durov diagram, can provide striking evidence for the contribution of fresh–saline water mixing to the overall evolution of groundwater chemistry within a given aquifer. For instance, Trend "B" on Figure 4.6 shows a clear mixing trend from fresh Ca-Mg-HCO_3 facies groundwater (typical of the aquifer recharge zone in this dolomitic aquifer) to Na-Cl facies (samples from wells near the coast).

4.4.4 Tracking mass balances along groundwater flow-paths

Given the wide range of processes which affect groundwater quality evolution (evaporation, mineral dissolution/precipitation, sorption, mixing, etc.), it might seem unlikely that we would be able to make much progress in unravelling the evolutionary history of a given groundwater. However, as the trends plotted on Figure 4.7 reveal, some trends in groundwater quality evolution can be relatively easy to identify. Even for more complex cases, it is possible to develop models which can quantitatively explain the differences in chemical composition between two water samples at either end of a given flow path within an aquifer. Given our knowledge of the chemical composition of the common rock-forming minerals, coupled with insights into the processes of cation exchange, it is possible to construct plausible explanations for changes in the molar concentrations of given solutes as water migrates from one point to another.

For instance, in a limestone aquifer, in which the principal rock-forming mineral is invariably calcite ($CaCO_3$), it is reasonable to infer that an observed increase in dissolved Ca^{2+} concentrations between an up-gradient and down-gradient well located along the same flow path can be ascribed to dissolution of calcite. Using molar concentrations and recalling that one mole of any substance has a mass in grams which is equal in numerical value to its relative molecular mass (Box 4.1), we can go on to calculate how many grams of calcite must have dissolved in each liter of groundwater to explain the observed rise in dissolved Ca^{2+} between the two wells. Such a model is known as a **geochemical mass balance** model.

One common application of geochemical mass balance modeling is the deduction of the processes which occur during the transformation of rainwater into groundwater (e.g. Chen et al. 1999; Younger 2004a). Such studies have a number of uses. For instance, they can help to elucidate the rate at which chemical erosion is occurring in the subsurface, which can be an issue of economic importance where it occurs sufficiently rapidly to affect ground stability (e.g. Lamont-Black et al. 2002; see also Chapter 8). They can also provide invaluable insights into the rates at which geochemical reactions occur under real field conditions. While much is known about how pure samples of individual minerals dissolve under laboratory conditions, relatively little is known about how the complex mixtures of different minerals found in undisturbed aquifers behave at ambient groundwater temperatures. By means of mass-balance modeling, we can obtain reaction rate information which can be crucial in determining whether a given aquifer has the potential to naturally clean up polluted waters which have accidentally entered the subsurface (e.g. Malmström et al. 2000; Banwart and Malmström 2001; Banwart et al. 2002; see also Chapter 11).

It is important to recognize that geochemical mass balance models are never *unique* explanations of observed changes in water chemistry; however, they are often the most credible explanations consistent with what we know of aquifer composition and the rates of dissolution of different minerals. Freely available software tools now exist which greatly facilitate the development of mass balance models for aquifers (e.g. Parkhurst and Appelo 1999). However, any such model can only be as good as the knowledge and experience of its developer will allow: considerable knowledge of mineralogy and dissolution rates is a necessary background for developing plausible geochemical mass balance models (Appelo and Postma 1993; Hounslow 1995; Parkhurst and Appelo 1999).

Endnotes

i A variety of classifications are used to distinguish "major", "minor," and "trace" constituents; in practice, setting the boundary between major and minor constituents at 1 mg/L works for most purposes (e.g. discussions in Sections 4.1.3, 4.2.3, and 4.2.4).

ii Or relative molecular mass, where we are dealing with a compound rather than a free ion of a single element.

iii Also known as "ionic balance" or simply as "electroneutrality"; it should be noted that some authors use slightly different definitions of this parameter, and this should be checked before coming to conclusions on values quoted.

iv The specific heat capacity of any solid or liquid is defined as the heat required to raise the temperature of a unit mass of that substance by one degree Celsius. The specific heat capacity of water (around 4.2 joule/gram/°C) is higher than that of any other common natural substance (which is why it is such a wonderful coolant).

v For the vast majority of natural waters, this holds true. Nevertheless, simple math tells us that for H^+ concentrations in excess of 1 mol/L, it is possible for *p*H to take on negative values. Values of *p*H as low as −3.5 have indeed been reported from abandoned underground mine workings at Iron Mountain, California, where waters initially acidified by dissolution of pyrite (FeS_2) have their acidity further increased by evaporation (Nordstrom et al. 2000).

vi Atmospheric pollution exacerbates, rather than originates, the phenomenon of "acid rain."

vii The terms "humic" and "fulvic" are used to describe the common apparently "soluble" components of ordinary soils. The distinction between the two is complex in detail; a *de minimis* definition is that fulvic substances are those which are soluble in water of any *p*H, whereas humic substances will only dissolve in alkaline waters.

viii For a detailed explanation of the nature and occurrence of ultrabasic, basic, and acidic igneous rocks, the reader is referred to standard igneous petrology texts such as that of Best and Christiansen (2000).

ix A similar development in groundwater tracing using chlorofluorocarbons (CFCs), the use of which was banned worldwide in 1996, is now underway (see Cook et al. 1995).

x It should be noted that a further class of reactions which commonly affect the chemistry of infiltrating waters (so-called "surface processes") are, for reasons of convenience, described in the following section.

5

Groundwater Discharge and Catchment Hydrology

From Your dwelling You water the hills; Earth drinks its fill of Your gift.
You make springs gush forth in the valleys: they flow in between the hills.
(Psalm 104)

Key questions

- What happens where the water table intersects the land surface?
- What's the difference between perennial, intermittent, and ephemeral groundwater discharges?
- What causes springs to occur in some places, and seepage faces in others?
- Wetlands, ponds, and lakes: are they just giant springs?
- How does groundwater make its way into streams?
- What can the study of streams tell us about groundwater characteristics?
- What controls groundwater discharge to streams at the catchment scale?

5.1 Groundwater discharge features

5.1.1 Dynamic outcrops of the water table

We have already seen that the spatial patterns of natural groundwater discharge are greatly influenced by geomorphology (see Section 3.3.1). Temporal patterns of groundwater discharge are also fundamentally controlled by the locations and physical nature of those geomorphological features which provide pathways to the surface for upwelling groundwaters. In this chapter, we will deepen our understanding of such features and the manner in which they determine the availability of discharging groundwaters in a wide range of surface environments. In passing, mention will also be made of some cases in which landscape features which normally act as points of groundwater discharge can temporarily switch roles, acting as pathways for indirect recharge (cf. Section 2.2.5).

In formal terms, all groundwater discharge features correspond to low points ("minima") in the head fields of the aquifers which they drain. In other words, if we were to trace a given groundwater flow path towards a groundwater discharge feature, we would find that head steadily decreases until it reaches a minimum value at the point of emergence at the ground surface. In essence, all groundwater discharge features amount to "outcrops" of the water table. However, unlike natural rock outcrops, which tend to remain static, the water table crops out in a rather dynamic manner, exhibiting marked changes over time in both elevation and consequent groundwater discharge rates. Many groundwater discharge features are **perennial**, which means that they emit water all year round, albeit the rate of emission can vary substantially over time. Such perennial discharge features tend to correspond to the lowest outcrop areas of major aquifer horizons. Other groundwater discharge features may be **intermittent**, functioning only for a few weeks or months in any one year, when the water table reaches seasonal highs following periods of sustained recharge. In some cases, groundwater discharge features are only **ephemeral**, which is to say that they operate for extremely limited periods when head reaches a temporary peak level. Ephemeral discharge features are often (but by no means invariably) associated with perched aquifers (see Section 1.3).

In the sections which follow we will examine the principal types of groundwater discharge features found in river catchments,[i] paying particular attention to the manner in which they transmit water from their parent aquifers to the surface environment. Figure 5.1 summarizes the typical occurrences of the most important types of groundwater discharge features found in the catchments of major freshwater streams and rivers.

5.1.2 Seepage faces

Where the water table intersects a hillslope, it is common for groundwater outflow to occur in a very diffuse manner throughout a large belt of ground running for tens or hundreds of meters along the hillside. Although the amounts of outflow occurring at any one point will generally be immeasurably small, the total discharge might well be substantial. Such a belt of diffuse outflow is known as a **seepage face**. Its upper boundary corresponds to the elevation of the water table within the ground immediately behind the hillslope. Recognition of seepage faces in the field is not always easy, for where the outflowing groundwater is of good quality and not too voluminous, the seepage face may be overgrown. However a seepage face may be a very obvious landscape feature in cases where the rate of outflow is so high, or the quality of the discharging groundwater so poor, that it prevents plant growth (e.g. Figure 5.2).

By nature, seepage faces are associated only with unconfined portions of aquifers. The development of a wide seepage face implies that groundwater is flowing to the surface fairly evenly through most of the aquifer material behind the hillslope. Where the bulk of the flow is focused in a small number of preferential pathways, groundwater discharge is more likely to occur in the form of discrete springs.

5.1.3 Springs

A **spring** is a natural opening in the Earth's surface from which groundwater flows. In order to distinguish springs from a range of other groundwater discharge features (such as wetlands), it is worth expanding this simple definition by adding that the emerging groundwater flows briskly away from the spring in an open channel. Springs provide the principal means of natural discharge from confined aquifers and are also important outflow features in many unconfined aquifers. Figure 5.1 summarizes the most important types of springs found in the majority of catchments. In practice, it is helpful to be able to distinguish on hydrostratigraphic grounds between the following four types of spring:

1 "**Depression** springs," which are formed simply by the intersection of the land surface by the water table.

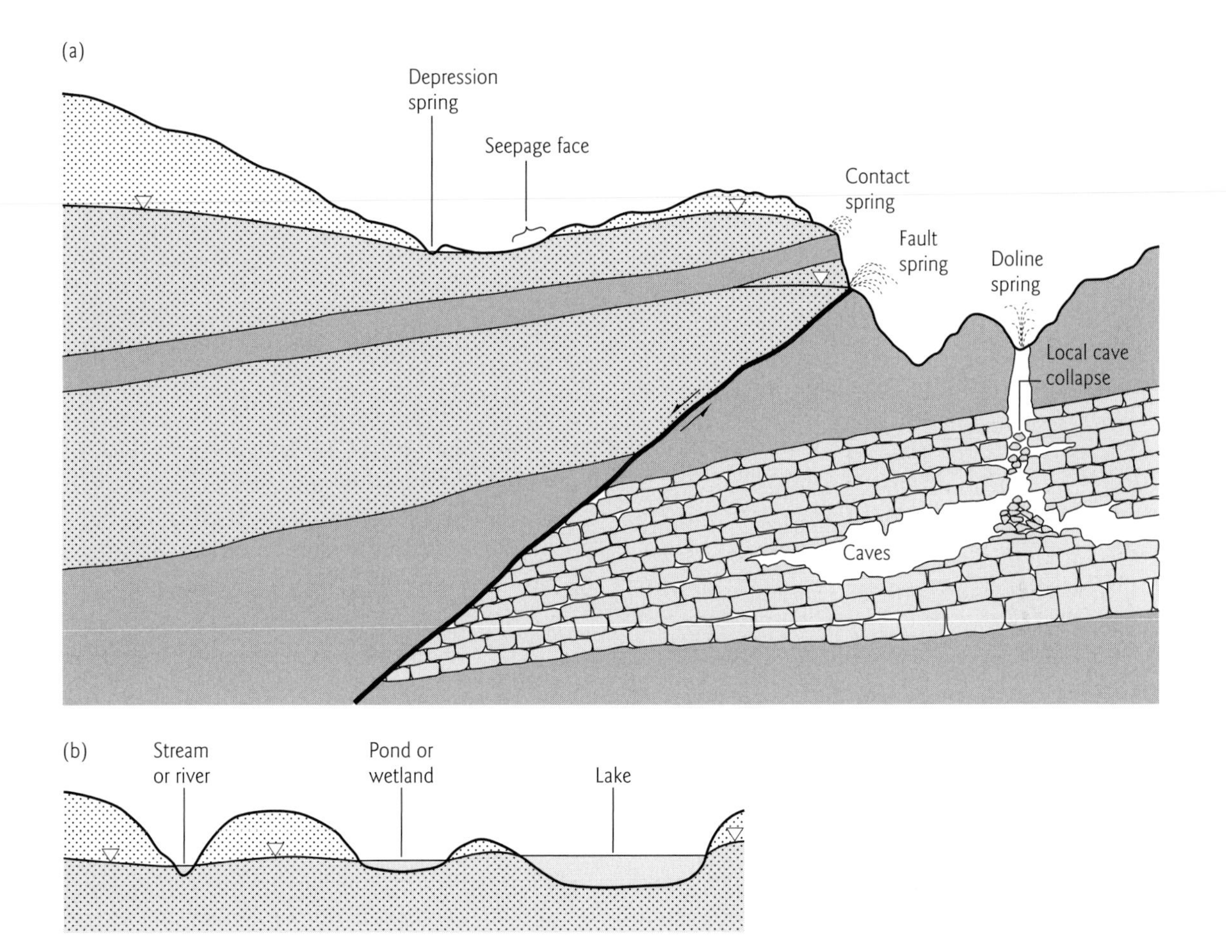

Fig. 5.1 Diagrammatic cross-sections illustrating the occurrence of major types of natural groundwater discharge features found in river catchments. (a) Springs (four major categories) and a seepage face. (b) Stream/river, wetland, and lake groundwater discharge zones.

2 "**Contact springs**," which typically arise at the lowest-lying point of outcrop of the stratigraphic contact between an aquifer and an underlying aquitard.
3 "**Fault springs**," which arise where a fault brings an aquifer into contact with an aquitard.
4 "**Doline springs**," which occur only in karst aquifers (cf. Section 1.5.2). Dolines are natural, vertical shafts (see Ford and Williams 1989), which can provide highly permeable pathways to the surface from deeply flooded caves and other conduits.

A general hierarchy of flow variability exists amongst these four different types of springs: Contact springs tend to display more variable flows than all other types (reflecting rapid response to shallow recharge), while the lowest variability in flow rates tends to be shown by fault and doline springs arising from confined aquifers (reflecting the generally slow response to recharge at depth). By nature, depression springs and contact springs are restricted to unconfined aquifers (or at least to the unconfined portions of aquifers which are elsewhere confined). Fault springs can occur in either unconfined or confined aquifer settings. Where they occur in confined aquifers, the fault zone in question must itself be permeable, thus providing a pathway to the surface for waters confined at depth. Doline springs can also discharge

Fig. 5.2 Example of a seepage face near Vryheid, KwaZulu, South Africa, in which a voluminous outflow of polluted (very acidic) groundwater results in an obvious impoverishment of vegetation below the line of the water table.

groundwater from both unconfined and confined aquifers. Some of the world's largest springs are doline springs fed by confined karst limestone aquifers.

Although these four categories cover most types of springs, it is important to note that even more specific terms are sometimes used to denote springs which display particular types of behavior. For instance, in terms of hydrochemical behavior, a spring which precipitates **travertine** (i.e. encrustations of mineral deposited by water emerging from the ground[ii]) may be termed a **travertine spring**. In terms of physical behavior, karst terrains give rise to an impressive array of spring types. For instance, certain karst springs tend to surge in flow every few hours, as a result of being fed by natural self-actuating siphons formed by natural caves and smaller conduits. One example is the Gihon Spring in Jerusalem, which feeds water into a basin called The Sheep Pool; in biblical times the stirring of the waters of this spring was considered to be due to the action of an unseen angel, and for this reason was regarded as having special healing powers (John 5:2–4). Some karst terrains also include a type of spring called an **estevelle** (Hardwick and Gunn 1995), which is a depression spring which discharges groundwater when the water table is high, but which can become a sink for surface water when the water table is low. Some estevelles are responsible for the dynamics of intermittent lakes or wetlands known as Turloughs (Hardwick and Gunn 1995).

Although it is usual to describe springs as a distinct category of groundwater discharge features, it is important to note that many groundwater discharge features which lie below the water lines of surface water bodies (i.e. wetlands, ponds, lakes, streams and rivers) are identical in form to surficial springs. Indeed, some authors refer to specific submerged groundwater discharge features as "underwater springs" or "subfluvial springs." However, because such features are normally inconspicuous and invariably difficult to observe directly, they are rarely documented.

5.1.4 Wetlands

The protection of wetlands has become one of the most notable environmental *causes célèbres* of modern times. While wetlands are informally referred to by a range of names with various shades of meaning (e.g. marshes, swamps, meres, fens, bogs; see Bullock and Acreman 2003), wetlands have also been the subject of numerous attempts at a more formal definition (Mitsch and Gosselink 2000). Perhaps the simplest of the more all-encompassing definitions is that adopted by the international Ramsar Convention, which defines wetlands as bodies of surface water less than 6 m in depth. The apparently arbitrary threshold of 6 m depth was chosen to ensure that the term "wetland" covers all circumstances in which plants rooted in the bed of the water body develop emergent stems and leaves above the water line. (In fact a threshold of only 1 m would achieve the necessary distinction in most freshwater systems, but brackish water wetlands of greater depth can host substantial emergent colonies of mangrove and allied species.) Using the 6 m depth criterion, the definition of wetland not only covers discrete bodies of water in which water depth is nowhere deeper than 6 m, it also includes the shallow margins of many larger, deeper (>6 m) lakes.

Ecology is replete with many big words denoting simple concepts. Thus depending on the principal source of water feeding them, wetlands are categorized as being **ombrotrophic** (fed solely by rainfall), **fluviotrophic** (fed largely by inflows of surface water), and **phreatotrophic** (groundwater fed). Although examples of each of these three "pure" types abound, it is important to bear in mind that many wetlands receive their waters from a combination of sources. In the context of this book we are principally concerned with understanding phreatotrophic wetlands.

Some phreatotrophic wetlands receive all of their inflow from subaerial springs and/or seepage faces which lie above the high water line. However, inconspicuous inflow of groundwater below the water line probably accounts for most groundwater discharge to phreatotrophic wetlands. Such subaqueous inflows may be localized (submerged springs) or diffuse (seepage flowing across a large area of the wetland bed). At the catchment scale, it is difficult (and rarely worthwhile) to distinguish phreatotrophic wetlands from large depression springs.

While many wetlands are thought of as being bodies of standing water, slow lateral movement of water is the norm. Wetlands transmit water from its point of entry to some outflow zone, either a point at which a stream leaves the wetland or a zone of infiltration to the subsurface. Lateral flow in most natural wetlands occurs by slow movement of surface waters between the stalks of emergent plants. However, the possibility of some lateral movement of groundwater within the bed sediments of the wetland should not be overlooked. Flow through the bed is commonly termed **hyporheic flow**. Another example of a "big" word from the ecological dictionary, hyporheic was originally coined by Orghidan (1959) by combining two Greek words: *hypo* (below) and *rheos* (flow). Hyporheic flow is a phenomenon more commonly associated with streams and rivers, as we will see in Section 5.1.6.

It should be noted that most **oases** in desert climates can be classed as phreatotrophic wetlands, in that all of the water entering the shallow surface depression is groundwater, which leaves naturally again either by onward groundwater flow or by evaporation. Large oases might fall within the classification of "through-flow" ponds or lakes (see next section).

5.1.5 Ponds and lakes

From a hydrogeological perspective, there is no fundamental difference between a pond and a lake.[iii] In terms of modes of groundwater discharge into and through ponds and lakes, most of the prior comments concerning wetlands are equally applicable here: groundwater can enter ponds and lakes from seepage faces or springs above the water line or, more commonly, by preferential or diffuse upflow through the bed. Given the substantial depth of water, hyporheic flow is even less likely to occur in ponds and lakes than in wetlands, although diffusional exchange of solutes between the water column and the bed sediment is commonplace and geochemically important.

Given the great size of many lakes, they often play significant roles in catchment water balances. A clear understanding of lake–groundwater interactions is clearly an important issue in many cases (see Winter et al. 1998). Lake–groundwater interactions in humid, temperate areas can typically be ascribed to one of four main scenarios (Figure 5.3):

1 Lakes fed by groundwater which deliver their entire outflow to surface streams (Figure 5.3a).
2 Lakes which are fed principally by surface water inflow, but which release most (if not all) of their water to the subsurface, thereby acting as major foci of indirect recharge (Figure 5.3b).
3 "Through-flow" lakes, which receive all of their water from an adjoining aquifer, and release all of their outflow to the same (or another) aquifer (Figure 5.3c).
4 Hybrid systems, in which lakes are fed by both groundwater and surface water inflows, but release their waters only via subsurface outflow.

Besides considering the balance of inflows and outflows of ground and surface waters, it is essential to note that all lakes are subject to significant losses of water to the atmosphere by evaporation. Indeed, in arid and semi-arid areas, many lakes are fed solely by groundwater inflow, all of which is lost to evaporation. It is therefore essential that lake evaporation be adequately quantified when evaluating the role of groundwater in the overall behavior of river catchments.

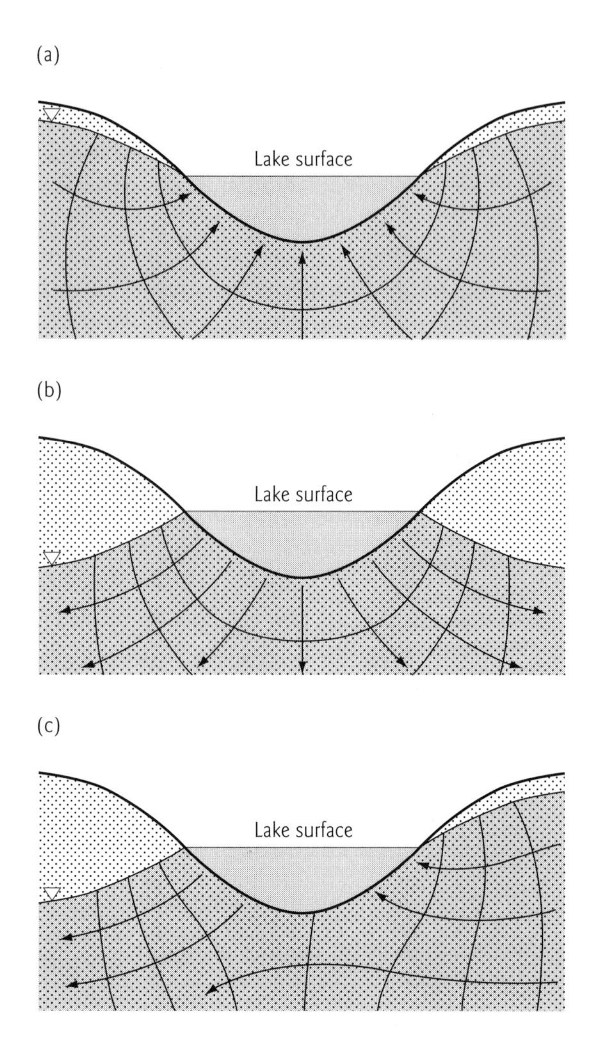

Fig. 5.3 Lake–groundwater interactions: three major scenarios (the same scenarios apply also to wetlands). (a) "Discharge lakes," which only *receive* groundwater discharge (releasing it all to surface outflow and/or evaporation). (b) "Recharge lakes," which are fed principally by surface water inflow, but which release most (if not all) of the water to the subsurface. (c) "Through-flow lakes," which receive all of their water from an adjoining aquifer, and release all of their outflow to the same (or another) aquifer. (Adapted after Winter et al. 1998.)

5.1.6 Streams, creeks, and rivers

Natural open channels come in all sorts of shapes and sizes, from the humblest brook to the mighty Amazon. Various terms are used in the English language to denote channels of varying size: brooks, burns, gills, becks, creeks, streams, rivers, etc. From the hydrogeological perspective, the size of a river does not fundamentally alter the manner in which it interacts with surrounding groundwater systems, and it is therefore possible to discuss the groundwater interactions of all natural open channels under a single category. Accordingly, the word "stream" is used throughout this book to denote a natural open channel of any size; the other terms will only be used where they form parts of names of real streams.

Groundwater enters stream channels from seepage faces and/or springs present above the water's edge, and/or by direct upflow through the stream bed, which (as in the cases of wetlands and lakes) can be highly localized or very diffuse. As we have already seen in Chapter 2, many streams lose water to underlying aquifers through their beds. Two factors control whether a given stream (or reach of a stream) gains or loses water by interaction with an adjoining aquifer (Figure 5.4):

- The difference in water level between the stream channel and the water table in the adjoining aquifer.
- The hydraulic conductivity of the channel perimeter.

Let us explore the implications of these two factors a little further. The water level at any one point in a stream at any specified time is called its **stage**. Unless the head in the adjoining aquifer exceeds the stage, it is impossible for the stream to gain water from groundwater discharge. That groundwater head should exceed stream stage is thus a *necessary* condition for groundwater discharge to a stream. However, it is not in itself a *sufficient* condition to guarantee a substantial discharge: the permeability of the channel perimeter is the final deciding factor. In many geological settings, the sediments which immediately flank the stream channel have a lower hydraulic conductivity than the adjoining aquifer materials (see Box 5.1 for an explanation of why this is often the case). Recalling Darcy's Law (Section 3.2), given that the cross-sectional area of contact between a stream and an adjoining aquifer is fixed by the channel dimensions, then, for a given hydraulic gradient between an aquifer and a stream channel, the rate of groundwater discharge to the river is wholly governed by the hydraulic conductivity of the channel perimeter. Of course this works both ways: if the stream bed is of sufficiently low hydraulic conductivity, then streamflow can occur without significant loss to the subsurface even where the head in the adjoining aquifer lies well below the elevation of the stream bed.

Streambed sediment comes in a wide range of particle sizes. In many cases it is silty, but in others it can consist of coarse sands, gravels, and even cobbles. Coarse bed sediments are common in streams which drain mountainous areas. Given the steep topographic gradients which are also typical of such streams, it is not surprising to learn that coarse-grained bed sediment is often a zone of intense mixing between groundwater upwelling from an adjoining aquifer and stream water which locally enters the bed sediment in response to local hydraulic gradients. The bed sediments and underlying aquifer materials within which this hyporheic flow occurs is termed the **hyporheic zone** (Figure 5.4f). At the catchment scale, the processes of water storage in the hyporheic zone tend to have a negligible impact on water resource availability. However (as we shall see in Sections 6.3 and 6.4), the dynamics of the hyporheic zone are of great importance to the ecological quality of many streams which gain water from adjoining aquifers (see Gibert et al. 1994; Brunke and Gonser 1997; Griebler et al. 2001; Malcolm et al. 2002).

In our discussion so far, the words "gain" and "loss" have been used to refer respectively to groundwater discharge to a stream and water migration from the channel to an underlying aquifer. Using this terminology, it is possible to talk of **gaining streams**, i.e. those which receive water from an adjoining aquifer, and **losing streams**, i.e. those which lose water to an underlying aquifer. These concepts of "gaining" and "losing" can be applied to entire streams or to individual reaches of streams. Indeed, it is often found that a particular stream will be gaining in some of its reaches and losing in others. The distribution of gaining and losing reaches often shifts seasonally: where a reach is always gaining, streamflow will be perennial, whereas intermittent streamflow is typical of reaches in which a switch from gaining to losing conditions occurs during the seasonal cycle (cf. Section 5.1.1). One of the most celebrated examples of intermittent, groundwater-fed stream reaches are the so-called

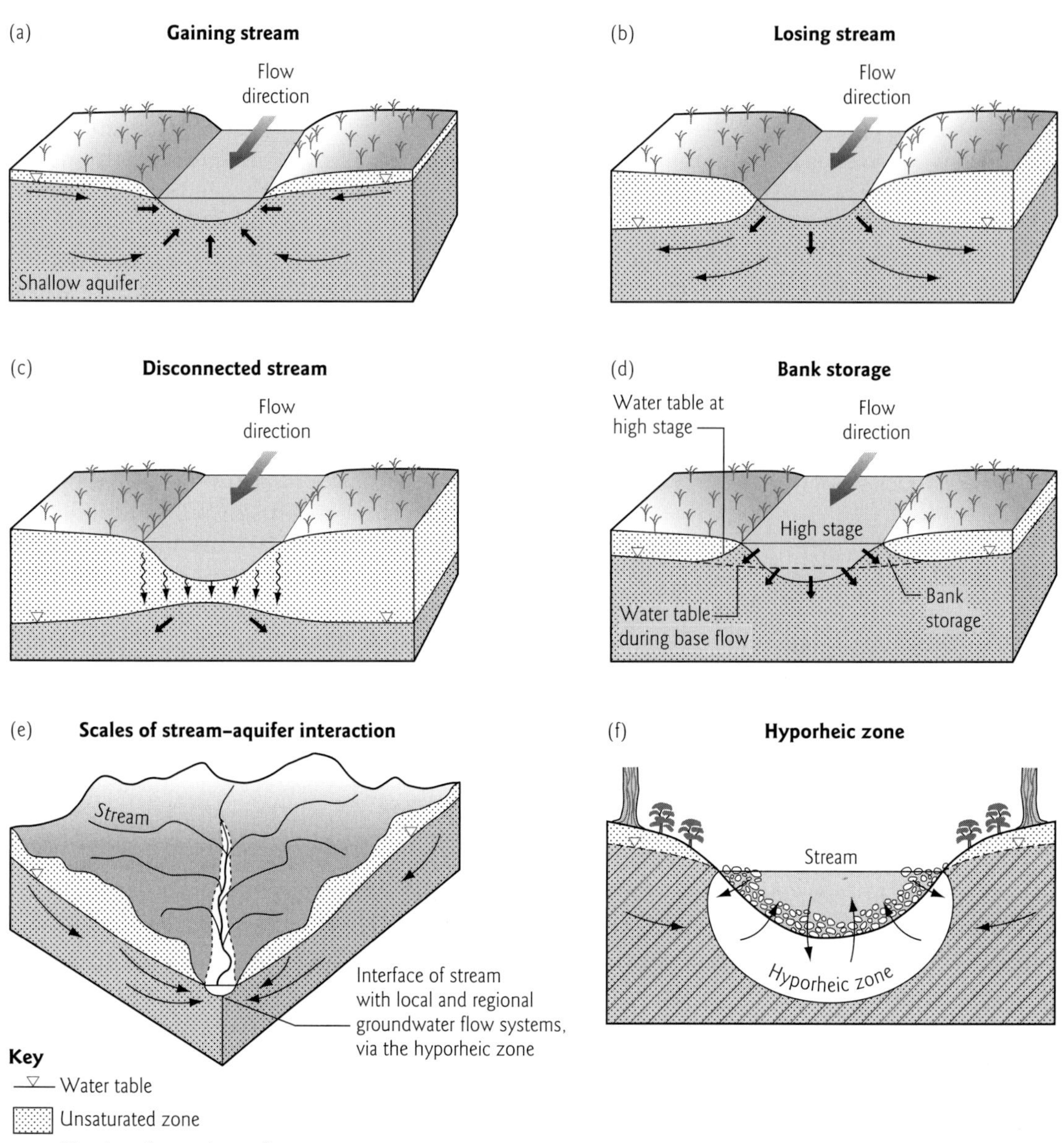

Fig. 5.4 Stream–aquifer interactions. (a) A gaining stream draining an aquifer. (b) A losing stream, recharging an aquifer. (c) A hydraulically disconnected stream: an example of a losing stream in which the water table lies far below the stream bed. (d) The bank storage phenomenon, whereby water enters the channel bank aquifer materials during periods of high stage (storm runoff), creating a "wedge" of stored water which returns to the channel after stage declines once more (cf. Figure 5.5). (e) Representation of the multiple spatial scales over which stream–aquifer interactions occur in real catchments. (f) Zooming in on the small-scale zone of interactions: the hyporheic zone. (Adapted after Winter et al. 1998.)

"winterbournes" of the Chalk downland in the countryside surrounding London (UK), which flow only in late winter and early spring when the water table is at its highest.

Complicated patterns of stream–aquifer exchange can arise where a previously gaining stream reach is suddenly subjected to flood flows due to surface runoff generated upstream. The possible sequence of events is illustrated in Figure 5.5. As the stage rises, the hydraulic gradient from the aquifer into the stream declines until it is actually reversed. At this point, flow can occur from the stream through the channel perimeter into the aquifer. Under these circumstances, the quantity of water in the channel is reduced; stream water is said to have entered **bank storage** (Figures 5.4d, Figure 5.5). Where the flood remains within the regular stream banks (Figure 5.5a,b), the

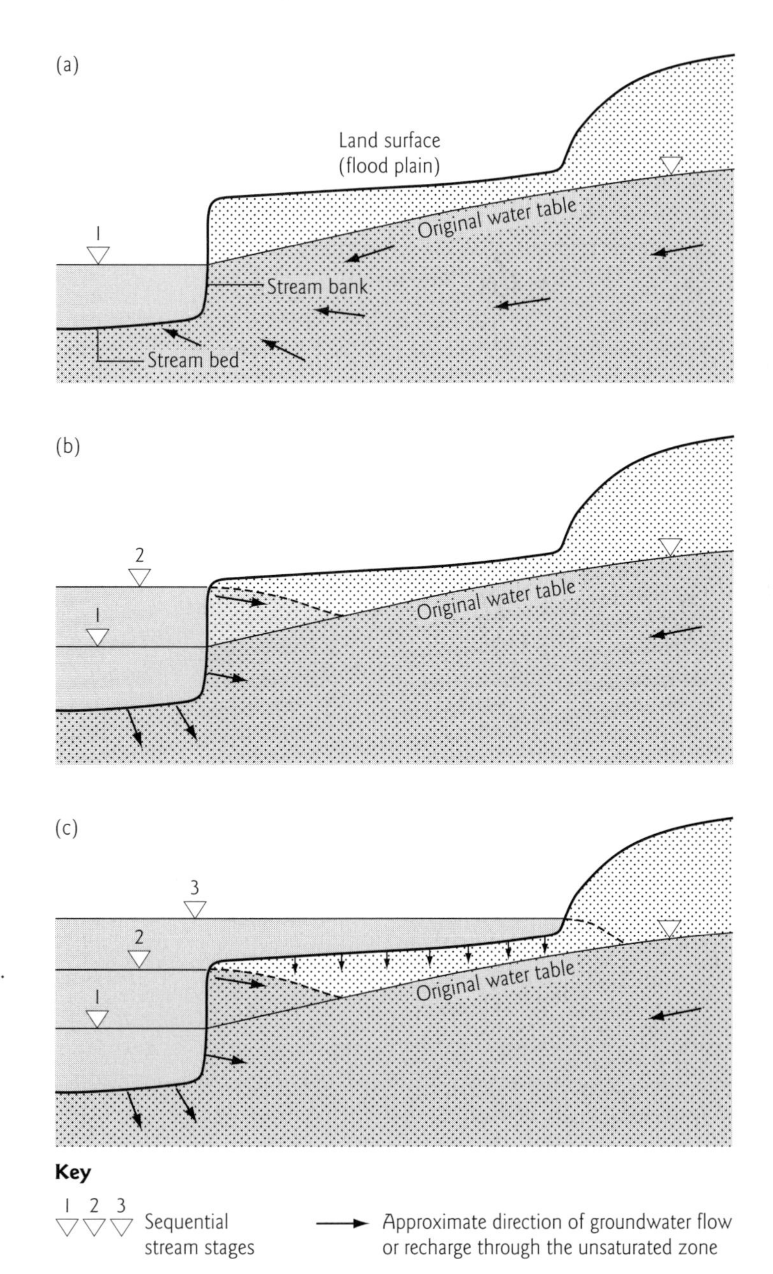

Fig. 5.5 The bank storage phenomenon: typical sequence of events during the creation of a wedge of stored water. (a) Pre-storm conditions: stage in stream (level 1) at typical base flow level, with groundwater draining into the stream. (b) Initial stage rise during a storm (level 2), although with stage still below bankfull channel capacity; the creation of a bank storage wedge is now underway. (c) Peak flood stage (level 3), with overbank flooding inundating the floodplain, and downward flow of water into storage. (Adapted after Winter et al. 1998.)

bank storage effect is usually rather subtle. However, if the flood is sufficiently large to exceed the bank-full capacity of the channel, then the floodplains will be submerged (Figure 5.5c), and downward infiltration of flood waters into the adjoining aquifer can occur. Bank storage can be far more marked under these conditions (Figure 5.5c), as the area through which flow can pass to the subsurface is much greater than is the case while flow remains within the stream banks (Figure 5.5b). (Remember Darcy's Law). When the flood wanes and stage declines, the "wedge" of water which was added to the saturated zone by bank storage now forms a body of riverside groundwater with an unusually steep hydraulic gradient back towards the channel. Drainage of the stored water into the channel therefore proceeds briskly. The overall hydrological effect of bank storage is to lessen the magnitude of peak flood flows, by storing flood runoff to be released slowly after the end of the storm (thus prolonging the length of the runoff event). As such bank storage offers considerable natural mitigation of flood flows.

Because many gaining streams receive substantial quantities of groundwater discharge, they often form the lowermost boundaries of many aquifer flow systems. When attempting to quantify the available water resources in an aquifer, it is sometimes convenient to regard streams as boundaries along which head is fixed equal to stream stage. In reality, if a stream is to absolutely determine the head in the adjoining aquifer, it is necessary that the stream perimeter offers absolutely no resistance to groundwater flow, which is rare in practice, due to the common presence of low-permeability alluvium and/or bed sediments (Box 5.1). Nevertheless, the temptation to assume that a stream behaves like a fixed head boundary to a groundwater flow system is very tempting when it comes to groundwater modeling (see Section 10.2). However, besides overestimating the ability of the stream to dictate heads in the adjoining aquifer, the fixed-head assumption also implies that the stream completely penetrates the entire saturated thickness of the aquifer. In nature, streams never fully penetrate aquifers of any size: partially penetrating streams are the norm, and their perimeters are very often lined with low-permeability sediment.

5.2 The role of groundwater in generating surface runoff

5.2.1 Baseflow

The principal role which groundwater discharge plays in catchment hydrology is the sustenance of stream flows during dry periods. In most river basins of the world, flow would tend to cease altogether within a week of a major rainfall event were it not for the ongoing release of water from stored sources. Where a lake or manmade reservoir exists in a catchment, such surface water storages can help sustain flows during dry periods. However, in catchments where no such surface storages exist, the persistence of river flows during dry periods is a sure sign of the continuance of groundwater discharge, which slowly depletes aquifer storage long after the end of the most recent rainfall event. The overall effect of groundwater discharge is therefore to provide a background level of stream flow, known as **baseflow**, upon which surface runoff peaks are superimposed.

Baseflow represents the natural drainage of aquifers. In the absence of recharge, sustained baseflow will result in a gradual lowering of heads in its source aquifer. As the head declines, so will the hydraulic gradient from the aquifer to the stream (the bed elevation of which is usually fixed). Given Darcy's Law, it is clear that as hydraulic gradient declines, so must the rate of baseflow. There is thus a natural tendency for the rate of baseflow to decline over time, following a period of recharge. The natural patterns of decrease in baseflow rates are known as **baseflow recessions**. The recession of baseflow in some streams can continue all the way to zero, so that the stream displays an intermittent flow regime. Where baseflow never ceases, the stream is perennial.

Various means of investigating baseflow rates and recessional behavior are described in Section 5.3.

Box 5.1 Why should sediments near streams be less permeable than the adjoining aquifers?

While major gravel-bed rivers are lined with sediments more permeable than the adjoining aquifers, many other types of river are lined with sediments which are rather fine-grained in nature, and are thus often less permeable than the adjoining aquifers. The reason that this is so often found to be the case lies in processes of global change: many streamflow regimes have been far more gentle over the last few thousand years than they were previously. Streams that now flow at fairly gentle rates, carrying and depositing only silts and clays, are often the "poor descendants" of raging torrents which formerly flowed in the same valleys. These ancient rivers deposited sand and gravel, which is now buried beneath recent fine-grained alluvial deposits associated with today's relatively gentle streams.

But why has the flow regime of so many streams become so much more gentle in the recent geological past? The answer lies in the periods of profound climatic change referred to as the "ice ages." The most recent "ice age" ended around 10,000 years ago. Before this, glacial conditions held sway at latitudes greater than about 50° in the northern and southern hemispheres (and at even lower latitudes in mountainous areas). Thus all of Canada, the northern parts of the USA, Europe and Asia, and the southernmost areas of Australasia and Latin America were all heavily glaciated until about 10,000 years ago. Glacial erosion generates large quantities of coarse sediment scoured from the underlying bedrock: meltwater streams typically flow at high rates and deposit sands and gravels. Even outside of the glaciated areas, in lower latitude areas the climate was decidedly wetter than at present. For instance, geologists working in north Africa and the Middle East have long recognized the existence of "pluvial" periods (periods of far greater rainfall than at present) which were synchronous with the glacial periods at higher latitudes. As in the glaciated areas, runoff regimes were far more energetic during pluvial periods than at present. Since the end of the glacial-pluvial episodes, less extreme flow regimes have generally obtained, resulting in the accumulation of much finer-grained sediment in close proximity to the river channels.

5.2.2 Rainfall-runoff responses

Although the percentage of streamflow coming from groundwater sources is usually greatest during periods of sustained baseflow, it should not be thought that groundwater plays no part in runoff generation at any other time. In many cases, baseflow will continue throughout wet periods. Also, groundwater plays a role to varying degrees in the generation of surface runoff. Most surface runoff originates as overland flow, which literally flows over the land surface until it finally falls into a stream channel. Two principal concepts of overland flow generation have been championed over the years, both of which depend to some degree on subsurface flow processes.

The earliest scientific theory of surface runoff generation was proposed by Horton (1933), who hypothesized that surface runoff occurs only when the rate of rainfall (or provision of free water by snowmelt) exceeds the rate at which the soil is capable of absorbing water. In other words, once a soil is sufficiently wetted that infiltration is occurring at the maximum possible rate for that soil, any further rainfall landing on the soil surface will be unable to enter the subsurface and will thus become overland flow. Hortonian runoff generation is nowadays usually referred to as

infiltration-excess overland flow, though the older term **Hortonian overland flow** is still occasionally used. Detailed field studies have now established that, except in areas of extremely high rainfall, this mode of runoff generation is actually rather rare (Freeze and Cherry 1979, p. 219), save on very low permeability soils and paved areas.

It should be noted that the basic theory of Horton (1933) makes no reference to the depth of the water table. As long as the maximum infiltration rate is attained in the surface soil layer, then infiltration-excess overland flow will be generated, irrespective of the depth of the water table. But what if the water table is near surface? In such circumstances, a second runoff generation mechanism is more likely to occur. This occurs wherever the water table rises so rapidly during a storm event that groundwater discharges at the ground surface. When this happens, any further rainfall cannot infiltrate and also joins the overland flow. This mode of runoff is termed **saturation-excess overland flow**. The water table is naturally shallow around the headwaters of many streams, particularly where the streams drain low permeability bedrock, so that wetland areas of various sizes flank the hillsides and channel margins. Often only a few millimeters of water table rise is needed to turn a previously stagnant wetland area into a dynamic source of overland flow. As periods of wet weather come and go, such wetlands wax and wane substantially in surface area and water depth. For this reason, these runoff-generating wetland areas have come to be known as **variable source areas**. The propensity of variable source areas to generate runoff depends on antecedent weather conditions: after a long dry spell, several rain storms may be required before the water table will have risen sufficiently to give rise to saturation-excess overland flow. Conversely, after prolonged periods of wet weather, virtually all additional rainfall arriving on or near these wetlands will give rise to overland flow.

In some cases, the rate at which the water table rises in response to rainfall can be truly impressive. Relatively modest quantities of infiltrating rainwater have been shown to give rise to substantial water table rises, far in excess of what would be expected by dividing the depth of water added by the specific yield of the aquifer (cf. Box 2.1). How can this be explained? The answer to the mystery lies in the capillary fringe, which we first met in Section 1.2. Until the early 1980s, hydrologists ascribed little importance to the capillary fringe (beyond its value for trapping unwary students in exam questions concerning the definition of "saturated zone"!). However, by the mid 1980s, it was becoming clear that rapid water table rise during periods of storm runoff generation was explicable by the sudden conversion of the capillary fringe into pressure-saturated groundwater (e.g. Gillham 1984). In other words, while the capillary fringe effectively sits on the water table as a tension-saturated mantle (in which all pores are 100% filled with water, but the water is held in tension with a pressure less than atmospheric pressure), a sudden change in water pressure can almost instantaneously convert the capillary fringe into "true" groundwater (i.e. with pore water pressure > atmospheric pressure). As this happens, the water table effectively jumps upwards by a height equal to the prior thickness of the capillary fringe. This phenomenon of **capillary fringe conversion** can result in the rapid establishment of steep hydraulic gradients in aquifers adjoining deeply incised streams, leading in turn to very rapid runoff of groundwater into the stream channel (Figure 5.6e; see Sklash and Farvolden 1979). Where capillary fringe conversion is operative, a very high proportion of the total flow in a stream in spate can be groundwater, with values as high as 90% quoted by some authors (see Buttle 1994, for a review).

5.3 Estimating the groundwater component of catchment runoff

5.3.1 Deconstructing runoff: hydrograph separation and the baseflow index

A stream hydrograph is simply a graph of stream flow rates (on the *y*-axis) plotted against time (on

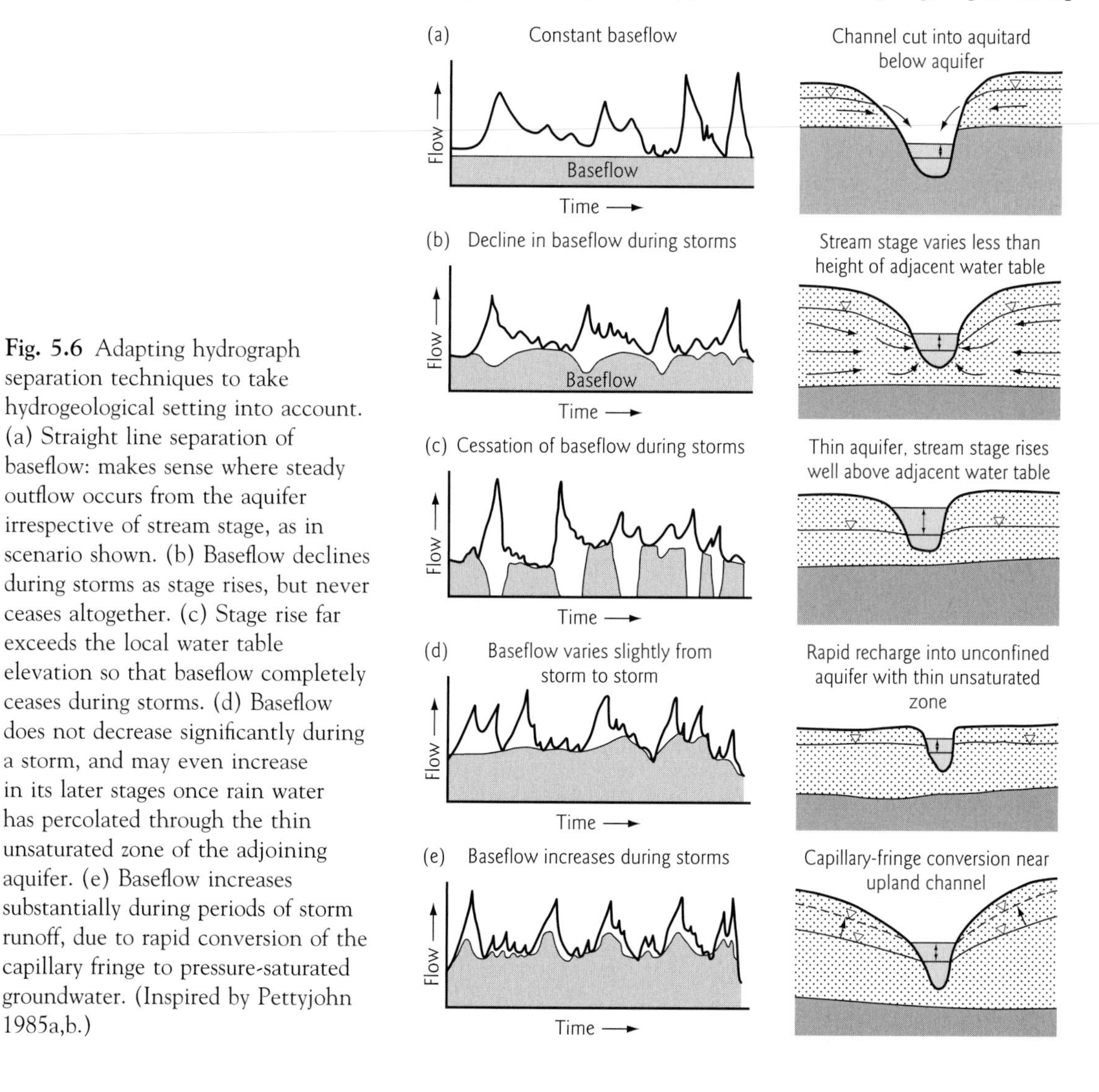

Fig. 5.6 Adapting hydrograph separation techniques to take hydrogeological setting into account. (a) Straight line separation of baseflow: makes sense where steady outflow occurs from the aquifer irrespective of stream stage, as in scenario shown. (b) Baseflow declines during storms as stage rises, but never ceases altogether. (c) Stage rise far exceeds the local water table elevation so that baseflow completely ceases during storms. (d) Baseflow does not decrease significantly during a storm, and may even increase in its later stages once rain water has percolated through the thin unsaturated zone of the adjoining aquifer. (e) Baseflow increases substantially during periods of storm runoff, due to rapid conversion of the capillary fringe to pressure-saturated groundwater. (Inspired by Pettyjohn 1985a,b.)

the *x*-axis). The most common way of trying to interpret baseflow rates is to separate a stream hydrograph into the surface runoff and baseflow components. The most straightforward approach to hydrograph separation is simply to join all of the low turning-points on the hydrograph by straight lines (Figure 5.6a), on the assumption that all of the flow below the resultant chain of straight lines must be groundwater-fed baseflow. However, better results will normally be obtained if the method of hydrograph separation is customized to take into account the hydrogeology of the catchment in question (e.g. Pettyjohn 1985a). Figure 5.6 illustrates alternative hydrograph separation approaches appropriate to specific stream–aquifer settings. In some cases, baseflow can be expected to decline (Figure 5.6b), or even cease altogether (Figure 5.6c), during surface runoff events, as the stage in the channel reverses the hydraulic gradient and prevents further groundwater inflow through the channel perimeter. In other settings, baseflow may remain

(i) Take a stream hydrograph and trace all of the falling limbs of flood peaks onto a single plot

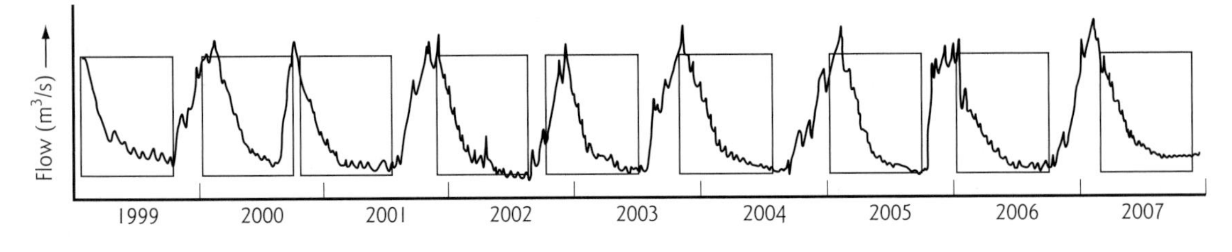

(ii) From multiple plot of recession curves, identify medial line and "envelope curves"

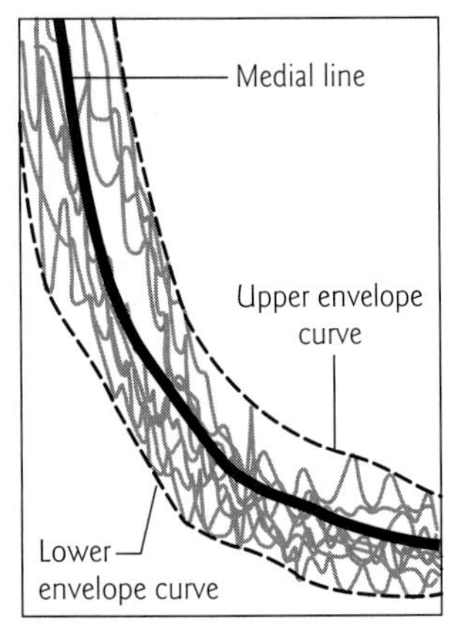

(iii) Adopt medial line as "master baseflow recession curve", and use it to separate other hydrograph periods

(iv) Fit exponential equations to the curves ($Q = Q_0e^{-at}$) and use to predict flows in dry periods

Fig. 5.7 Derivation of a "master baseflow recession curve" from the declining limbs of hydrographs for a series of storm runoff events. (Inspired by Pettyjohn 1985a,b.)

constant (Figure 5.6d), or even (where capillary fringe conversion is operative; Figure 5.6e) increase significantly during surface runoff events.

One approach to hydrograph separation which lends itself to routine processing of large numbers of hydrograph records is the establishment of standardized **baseflow recession curves**[iv] for specific flow gauging stations on major rivers. A baseflow recession curve is compiled from a series of successive periods of flow recession following storm runoff, simply by tracing the declining limb of the hydrograph for each flood event (Figure 5.7), and superimposing the individual traces for a large number of individual events (ideally spanning more than one full year's worth of flow gauging records) until a "master" baseflow recession curve for the gauging station in question has been identified. Baseflow recession curves obtained in this manner have two principal uses:

1. They provide an observationally based model for likely future storm flow/baseflow recession patterns in the catchment in question, which can be helpful in devising a customized hydrograph separation routine.
2. They make prediction of future baseflows possible. This is because baseflow recession curves usually turn out to have a concave form which conforms to the pattern defined by an exponential relationship between flow rate and time. If an exponential equation can be fitted to a given baseflow recession curve, then this equation can be used to predict future baseflow rates, prior to the onset of the next surface runoff event. Such predictions can be very useful in the context of developing drought management policies for rivers subject to artificial water abstractions.

It should be noted that both of these applications of baseflow recession analysis should be used with caution, bearing in mind the hydrostratigraphy of the catchment in question. The coefficients used to define exponential baseflow recession equations are sometimes taken as indices of individual drainage basins. However, since exponential equations are normally derived from absolute values of flow rather than flow rates normalized against total annual runoff or catchment area, it is rarely possible to rationally compare the coefficients obtained from analysis of one basin with those from another. More useful comparative indices are discussed below.

Besides separating hydrographs using graphical techniques (Figures 5.6, 5.7), it is also possible to achieve the same end by using hydrochemical calculations based on concepts of mixing two waters of different water quality. For this to be applicable, it is important that the discharging groundwater is significantly different in chemical composition from overland flow generated in the same catchment. Bearing in mind the points made in Chapter 4 concerning groundwater chemistry, it is often reasonable to anticipate that overland flow will have a composition close to that of rainwater, whereas groundwaters will be considerably more mineralized. Ideally, two ideal "end-member" compositions will be identifiable: pure groundwater and pure overland flow. End-member compositions can be defined by directly sampling overland flow during storms and by sampling near-channel groundwaters using boreholes. In other cases, it may be sufficient to sample only stream waters and simply assume that samples obtained during periods of peak runoff represent 100% overland flow, whereas those sampled during low-flow periods represent 100% groundwater. (Whether such assumptions are justified depends on the local hydrogeological setting; cf. Figure 5.6.) Once the two end-members have been characterized (and shown to be significantly different from one another using statistical techniques), the proportion of groundwater in the total stream flow at any one time can be calculated using a series of calculations which have been dubbed **end-member mixing analysis (EMMA)**. Best practice is to implement EMMA using a range of natural tracers and compare the results obtained. Relatively nonreactive solutes (such as Cl, SO_4, Na, and Mg, which are not generally incorporated into fresh mineral precipitates in most natural stream–aquifer systems, and are not especially prone to sorption) are most useful for calculating mixing of groundwater and overland flow. Where gross differences in these solutes are manifest between groundwaters and overland flow, EMMA should be easy to implement. It is still feasible to use EMMA where the differences between overland flow and groundwater are more modest, for instance restricted to contrasting ratios of environmental isotopes (see Buttle 1994). Having identified the relative proportions of groundwater and overland flow in the channel at any one time, EMMA can be extended to investigate the aquifer-to-river transfer rates of reactive solutes (such as NO_3 and PO_4); by comparing their apparent rates of transfer in comparison with those of nonreactive solutes, it is possible to quantify the degree to which biogeochemical reactions in the hyporheic zone alter groundwater composition during discharge to streams. Further explanation of EMMA and related mass-balance mixing calculations are beyond the scope of this text. The interested reader is referred to Appelo and Postma (1993) for general guidance, and to Soulsby et al. (2003) for a useful exposition of some of the pitfalls of applying EMMA in complex natural stream–aquifer systems.

Having separated a hydrograph, it is relatively simple to calculate the proportion of the total stream flow which is accounted for by baseflow contributions. This is achieved by bearing in mind that the areas under the curves on the hydrograph plot amount to volumes.[v] If the area under the baseflow separation curve is divided by the area under the original total flow curve, a ratio is obtained which is termed the **baseflow index (BFI)**. The BFI represents the proportion of total stream flow, over the entire period of records analyzed, which is derived from *all stored sources* (Institute of Hydrology 1980). In other words, where surface lakes or man-made reservoirs exist in the catchment, or where large man-made discharges of sewage or industrial effluents are discharged into the channel, the BFI will include

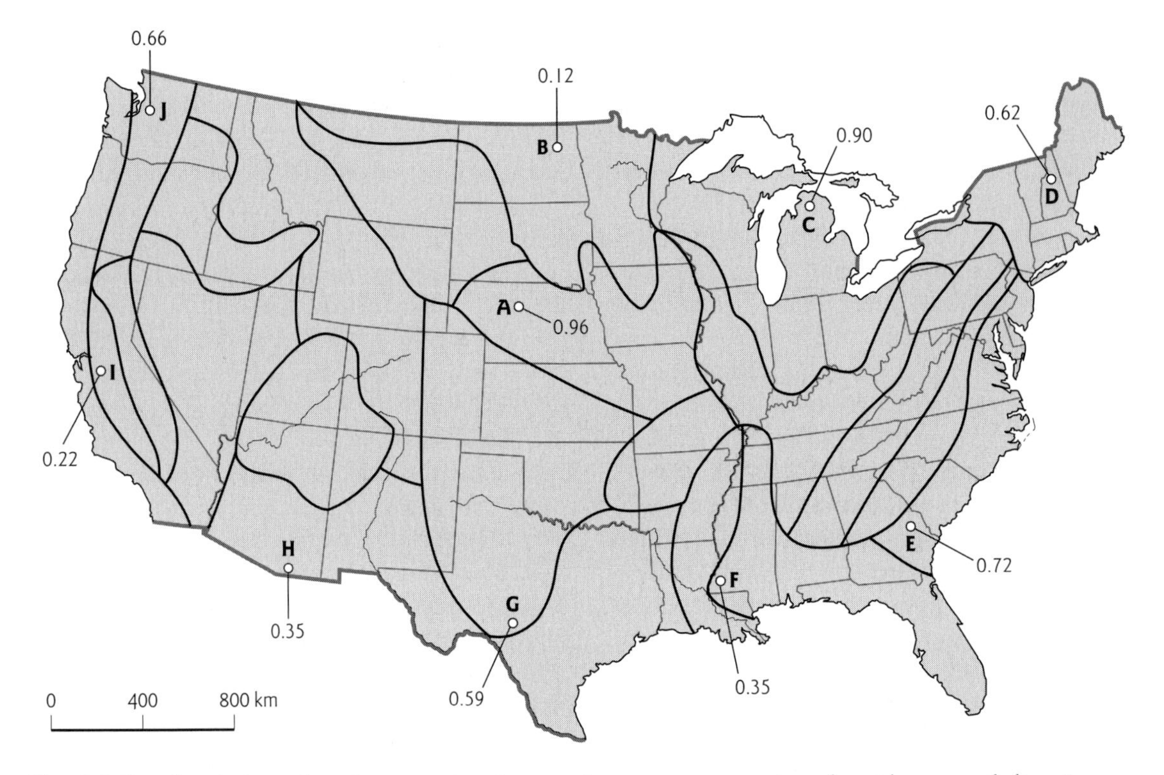

Fig. 5.8 Baseflow index values for ten major river catchments representative of a wide range of climatic regions within the contiguous United States of America. The figures shown are those proportions of total runoff (= 1) accounted for by baseflow. Key to rivers shown: A, Dismal River, Nebraska; B, Forest River, North Dakota; C, Sturgeon River, Michigan; D, Ammonoosuc River, New Hampshire; E, Brushy Creek, Georgia; F, Homochitto River, Mississippi; G, Dry Frio River, Texas; H, Santa Cruz River, Arizona; I, Orestimba Creek, California; J, Duckabush River, Washington. The heavy black lines delineate 24 regions within which stream–aquifer interaction processes at the catchment scale display relatively consistent patterns. (Adapted from data presented by Winter et al. 1998.)

these components in addition to any natural groundwater discharges. Once these surficial and artificial stored sources of water have been taken into account, however, the BFI is a good measure of the groundwater component of stream discharge. Figure 5.8 illustrates some values of BFI for various catchments around the USA. A substantial range of values is displayed, reflecting the contrasting geology of the different catchments. It should be noted that BFI does not vary coherently with climate, as it is a ratio rather than an indicator of absolute flow rates. However, care must be taken when applying such indices to ephemeral or intermittent stream reaches, for BFI is really only a logical index in relation to perennial streams.

Table 5.1 summarizes some typical values of BFI for catchments underlain by various rock types. Catchments underlain by prolific aquifers typically have BFIs in excess of 0.70, with values in excess of 0.95 in the most permeable catchments. It is interesting to note that BFI values tend to exceed 0.15 even in catchments underlain wholly by rocks normally regarded as aquitards (e.g. mudstones), with values as high as 0.50 being displayed by many catchments with such geology. One logical explanation for this is that soils developed above aquitard bedrocks tend to have significantly greater permeability and specific yield than their parent materials. This ensures that groundwater discharge remains a non-negligible component of total stream flow even where the

Table 5.1 Typical ranges of values of base flow index (BFI) for perennial stream catchments underlain by a range of different rock types. Based in part on Institute of Hydrology (1980, table 3.3), modified in the light of a range of other more recent publications (cited in this text) from numerous catchments around the world.

Aquifer lithology	General *T* and *S* characteristics of aquifer	BFI (typical range)
i: Unconsolidated sedimentary aquifers (permeability almost wholly intergranular)		
Gravel (well sorted)	*T* and *S* both very high	0.95–1.00
Sand (well sorted)	*T* moderate, *S* very high	0.85–0.95
Sands and gravels (poorly sorted)	*T* moderate, *S* moderate	0.60–0.80
Silt or mud	*T* and *S* both low/very low	0.15–0.45
ii: Bedrock aquifers with fracture permeability only		
Karstified rocks (limestone/gypsum)	*T* very high, *S* low	0.75–1.00
Thoroughly jointed limestones (negligible drainable intergranular porosity; not heavily karstified)	*T* high to moderate, *S* low	0.80–0.98
Basaltic volcanics containing unfilled lava tubes and open joints	*T* high, *S* low	0.60–0.90
Nonbasaltic volcanics, and cemented basaltic volcanics	*T* moderate to low, *S* low	0.40–0.70
Heavily cemented sandstones/limestones/volcanic rocks	*T* and *S* both low	0.30–0.55
Mudstones and siltstones	*T* and *S* both very low	0.15–0.40
Plutonic and metamorphic rocks	*T* and *S* both very low	0.30–0.50
iii: Bedrock aquifers with both intergranular and fracture permeability		
Coarse-grained sandstones (pores not occluded by cements)	*T* moderate, *S* moderate to high	0.70–0.80
Oolitic limestones (nonkarstified) and other limestones with large primary pores	*T* moderate to high, *S* moderate	0.85–0.95

bedrock geology includes few major aquifers (see Section 5.4.3 for further discussion).

5.3.2 Analyzing annual variations in streamflow rates

From many practical perspectives the most important question to be asked about groundwater discharge to streams is: what level of stream flow will it sustain during the dry season? To determine this, it is necessary to examine how stream flow varies over entire annual (or, better still, multi-annual) cycles. The easiest way to make such an examination is to prepare a **flow–duration curve** (Figure 5.9), which is simply a cumulative frequency curve (a type of graph widely found in basic statistical studies) showing flow rate versus the percentage of time a given flow is equalled or exceeded. Box 5.2 explains how to prepare a flow–duration curve from records of mean daily flows at a gaging station. Having developed a flow–duration curve for a particular catchment, it is easy to read off different percentile values indicative of a particular flow condition.

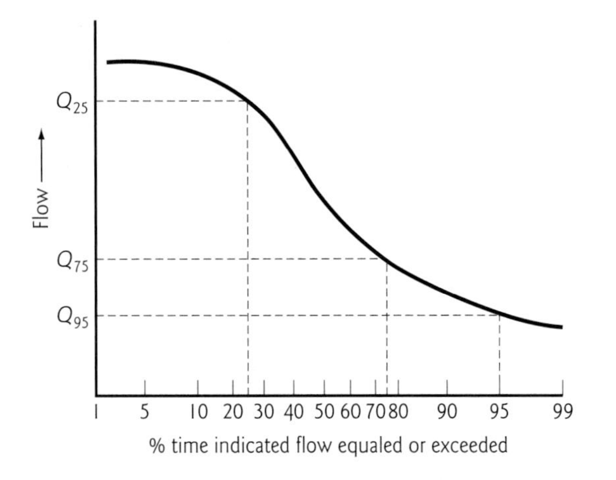

Fig. 5.9 An example of a flow duration curve. Three specific percentile flow values are shown (Q_{25}, Q_{75}, and Q_{95}), as discussed on Section 5.3.2 of the text. Of these three Q_{95} is the most widely used index of the absolute magnitude of baseflow in rivers.

Obviously, for low-flow conditions which are likely to be sustained solely by groundwater discharge, it is the tail-end of the flow–duration curve which is of primary interest. The most common index of low flows derived from scrutiny of flow duration curves is the so-called $\mathbf{Q_{95}}$ value, which is simply the flow equalled or exceeded 95% of the time. The popularity of Q_{95} as a low-flow index lies in the fact that most flow duration curves will be supported by a sufficient spread of data points to allow accurate reading of Q_{95}, whereas higher percentile values (Q_{97}, Q_{99} etc) will be far less accurately defined.

If it desired to compare the baseflow characteristics of one catchment with another, it will normally be best to prepare an area-normalized flow–duration curve. This is achieved simply by dividing all flow records by the catchment area before preparing the cumulative frequency curve, so that the flow rate axis of the resultant flow–duration plot will be in units of "m^3/s per km catchment area" or similar. A further useful comparator statistic, which is useful for comparing the degree of groundwater dominance of stream runoff regime between catchments, is to calculate the ratio of Q_{25}/Q_{75}. Low values of this ratio (i.e. $1 < Q_{25}/Q_{75} \leq 2$) are typical of very

Box 5.2 Preparing a flow–duration curve.

To prepare a flow–duration curve you need at least a single full year of flow records (preferable daily mean flows) for a stream flow gaging station. It is even better if you have several years' worth of records, but only *complete* years of data should be included in the analysis. The first step is to examine the flow records and determine the maximum and minimum values. Between these two extremes, subdivide the entire range of flows into a number (>10) of arbitrary intervals. For instance, if the minimum flow was 0.12 m^3/s and the maximum was 2.14 m^3/s, I would probably subdivide the range into intervals of approximately 0.1 m^3/s, as follows: 0.10–0.19; 0.20–0.29; 0.30–0.39; 0.40–0.49; and so on up to 2.11–2.19. Now go through the flow records and make a tick for each record in the interval to which it belongs. For instance, a value of 1.75 m^3/s would give a tick in the 1.70–1.79 interval, and so on for all other records. In the end, we will be able to count the number of records allocated to each of the intervals. The percentage of time in which flow fell within each interval can then be simply calculated by dividing the number of "ticks" in each interval by the total number of records used in the analysis (i.e. 365 if we used 1 year's worth of records). Then, starting from the highest interval, we calculate the cumulative frequency for each interval until we reach the lowest interval. For the lowermost value of each interval, we can now specify a percentage of time for which that particular flow is equalled or exceeded. By plotting each such value against flow on a graph, we obtain our final flow–duration curve, from which values such as Q_{95}, Q_{75}, and Q_{25} can be easily read by interpolation (Figure 5.9).

permeable catchments in which almost all of the annual flow is supplied by groundwater discharge, whereas increasingly higher values typify increasing dominance of surface runoff. Obviously indices such as this can be expected to correlate with BFI values. For instance, Smakhtin and Toulouse (1998) were able to develop reasonable positive correlations between BFI and the ratio of Q_{75} to average daily flow for a number of stream systems in South Africa. It should be noted, however, that such correlations tend to be specific to a particular region, or even to the tributaries of a single large catchment.

5.3.3 Dry weather stream surveys

By now you're likely to have understood the message that there's a vast amount of groundwater making its way into the world's streams. To quote a favorite adage of my own "Apprentice-Master," Professor Wayne Pettyjohn, as far as a hydrogeologist is concerned "a gaining stream is really little more than a very long, very shallow well." As such, there is much to be learned about aquifers from an investigation of the streams which drain them. Indeed, we have already seen that catchment-scale flow indices (such as BFI, Q_{95}, and Q_{25}/Q_{75}) shed light on the permeability of the aquifers drained by gaining streams. At a much smaller scale, it is possible to learn even more about the aquifers in a catchment by means of careful studies of the quantity and quality of stream water during dry weather periods (when it is reasonable to expect virtually all of the flow in the stream to originate as groundwater, except for any flow supported by surface reservoirs).

A wide range of dry weather survey techniques has been devised.[vi] Perhaps the simplest is to wade barefoot into a stream on a hot summer afternoon (cf. Holmes 2000). After many hours of sunlight, water that has been in the channel for some time will have become pleasantly warm. By contrast, groundwater that is just emerging into the channel will be much cooler (typically with a temperature close to the local mean annual air temperature; see Section 4.2.1). After a few minutes of strolling around a given stream reach, you should have been able to identify the cool spots in the water at the streambed corresponding to localized zones of groundwater discharge. On large scales, similar surveys can be made using thermometers trailed from boats, or even by airborne thermal imaging. If you can't wait until summer to identify groundwater discharge zones, you can always take a walk on a cold winter's day and identify groundwater discharge zones from the occurrence of unfrozen patches (sometimes emitting steam into the frosty air) along otherwise iced-up stream channels. Airborne visual surveys can also be useful under such winter conditions.

At the next level of sophistication, it is possible to quantify groundwater discharge rates by making a series of "spot gagings" (i.e. one-off measurements) of stream flow at a number of points along a channel (e.g. Pettyjohn 1985a). The increase in flow between each gaging point and its downstream neighbor is called the **transmission gain** (G_T), and it is calculated by subtracting the upstream flow (Q_u) from the downstream flow (Q_d). In the absence of stream tributary inflows or artificial discharges, G_T simply equals the amount of groundwater entering the channel between the two points. If the transmission gain is divided by the distance between the two measurement points, a **specific gain** value is obtained, typically in units of m^3/s per km length of stream channel. Alternatively, specific gain can be defined as the transmission gain divided by the area of the streambed (A_s) between the two stations, so that the value obtained will be in units of m^3/s per m^2 of streambed (or similar). (A_s can be readily estimated by multiplying the distance between neighboring gauging points by the average channel width in that reach.) If observation boreholes are present in the adjoining aquifer, so that the gradient (i) of the water table towards the river can be determined, then it is even possible to estimate the mean transmissivity (T) of that aquifer using the following simple formula, derived from Darcy's Law:

$$T = G_T / (i \cdot A_s) \tag{5.1}$$

Comparison of specific gain values for successive downstream reaches facilitates identification of more or less prolific gaining and losing reaches. Such variations in specific gain can be checked for correlation with known geological features (e.g. intersection of aquifer outcrops with the river bed, traces of faults, etc.), or, indeed, used as evidence for the likely positions of such features during a priori hydrostratigraphic mapping.

Hydrochemical samples collected during dry weather surveys can yield important information about the chemistry of local aquifers (Pettyjohn 1985a,b). Indeed, by such means it is possible to obtain considerable insight into variations in groundwater quality within a catchment, before a single borehole is drilled (Pettyjohn 1985b). This can be especially helpful in developing countries, where resources for drilling observation boreholes are often severely limited. For instance, Kuma and Younger (2004) successfully used dry weather stream surveys, coupled with opportunistic sampling of village hand-pump wells, to delineate distinctive bodies of groundwater in the goldfields of the Tarkwa area, Ghana. In some cases, such dry weather hydrochemical surveys have been successfully used to identify bodies of contaminated groundwater (e.g. Younger and Bradley 1994), which have then been further investigated using more conventional site investigation techniques.

5.4 Physical controls on groundwater discharge at the catchment scale

5.4.1 The ubiquity of groundwater discharge

The biblical proverb "seek and ye shall find" has significant resonance in the context of contemporary research into groundwater–surface water interactions. The more one appreciates the many manifestations of groundwater discharge within the surface water environment, so it seems, the more apparent are groundwater discharge features across the full range of landscapes. Of course the interactions between regional-scale aquifers and major lowland rivers have been appreciated for many years (e.g. Theis 1940; Pinder and Sauer 1971; Freeze 1972a,b; Bredehoeft et al. 1982; Winter et al. 1998). Much more recent is the emerging appreciation of the degree to which groundwater participates in the generation of surface runoff in mountainous areas (Buttle 1994; Soulsby et al. 1998). With many of the world's mountain chains being composed of plutonic and metamorphic rocks of generally low permeability, the historic consensus used to be that groundwater was unlikely to play an important role in the hydrology of such areas. However, once scientific investigations of upland catchments began in earnest, a rather different picture soon emerged. The currently emerging consensus is that "shallow drift deposits and fracture flow in slowly permeable sedimentary and crystalline rocks [host] significant groundwater sources that can exert important controls on the hydrology and hydrogeochemistry of upland streams" (Soulsby et al. 1998).

So groundwater is pretty much "bursting out all over": it upwells abundantly in the floodplains of major rivers; it moves dynamically through weathered bedrock mantles of bedrock and stony soils in mountainous areas; it even plays a significant role in the hydrology of the driest places on Earth. For instance in the Atacama Desert of northern Chile, where centuries may pass between successive recharge events, groundwater discharge sustains the supply of moisture to saline lakes and salt flats (e.g. Houston and Hart 2004). The apparent ubiquity of groundwater discharge is not even restricted to this planet: many of the landforms of Mars are now believed to have been formed by prehistoric discharges of groundwater (see Coleman 2003).

5.4.2 Geological and geomorphological factors

In all of its manifestations, groundwater discharge is controlled by the interplay between subsurface geological structure and landscape. We have already considered the manner in which geological structure affects the distribution of permeable rocks in the subsurface (see Section 1.5) and have touched upon the interplay between landforms and groundwater discharge features (see Section 3.3.1). At the scale of entire catchments, the details of the interaction between geological structure and landscape critically determine patterns of groundwater discharge via springs, wetlands, and stream beds.

As Higgins et al. (1988) have noted: "The landscape consists of a series of interconnected hillslopes. The variability of hillslope length, steepness and shape are reflected in variable water-table configurations, which in turn translate into other aspects of ground-water conditions." Clearly the occurrence of different types of springs (Figure 5.1) depends upon the manner in which hillslopes intersect sedimentary contacts, intrusive contacts, faults, and other geological features. Similarly, where low-permeability bedrock reaches the surface in the axis of a stream valley that is elsewhere occupied by more permeable unconsolidated sediments, and the groundwater present in the latter meets the bedrock high, it will be forced to flow upwards through the stream bed (e.g. Malcolm et al. 2005).

Rather more subtly, the aspect of a slope (i.e. the direction in which it faces) can considerably influence its hydrogeological behavior: "In the Northern Hemisphere, southern and southwestern slopes receive greater insolation than northern and northeastern slopes. This leads to differences in soils, vegetation, and water retention and infiltration rates. Such factors create feedback systems that affect the amount of water that enters the ground-water zone" (Higgins et al. 1988).

We have already seen that the dissolution of minerals naturally present in rocks accounts for much of the dissolved solids content of groundwaters. At a more general level, the total dissolved solids contents of groundwaters are influenced by geomorphological factors: it is commonly found, for instance, that groundwater lower down hillslopes is considerably more mineralized than groundwater in upland zones (e.g. Soulsby et al. 1998). This is simply because "groundwater near hilltops and upper slopes is younger and has had a shorter time to dissolve minerals than the water near the base of the hills" (Higgins et al. 1988).

5.4.3 Soils and groundwater discharge patterns

Soils[vii] are complex biogeological phenomena. The structure and chemistry of natural soils reflect the net result of the operation of an array of biological and physicochemical processes, which naturally affect bedrock. Disruption of soils by a range of human activities further complicates the picture. Clearly, both soil properties and the patterns of occurrence of different soils within the landscape tend to be correlated with bedrock geology. Given that geological features are known to influence groundwater discharge patterns, it is logical that correlations should also exist between the base-flow behavior of streams and the properties of the soils into which their channels are incised. The physical characteristics of soils influence groundwater discharge patterns in three principal ways:

1. Soil properties determine how readily recharge will be transmitted to the saturated zone (cf. Section 2.2), and/or how readily incoming rainfall will be converted into surface runoff (Section 5.2.2).
2. The local configurations of soils within valley axes can exert an important "throttling" of the final stages of groundwater upwelling to the surface.
3. As was noted previously (Section 5.3.1), soils can function as aquifers in their own right, helping to sustain baseflows in rivers which are underlain by aquitard bedrocks. This is simply because the processes of bedrock weathering which lead to soil formation almost always tend to increase the permeability and intergranular porosity of the soil in comparison to its parent rock material.

Integrated studies of baseflow hydrology and soil properties are beginning to provide insights into how these influences operate in real catchments. For instance in the UK a soil classification system known as 'HOST' (**H**ydrology **O**f **S**oil **T**ypes) has been in use for more than a decade now to facilitate the prediction for ungauged catchments of values of BFI, Q_{95}, and other low-flow indices

(Gustard et al. 1992). More recently, patterns of groundwater discharge in the semi-arid Tarkwa area of southwestern Ghana have been found to correlate significantly with the distributions and measured hydraulic properties of the various types of soils developed in the district (Kuma and Younger 2001).

Endnotes

i The international term **catchment** is used throughout this book to denote the entire surface area feeding runoff to a given point on a surface water drainage system. As such, "catchment" is synonymous with the US term **watershed** (a term which has a rather narrower meaning in the English used in Commonwealth countries). It is also synonymous with the rather less formal term **river basin** which is favored in European Commission documents.

ii Travertines most commonly comprise calcite ($CaCO_3$) (Pentecost 1996), though other minerals such as barite ($BaSO_4$) can also form travertines (e.g. Younger et al. 1986). (The term "tufa" is a synonym for travertine, though its similarity to the word "tuff" (volcanic ash) gives rise to confusion, especially in translation; hence "travertine" is preferable).

iii Although the layperson would rarely struggle to tell a pond from a lake, there is no agreed scientific distinction between the two. In order to distinguish them from wetlands (as defined by the Ramsar Convention), ponds and lakes must both include areas of water exceeding 6 m in depth. Beyond that requirement, any further differentiation between ponds and lakes must come down to surface area. A threshold of around 2000 m^2 is suggested to provide a workable distinction between a large pond and a small lake.

iv Baseflow recession curves are also called "flow depletion curves" by some authors.

v If we multiply flow rate (i.e. unit volume of water per unit time) by time, we obtain a volume.

vi Dry weather stream surveys are not only instructive – since they must be carried out during periods of pleasant weather, they offer all kinds of opportunities for combining business with pleasure!

vii In this text, the term "soil" is used in the scientific sense, to mean a surficial sediment which has been subject to biological activity and ongoing interaction with the atmosphere. It should be noted that in civil engineering terminology "soil" denotes any unconsolidated deposit, including many ancient sediments.

6

Groundwater and Freshwater Ecosystems

What would the world be, once bereft of wet and of wildness? Let them be left.
O let them be left, wildness and wet; long live the weeds and the wilderness yet.
(Gerard Manley Hopkins (1844–1899), "Inversnaid")

Key questions

- What are the requirements for a healthy freshwater ecosystem?
- What role does groundwater play in wetlands?
- How do wetlands affect the quality of discharging groundwater?
- What are the ecological implications of groundwater/surface water interactions in the hyporheic zones of streams?
- What is meant by "groundwater ecology"?
- What governs the distribution of invertebrates in alluvial aquifers?
- What is known about the ecosystems present within cave systems below the water table?

6.1 Freshwater ecosystems

6.1.1 Definitions and conservation issues

Ecology is the scientific study of the interactions between living organisms and their environment. Ecologists use the term "**ecosystem**" to refer to any specific assemblage of organisms and their natural environmental surroundings, or **habitat**. As such, the "ecosystem" is the fundamental frame of reference for most ecological investigations, and the maintenance of suitable habitats is the key to ensuring the welfare of ecosystems. Given the diversity of landscapes on Earth and the hugely varying sizes and lifestyles of living organisms, it is not surprising that ecosystems come in a vast range of forms and sizes, ranging from colonies of microbes to forests thronged with birds and mammals. Ecosystems also adjoin one another, forming a continuous mosaic in the landscape/seascape. Boundaries between ecosystems tend not to be fortified frontiers: rather, one ecosystem gives way to its neighbor in a transitional manner, so that a zone often exists in which elements of both ecosystems are identifiable. Such a transitional zone is termed an **ecotone**. Good examples of ecotones include the margins of lakes and wetlands and the interface between the surface and subsurface water environments which occurs within permeable stream beds. Clearly groundwater discharge is a key element in these examples of ecotones. Indeed, the concept of ecotones is almost as important as that of ecosystems in many investigations concerning the interactions between groundwater and the ecology of freshwater systems.

Freshwater ecosystems physically coincide with bodies of standing or running water, which are conventionally regarded as "fresh" in ecological terms if their total dissolved solids content is less than about 10,000 mg/L. As such, freshwater ecosystems are usually identified with particular streams, rivers, wetlands, or lakes (see Baskin 2003). In recent years, many aquifers have come to be recognized as ecosystems in their own right (e.g. Gibert et al. 1994; Griebler et al. 2001), a point which is explored in greater detail in Section 6.4. As we saw in Chapter 1, nonmarine surface waters do not represent a very large fraction of the world's total water budget (see Figure 1.1). For this reason, the total area of the Earth's surface occupied by freshwater ecosystems is really very modest, at about 1% of the total.

Despite their diminutive extent, however, freshwater ecosystems contain a disproportionately large number of the world's faunal species, amounting to 12% of all animal species and 40% of all fish species (WRI-UNEP 1998). Freshwater fishes alone account for 25% of all living vertebrate species (WWF 1998). Table 6.1 summarizes the identities, characteristics, and ecological importance of the many different groups of organisms found in freshwater systems.

One of the reasons why freshwater ecosystems have developed such a wide range of species lies in the nature of catchment geography. For many aquatic animals, overland migration over distances of more than a few meters is impossible. Therefore the animals in one river often tend to evolve in isolation from their relatives in the next catchment. Evolutionary divergence between neighboring freshwater ecosystems is thus the rule rather than the exception. One important consequence of this is that many freshwater species are restricted to only one or two catchments. Species that are native only to a very small area are said to be "**endemic**" to that area. Unfortunately, endemism[i] makes freshwater species highly vulnerable to disturbance by human activity, for if an endemic organism is driven to extinction in its sole habitat, the population cannot be replenished by immigration.

The general aim of conservation activities is to minimize loss of irreplaceable biodiversity. The focus is by no means restricted to ensuring the welfare of known members of endangered species. A more robust conservation goal is the maintenance or restoration of habitats appropriate to the support of a diverse assemblage of species. However, before beginning to plan habitat protection/restoration measures, it is important to assess the diversity of organisms present in a given freshwater ecosystem, identifying those which are most important (or most irreplaceable). The **biodiversity** of an area can be evaluated in several different ways (Groombridge and Jenkins 1998). For instance, an area with a large number of species (high **species richness**) can be described as being more biodiverse than an area with fewer. Similarly, moving up the classificational hierarchy of lifeforms, an area with more representatives of higher taxa (i.e. groups higher than species level, such as genera, families, or classes) will be classified as being more biodiverse than an area containing a smaller number of higher taxa. The identities of the taxa in question are also an important consideration: an area with more "primitive" taxa (e.g. lungfishes or sturgeon) will be regarded as more biodiverse than areas with representatives only of more recent evolutionary

Table 6.1 Organisms of freshwater ecosystems: characteristics and significance. (Adapted after Groombridge and Jenkins 1998.)

Group of organisms	Characteristics	Significance in freshwater ecosystems
Viruses	Microscopic; can reproduce only within the cells of other organisms, but can disperse and persist without host	Cause disease in many aquatic organisms, and associated with certain water-borne diseases in humans (e.g. hepatitis)
Bacteria	Microscopic; although generally less abundant than in soils, can nevertheless occur at very high densities (e.g. $10^6/cm^3$). Most derive energy from involvement in chemical reactions and are thus intimately involved in cycling of C, H, O, N, S, P, Fe, and other elements	Responsible for decay of dead material. Present on all submerged detritus. Food source for aquatic invertebrates. Certain types cause disease in aquatic organisms and humans
Fungi	Microscopic. Recycle organic substances. Able to break down cellulose plant cell walls and chitinous insect exoskeletons	Tend to follow bacteria in decomposition of dead material; also serve as a food source for invertebrates. Some cause disease in aquatic organisms and humans
Algae	Microscopic and macroscopic; include varieties of unicellular and colonial photosynthetic organisms. All lack the leaves and vascular tissues typical of higher plants. Green algae (*Chlorophyta*) and red algae (*Rhodophyta*) include freshwater species; stoneworts (*Charophyta*) are mostly freshwater species	Responsible for bulk of primary production (biomass growth) in most aquatic ecosystems. Forms which attach to bed are important in streams and wetlands. Free-floating forms (phytoplankton) are main biomass producers in lakes and slow reaches of rivers
Plants	Photosynthetic organisms; mostly higher plants that possess leaves and vascular tissues. Mosses, quillworts, and ferns important in some freshwater habitats. Most plants are rooted forms restricted to shallow water (<1 m), though free-floating surface species are locally important (e.g. water fern *Salvinia*, duckweed *Lemna*)	Provide a substrate for other organisms and food for many. Trees are ecologically important in providing shade and organic debris (leaves, fruit) plus structural elements (fallen trunks and branches) that enhance vertebrate diversity, in promoting bank stabilization, and in restricting or modulating floodwaters
Protozoans	Microscopic mobile single-celled organisms. Tend to be widely distributed through passive dispersal of resting stages. Attached and free-living forms; many are filter-feeders	Found in virtually all freshwater habitats. Most abundant in waters rich in organic matter, bacteria, or algae. Feed on detritus, or consume other microscopic organisms; many are parasitic on algae, invertebrates, or vertebrates. Some protozoans are agents of water-borne diseases

Table 6.1 (*Continued*)

Group of organisms	Characteristics	Significance in freshwater ecosystems
Rotifers	Near-microscopic organisms; widely distributed; mostly attached filter-feeders, some predatory forms	Important in plankton communities in lakes and may dominate zooplankton in rivers
Myxozoans	Microscopic organisms with complex life cycles, some with macroscopic cysts. (Formerly classified with protozoa)	Important parasites in or on fishes
Flatworms	A large group of generally ribbon-like worms, including free-living benthic forms (*Turbellaria*) and parasitic forms (*Trematoda, Cestoda*)	*Turbellaria* include mobile bottom-living predatory flatworms. The trematodes includes various flukes, such as the tropical schistosome that causes bilharzia; cestodes are tapeworms: both these groups are important parasites of fishes and other vertebrates including humans. Molluscs are often important intermediate hosts
Nematodes	Generally microscopic or near-microscopic roundworms, typically inhabiting bed sediments	May be parasitic, herbivorous, or predatory; some parasitic forms reach considerable size. Poorly known; may well be more diverse than is currently recognized
Annelid worms	Two main groups in freshwaters; oligochaetes and leeches	Oligochaetes are bottom-living worms that graze on sediments; leeches are mainly parasitic on vertebrate animals, some are predatory
Molluscs	Two main groups in freshwaters: *Bivalvia* (attached bottom-living filter-feeders, including mussels etc) and *Gastropoda* (mobile grazers or predators; snails etc). Both groups include a very large range of species, and are highly prone to endemism[i]	Both bivalves and gastropods have speciated profusely in certain freshwater ecosystems. The larvae of many bivalves are parasitic on fishes. Because of their feeding mode, bivalves can help maintain water quality but tend to be rather vulnerable to pollution
Crustaceans	A very large Class of animals, all characterized by having a jointed exoskeleton, often hardened with calcium carbonate	Include larger bottom-living species such as shrimps, crayfish, and crabs of lake margins, streams, alluvial forests, and estuaries. Also larger plankton: filter-feeding *Cladocera* and filter-feeding or predatory *Copepoda*. Many isopods and copepods are important fish parasites
Insects	By far the largest Class of organisms known. Jointed exoskeleton typically made of chitin. Because they are air-breathing, the great majority of insects are terrestrial	In rivers and streams, grazing and predatory aquatic insects (especially larval stages of flying adults) dominate intermediate levels in food webs (between the microscopic producers,

Table 6.1 (*Continued*)

Group of organisms	Characteristics	Significance in freshwater ecosystems
		mainly algae, and fishes). Also important in lake communities. Fly larvae are numerically dominant in some situations (e.g. in Arctic streams or low-oxygen lake beds), and some are important vectors of human diseases (e.g. malaria, river blindness)
Fishes	More than half of all vertebrate species are fishes. These comprise four main groups: hagfishes (wholly marine), lampreys (some types are wholly freshwater, others ascend rivers to spawn), sharks and rays (almost entirely marine, with few exceptions such as the Nicaraguan Lake Shark), and ray-finned "typical" fishes (>8500 species in freshwaters, or 40% of all fishes). Many of the latter group are highly prone to endemism[i]	Medium- to high-level predators in many freshwater ecosystems, dependent on aquatic invertebrates for much of their diet. In terms of biomass, feeding ecology, and significance to humans, fishes are the dominant aquatic organisms, not only in freshwater habitats but also in marine. Certain freshwater systems, particularly in the tropics, are extremely rich in species. Important fisheries exist in inland waters in tropical and temperate zones
Amphibians	Frogs, toads, newts, salamanders, caecilians. Require freshwater habitats. Larvae of most species need water for development	While larvae are typically herbivorous grazers, adults are predatory. Some frogs, salamanders, and caecilians are entirely aquatic (generally in streams, small rivers, and pools). Sensitive to loss of water-margin habitats
Reptiles	Turtles, crocodiles, lizards, snakes. All crocodilians and many turtles inhabit freshwaters but nest on land. Many lizards and snakes occur along water margins; a few snakes are highly aquatic	Because of their large size, crocodiles can play an important role in aquatic systems, by nutrient enrichment and shaping habitat structure. They are all predators or scavengers, as are freshwater turtles and snakes
Birds	Many birds, including waders and herons, are closely associated with wetlands and water margins. Relatively few, including divers, grebes, and ducks, are restricted to river and lake systems	Top predators. Wetlands are often key feeding and staging areas for migratory species. Likely to assist passive dispersal of small aquatic organisms
Mammals	Relatively few groups are strictly aquatic (e.g. river dolphins, platypus), several species are largely aquatic but emerge onto water margins (e.g. otters, desmans, otter shrews, water voles, water oppossum, hippopotamus)	Top predators and grazers. Large species widely impacted by habitat modification and hunting. Through damming activities, beavers play an important role in shaping and creating aquatic habitats

off-shoots. Even where species or taxon richness is not especially high, an area may still be regarded as hosting significant biodiversity if it contains one or more endemic species.

Human disturbance of freshwater ecosystems is depressingly commonplace. This is because most human settlements have developed along the banks of rivers, precisely in order to take advantage of water resources. With mounting pressures from habitat destruction, fishing, and pollution, it is not surprizing that ecologists now regard freshwater species as being the most endangered group of species in the world (WWF 2004). Indeed, freshwater ecosystems are already known to "have lost a greater proportion of their species and habitat than ecosystems on land or in the oceans" (WRI-UNEP 1998). The World Conservation Monitoring Centre has undertaken a preliminary evaluation of global freshwater biodiversity, which culminated in the publication of a list of 136 locations worldwide which are identified as "important areas for freshwater biodiversity" (Groombridge and Jenkins 1998). Although only eight of these locations are actually "springs and underground aquifers" (listed mainly on account of the rare gastropods which they support), many of the 128 other ecosystems are dependent to a significant extent on sustained groundwater discharge.

6.1.2 Physical requirements for a healthy freshwater ecosystem

Through integration of the findings of many previous ecological investigations on freshwater systems, the Ecological Society of America has identified five "dynamic environmental factors" which interact in various ways, in time and space, to regulate much of the structure and functioning of freshwater ecosystems (Baskin 2003). The five factors are:

1 **Flow patterns**, which define the pathways and rates of water movement through all types of freshwater ecosystem; in turn, these also determine **hydraulic retention times**, i.e. the rate at which the water content of a given freshwater ecosystem "turns over" (with old water being replaced by new), a factor of fundamental importance to nutrient cycling and the removal of toxins.
2 **Sediment and organic matter inputs**, which both supply and store nutrients that sustain aquatic plants and animals and, crucially, provide the raw materials from which key physical elements of habitat structure are constructed, such as substrates, spawning grounds, and **refugia** (i.e. microhabitats within which organisms can survive periods of stress, such as drought or freezing).
3 **Temperature and light penetration**, which regulate the metabolic rates, and thus the biological productivity, of aquatic organisms. Temperature exerts a direct control on all metabolic processes. Light penetration directly affects the viability of photosynthesis, and is thus a *sine qua non* for the growth of algae and plants containing chlorophyll. Given that photosynthesis accounts for virtually all of the primary productivity in surface freshwater ecosystems, light penetration indirectly influences the fauna also.
4 **Chemical and nutrient conditions**, which besides providing the basic building blocks for plant and animal productivity, also regulate pH and other aspects of water chemistry which affect the suitability of the habitat to support specific species/groups of organisms.
5 **The plant and animal assemblage itself**, which directly influences ecosystem process rates and community structure.

Focusing on each of these factors individually will not yield a true picture of ecosystem functioning. Rather, in order to obtain a valid evaluation of freshwater ecosystem integrity, it is necessary to jointly consider all five factors (Baskin 2003). All of these factors vary within defined ranges throughout the year, responding to short-term weather patterns and seasonal changes in temperature, precipitation, and day length. Individual species have adapted over the course of evolution to cope with these changes; so too have the ecosystems which they occupy. Emergent patterns of response by entire ecosystems often show an extraordinary capacity to survive periodic extreme hydrological events (floods and droughts), which have magnitudes far in excess of the normal annual highs and lows in flows, temperature, and other factors. Indeed, more than merely surviving such extreme events, many freshwater ecosystems seem to actually *require* periodic flooding, or periodic desiccation, in order for their long-term health to be maintained (e.g. Hammer 1992). For instance, occasional inundation of natural wetlands gives rise to differential patterns of erosion and sediment deposition, dynamically shifting the distribution of shallow and deep water areas for the coming years, and thus helping to maintain a diverse array of niches for re-colonization by wetland plants and insects; this in turn ensures that the wetland as a whole retains high species diversity and richness.

But what of groundwater? How does it fit into the promotion of these five dynamic environmental factors? Table 6.2 summarizes the key roles of groundwater in relation to all five factors.

Table 6.2 Role of groundwater systems in the operation of the five dynamic environmental factors which govern the sustained healthy functioning of all freshwater ecosystems.

Factor	Role of groundwater
(1) Flow patterns	Sustaining flows during periods of dry weather; giving rise to downstream increases in flow in the absence of surface tributaries; locally perturbing surface water flows where discharge pathways from the subsurface are highly localized; contributing to total flow rate in surface ecosystems, and to that extent helping to shorten retention times within them
(2) Sediment and organic matter inputs	Vigorous groundwater discharge can hinder deposition of sediments from suspension; some groundwaters can give rise to the precipitation of new chemical sediments (especially $Fe(OH)_3$ and $CaCO_3$) upon exposure to the atmoshpere; many groundwaters (especially those in peaty catchments) contain more dissolved organic matter than surface runoff, and can be a major source of organic carbon for the benthic ecosystem
(3) Temperature and light penetration	*Temperature*: Providing inputs of water of relatively constant temperature all year round, given that groundwater temperature year round tends to approximate closely to the local mean annual air temperature; as such, groundwater offers cooling in the summer, and warming in the winter. *Light penetration*: Since groundwaters do not carry appreciable plankton, they principally affect light penetration by changing the turbidity of receiving surface waters. Except where they carry excessive dissolved loads of certain metals (most notably Fe and Ca) which tend to precipitate and turn the water cloudy where they meet the atmosphere, most discharging groundwaters are essentially free of suspended solids; they will thus tend to reduce the turbidity of surface waters as they mix. However, where groundwater discharge is very vigorous and highly localized, it can locally lead to resuspension of bed sediment within the water column (though this is a rare effect)
(4) Chemical and nutrient conditions	Groundwater is the key source of several major nutrients (C, N, S, P) and many essential micronutrients (principally metals) in many freshwater ecosystems. On the other hand, most groundwaters are rather depleted in dissolved oxygen compared to surface waters, and where they upwell vigorously through the hyporheic zone they can jeopardize the survival of fish eggs and other fauna
(5) Plant and animal assemblage	Provision of permanently saturated refugia for invertebrates and small fish, both within and below the hyporheic zone; gene pool for numerous freshwater microbes which move from groundwater systems to colonize connected surface waters during wet periods

In the following sections of this chapter, these roles are described in greater detail, albeit the presentation is ordered by freshwater ecosystem type (rather than factor by factor as in Table 6.2).

6.2 Groundwater-fed wetland ecosystems

6.2.1 Groundwater quantity issues in wetland conservation

We have already briefly considered wetlands as groundwater discharge features in Section 5.1.4; here we will examine rather more closely the interconnectedness of wetlands and groundwater systems. Wetlands began to assume their present emblematic status for environmental campaigners in the 1980s. Many wetlands are ecosystems of global value, providing essential habitat for a very diverse range of organisms, of which the most prominent and emblematic are migratory waterfowl. It was precisely with the protection of wetland-dependent birds that the most important international measure for the protection of wetlands, the **Ramsar Convention**, was adopted in 1971. Those wetlands which have been formally adopted by the International Union for the Conservation of Nature and Natural Resources (IUCN) as falling under the protection of the Ramsar Convention are colloquially referred to as **Ramsar Sites**. The roll-call of Ramsar Sites is a "Who's Who" of the world's most important avian wetland habitats. For further insights into the ecology of wetlands and into the initiatives associated with the Ramsar Convention the reader is referred to Mitsch and Gosselink (2000).

Amidst the laudable enthusiasm of conservationists for compiling ever more reasons for protecting and restoring wetland habitats, much hyperbole has been expended on the supposed hydrological functions of wetlands. For instance, wetlands have been claimed to function as both recharge and discharge areas for groundwater systems and to reduce flood flows in rivers by storing water during storms and releasing it slowly during dry periods. How realistic are such claims? Bullock and Acreman (2003) have collated the results of a wide range of investigations concerning the role of wetlands in catchment hydrology. We have already seen that many wetlands owe their very existence to discharging groundwater, and this is amply borne out by studies worldwide (Fornés and Llamas 2001; Bullock and Acreman 2003). But what of the supposed role of wetlands as groundwater recharge zones? It is safe to say that relatively few wetlands are likely to act as recharge sources over extended periods of time. If the base of a wetland is sufficiently permeable to allow it to lose water freely to the subsurface, it is unlikely to be able to hold back enough water in the wetland to keep it saturated throughout the dry season. There are, of course, some wetlands that give rise to streams, which flow away onto aquifer outcrop areas where they then act as sources of indirect recharge. Even wetlands in this category are likely to support only intermittent recharge. What about the frequently mooted claim that wetlands can regulate flows in rivers by reducing peaks flows and sustaining baseflows? The baseflow arising from groundwater-fed wetlands originates in the adjoining aquifers, and is not attributable to the wetland *per se*. In relation to other types of wetlands, Burt (1995) has correctly noted that "most wetlands make very poor aquifers; . . . accordingly, they yield little base flow, but in contrast, generate very large quantities of flood runoff. Far from regulating river flow, wetlands usually provide a very flashy runoff regime."

Thus, when the scientific evidence is carefully examined (Bullock and Acreman 2003), the one claim about wetland–groundwater relations which is readily substantiated in many cases is that they tend to act as foci for groundwater discharge. Some major examples of wetlands fed by discharging groundwater are given in Box 6.1. The circumstances in which groundwater discharge can give rise to wetlands (or at least to partially feed wetlands which are also fed by surface runoff) are not too difficult to define: essentially, groundwater-fed wetlands arise in much the same manner as springs (see Figure 5.1). As such, the very existence of groundwater-fed

Box 6.1 Groundwater discharge and the major wetlands of Spain.

Nowhere is the importance of sustained, natural, groundwater discharge to the health of wetland ecosystems more closely studied (and at times more hotly debated) than in Spain. Lying closer to arid North Africa than any other European country, Spain is a land of enormous climatic contrasts, from the green and rain-soaked Cantabrian Cordillera in the north to the stereotypical sandy deserts of southeastern Andalucía. Many wetlands in Spain owe their existence, at least in part, to the natural discharge of groundwater. Two of the most important complexes of groundwater-fed wetlands in Spain are:

- **Doñana**, a large area of wetlands in the extreme SW of Spain, on the coastal plain of Andalucía some 50 km southwest of the city of Seville. Although the Doñana wetlands receive important inputs of both fresh surface waters and marine waters, groundwater discharge forms an important element of the water budget year round, and in the dry season it is crucial in the maintenance of residual flooded areas which act as important refugia for wildfowl. The groundwater in question emerges from a deep, regional aquifer which is recharged in a hilly area some 60 km to the north.
- "**La Mancha Húmeda**," a complex of wetlands in a semi-arid area in the central "meseta" of Spain (some 120 km to the south of Madrid). The wetlands of La Mancha Húmeda are fed by a sequence of aquifers (comprising clastic and carbonate rocks of various ages, from Triassic to Tertiary). Groundwater is forced to discharge at the surface where these aquifers adjoin aquitards of Precambrian age (schists, gneisses, and granites), and where the local rivers incise deeply into the aquifer materials. The most famous of the wetlands in La Mancha Húmeda are the Tablas de Damiel, a designated National Park covering an area of around 20 km^2 (or about 8% of the total wetland area in the district).

Both of the above wetland complexes are of such great wildlife value that they have been designated as UNESCO Biosphere Reserves. This designation has not spared either of the wetland complexes from experiencing significant stresses due to mismanagement of water resources. The wetlands of La Mancha Húmeda offer unequivocal evidence of the damage which can be caused to humid ecosystems by reckless interference with natural aquifer dynamics (Llamas 1988; Fornés and Llamas 2001). Originally, the total area of wetlands in La Mancha Húmeda exceeded 250 km^2. During the twentieth century, intensive agricultural development in areas underlain by the aquifers which feed La Mancha Húmeda gave rise to a burgeoning demand for irrigation water, almost all of which was obtained by pumping the same aquifers. The expansion of aquifer exploitation for irrigation water was particularly vigorous between 1970 and 1990, with the area of irrigated land in this region growing from 200 to 1400 km^2 over the period. This dramatic growth in groundwater use has been matched by a concomitant contraction of the wetlands in La Mancha Húmeda, with only 70 km^2 now remaining (i.e. a 70% decrease in wetland area). Governmental agencies have made a number of interventions to attempt to halt (if not reverse) further decline. Water transfers from adjoining river basins have helped to maintain flooded areas in the Tablas de Damiel wetlands. However, some of these transfers have themselves resulted in the depletion of other important wetlands (Fornés and Llamas 2001). Hydrogeological investigations have identified target levels of groundwater abstraction to which present irrigation pumping rates should be reduced in order to restore at least some of the natural functionality to La Mancha Húmeda (Fornés and Llamas 2001). It is as yet far from clear whether the political will exists to enforce any such reduction.

wetlands depends utterly on whether the water table is sustained above the local ground level (= wetland bed) for a substantial proportion of the year. Consequently, many wetlands are vulnerable to impoverishment, or even total destruction, in the event that the local water table is permanently lowered in response to pumping operations elsewhere in their feeding aquifers.

Although groundwater hydrologists have understood the relationship between water table lowering and wetland depletion for many decades (see Theis 1940), only recently have many verifiable instances of wetland damage due to aquifer pumping been documented. The scarcity of documented cases is likely explicable by two factors:

- Many wetlands had already been damaged by groundwater abstraction long before the advent of systematic studies of such phenomena in the late twentieth century.
- Few systematic studies of the hydrogeological settings of major wetlands were initiated prior to 1980, and only a small subset of such studies (e.g. Fornés and Llamas 2001; Burgess 2002) have been sustained over periods of time long enough (i.e. several decades) to definitively establish a causal link between long-term changes in wetland ecosystems and a gradual lowering of the water table.

In relation to the first of these factors, a certain amount of "forensic hydrology" can reveal the extent of wetland damage due to sustained pumping of aquifers. For instance, in an area near the Anglo–Scottish border that has been the locus of increasing groundwater abstraction since the 1930s, the perception of the present generation of regulatory authorities was that there were no ecological drawbacks to the sustained pumping of wells. However, using a combination of placename evidence and the collective memories of local families long resident in the area (who through their engagement in farming hold a keen appreciation of local land drainage conditions), it was demonstrated that groundwater pumping has in fact dried up former wetlands which were previously fed by natural discharge from the local aquifers (Younger 1998).

The task of investigating the impact of groundwater pumping on wetlands is not made any easier by the fact that many wetlands tend naturally to evolve into "dry lands" over time. This natural process of change, which ecologists term **hydroseral succession,** is a response to the gradual rise of the bed of the wetland due to the accumulation of undecayed plant debris and other sediments. Eventually the bed of the wetland will be so shallow that land plants can gain a foothold and begin to outcompete true wetland plant species. The first step in hydro-seral succession is typically for reedbeds to become invaded at the margins by willow scrub. The willows gradually spread throughout the former wetland area, shading the shorter reed plants and hindering their growth.

Eventually, an area of wet ground thoroughly colonized by willow trees develops, which is known as a **willow carr**. Eventually, willow carr will itself evolve into dry woodland. This process of habitat change can occur where the water table is static. Where human activities lead to falling water table levels, hydro-seral succession can be greatly accelerated.

Hydro-seral succession is only likely to result from an artificial lowering of the water table where this occurs very slowly. In cases where water table decline is more abrupt, a more likely sequence of events is:

1. Shallowing of the wetland, allowing invasion by terrestrial grasses.
2. Channelization of flow through the wetland, so that water moves through the area in the form of small channels rather than by gentle lateral flow through a generally inundated area.
3. Down-cutting of the bed sediments of the former wetland, and establishment of riparian grasslands/scrub.

By the time point 3 is reached, artificial changes in land use may soon follow, as humans decide to use the recently emerged dry land for grazing of animals or for arable farming.

Thus far this section has focused rather negatively on what can go wrong with wetland systems. On reflection, however, the same insights can be used to positively identify the role of groundwater discharge in supporting flourishing wetland ecosystems. In essence, these correspond to the functions summarized in Table 6.2. One crucial role of groundwater discharge which is not explicit in Table 6.2, but which is of overwhelming importance in the wetland context, is the maintenance of adequate water *levels*. A water depth in the range of 0.15–0.5 m provides ideal conditions for reeds to flourish without suffering competition from invading terrestrial grasses. Even if the bulk of the flow through a wetland is provided by seasonal inputs of surface runoff, therefore, as long as the wetland bed lies more than about 0.1 m below the local water table level then the groundwater system will ensure the maintenance of optimal water levels to maintain a wetland (as opposed to wet grassland) habitat.

6.2.2 Plant–water interactions in groundwater-fed wetlands

The dynamics of plant–water interactions in wetlands has already been alluded to above, in the discussion of the processes inherent in hydro-seral succession. Irrespective of whether they are fed by groundwater or surface water, wetlands are a challenging habitat for most plants. This is because, like ourselves, plants need to breathe air: they depend on oxygen for respiration. Because the solubility of oxygen in water is not very high (about 12–15 mg/L maximum at temperatures typical of the Earth's surface), submergence of plant roots amounts to oxygen deprivation. If the roots are devoid of oxygen (in a state of "**anoxia**"), then aerobic respiration cannot occur. Once anoxia sets in, most plants will experience a sequence of anaerobic biochemical changes that will shortly lead to the death of the plant. Wetland plants are specially adapted to counteract the establishment of anoxia, which they achieve by sustaining a small pocket of aeration around their roots. Effectively, this is achieved through "pumping" of air from the leaves above the water line, down through the stem, and into the roots, from which it leaks out into the surrounding water, creating an oxygen-rich haven. For a formal explanation of the plant adaptations that make this possible, the reader should refer to Mitsch and Gosselink (2000, pp. 209–224). This transfer of oxygen to the root zone requires a significant expenditure of energy. If the release of oxygen from the roots is to proceed sufficiently briskly to prevent the development of anoxia, then the resistance offered by the head of water above the roots must be overcome. If the head is too great, then the plant will struggle to prevent anoxia. It is for this reason that many wetland plants begin to display signs of stress where water depths exceed 0.5 m; where water depths extend to a meter or more, drowning and death will eventually follow.

The above considerations apply whether the wetland is fed by surface water or groundwater. However, groundwater-fed wetlands represent a particularly extreme case, for two reasons:

- The incoming groundwater will often be utterly devoid of dissolved oxygen, so that there is a greater oxygen deficit to be overcome before aerobic conditions can be established around the roots.
- Groundwater is usually continuously on the move within wetland bed sediments, thus tending to carry oxygen away from plant roots more briskly than will be the case in the rather stagnant bed sediments of many surface-fed wetlands.

Especially vigorous oxygen transfer to and through the roots is therefore a *sine qua non* for the maintenance of healthy plant communities (and therefore of invertebrates, birds, and other species) in groundwater-fed wetlands.

6.2.3 Water quality evolution during flow through wetlands

Given that many groundwaters are naturally devoid of dissolved oxygen, the release of oxygen from plant roots results in a marked increase in dissolved oxygen in the water as it flows through the wetland. Figure 6.1 clearly illustrates this phenomenon. In the wetland from which the data of Figure 6.1 were collected, flow measurements demonstrated that virtually all of the groundwater flowing out of the wetlands could be accounted

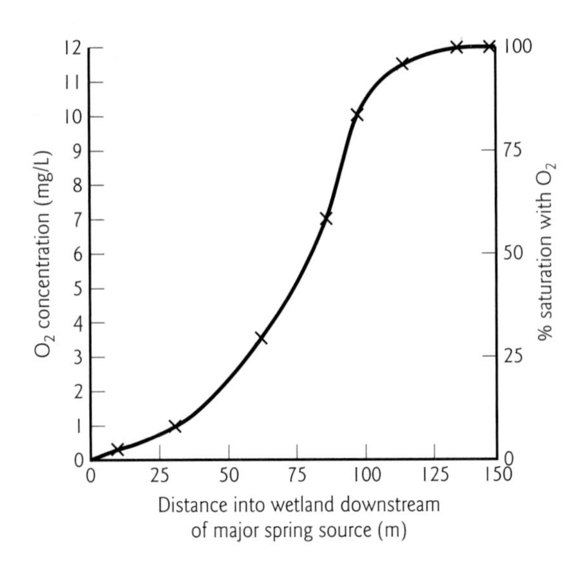

Fig. 6.1 Observed increase in dissolved oxygen in the water of a wetland in county Durham (UK) fed by a single large spring (which discharges completely anoxic groundwater).

for by inflow from a single spring source. In other wetlands in which groundwater inflows are rather more diffuse (upflowing through much of the bed sediment) the patterns of increase in oxygen will not correlate so neatly with distance from the upstream end of the wetland.

Many other geochemical processes which occur in groundwater-fed wetlands reflect the general tendency to increasing oxygenation. In general, as water flows through a wetland it will tend to become more oxidized, and geochemically reduced solutes will also be oxidized. For instance, many groundwaters contain soluble ferrous iron (Fe^{2+}), which is usually oxidized to the far less soluble ferric form (Fe^{3+}) during flow through wetlands. Solid precipitates of ferric hydroxide ($Fe(OH)_3$) consequently accumulate in the wetland sediments, and also on and within root tissues of wetland plants (see Batty and Younger 2002). Similar transformations affect other redox-sensitive metals, including manganese (which is oxidized from soluble Mn^{2+} to insoluble Mn^{4+}) and copper (soluble Cu^{2+} to insoluble Cu^{3+}). Amongst nonmetals, oxidation reactions also affect important nutrient species such as sulfur (reduced form HS^- to oxidized form SO_4^{2-}) and nitrogen (reduced form NH_4^+ to oxidized form NO_3^-), although in these cases both the reduced and oxidized forms are soluble.

Reactions with atmospheric oxygen are not the only processes affecting the quality of groundwaters discharging through wetlands. For instance, as the dissolved CO_2 concentration in discharging groundwaters comes to equilibrium with the atmosphere (usually by de-gassing (e.g. Hem 1985), though in some cases also by dissolution of atmospheric CO_2 (Khoury et al. 1985)), precipitation of calcite commonly occurs, resulting in a marked decline in dissolved Ca^{2+}. Many other solutes are at least temporarily retarded in their transport through wetlands by sorption onto solid surfaces. The most powerful sorbents in wetlands are living and dead plant materials, but ferric hydroxide precipitates and clay minerals are also important providers of electrostatic exchange sites. Most metals and many organic contaminants are prone to attenuation in this manner.

Biological processes occurring within wetland sediments can also significantly affect water quality. For instance, bacterial sulfate reduction can result in the precipitation of metallic sulfide minerals within the sediment body (away from the anoxic havens around roots). Simple organic molecules exuded by plant roots can in some cases provide energy sources for the sulfate-reducing bacteria. Other microbes can use organic contaminants as their energy source, leading to "biodegradation" of these contaminants. However, in some cases one organic contaminant may be biodegraded to release an even more toxic "daughter product" to solution. Similarly, mercury is rendered far more toxic by the attachment of methyl complexes, a process for which sulfate-reducing bacteria present in many wetland substrates are now known to be responsible (King et al. 2001). Given the array of potential contaminants, active microbes, and daughter-products, it is difficult to generalize about the role of wetlands in providing "biodegradational services" which improve water quality, and case-specific investigations are almost always warranted.

6.3 Fluvial ecosystems and the hyporheic zone

6.3.1 Spatial and temporal hierarchies of groundwater effects on in-stream ecology

A hierarchy of scale governs the degree to which groundwater affects in-stream ecology (Cannan and Armitage 1999): this hierarchy is governed by the magnitude of the baseflow index (BFI). For instance, where BFI is high (so that the flow regime is volumetrically dominated by groundwater discharge), the ecology of the stream is obviously dependent on groundwater discharge at every feasible scale of observation, from a small patch of streambed to the catchment as a whole (cf. Sear et al. 1999). It hardly needs stating that if virtually all of the water in the channel is groundwater, the in-stream ecology owes its very existence to aquifer outflows. Even in streams where BFI is more modest, in-stream ecology can still be powerfully influenced by groundwater discharge at the scales of individual reaches and/or channel perimeters. For instance, if groundwater inflows are restricted to one or two localized patches on the streambed, the ecological niches which correspond to those patches will tend to be strongly influenced by small-scale mixing of groundwaters and surface waters (e.g. Jones and Mulholland 2000; Malcolm et al. 2003).

This hierarchy can equally be expressed in terms of time-scales. Where BFI is high, groundwater discharge will be the predominant influence on in-stream ecology at almost all times (with the possible exceptions occurring during periods of flooding). In medium- to low-BFI streams, dynamic changes in groundwater/stream water exchanges might well occur over very short time-scales (e.g. Maddock et al. 1995; Alden and Munster 1997; Malcolm et al. 2003), such that relatively modest fluctuations in stream stage give rise to profound changes in the direction and velocity of groundwater movement near the channel (e.g. Alden and Munster 1997). Nowhere are the interactions between surface and subsurface waters more dynamic than in the hyporheic zone, in which various hydrogeochemical processes wax and wane in response to changing flow patterns. As we shall now see, these hydrogeochemical dynamics are often of critical importance for the ecology of the adjoining stream ecosystems.

6.3.2 Ecologically critical physicochemical dynamics in the hyporheic zone

The concept of the hyporheic zone was introduced in Section 5.1.6, where it was noted that it is a zone of intense mixing between discharging groundwaters and waters which have until very recently been in the open channel. On the rare occasions in which hyporheic zone processes have been investigated at the scale of a short stream reach (i.e. one which is little longer than the stream is wide), it is possible to distinguish between the following three modes of surface–subsurface hydrological exchange (Figure 6.2):

- Shallow exchange, in which surface waters locally enter the bed sediment and exit back into the channel a short distance away. Typically, water enters the subsurface through the upstream face of a stony riffle, re-surfacing through the downstream edge of the riffle, flowing into the adjoining pool.
- Upwelling, in which discharging groundwater (which entered the aquifer at some distance from the stream channel) flows upwards through the bed sediment and out into the open channel, and
- Downwelling, in which water from the open channel flows down through the stream bed sediment into the underlying aquifer.

A switch between upwelling and downwelling conditions can occur almost instantaneously when a flood wave passes through a stream (see Figure 5.5 and Section 5.1.6). For this reason the water present at any one point in the hyporheic zone, at any single moment in time, will closely reflect the recent history of stage fluctuations in the stream. A given sample of hyporheic water will thus amount to some mixture of waters of all three origins. As we have seen, groundwaters are typically depleted in dissolved oxygen, but relatively enriched in dissolved solids (see Chapter 4), whereas the opposite is the case for surface

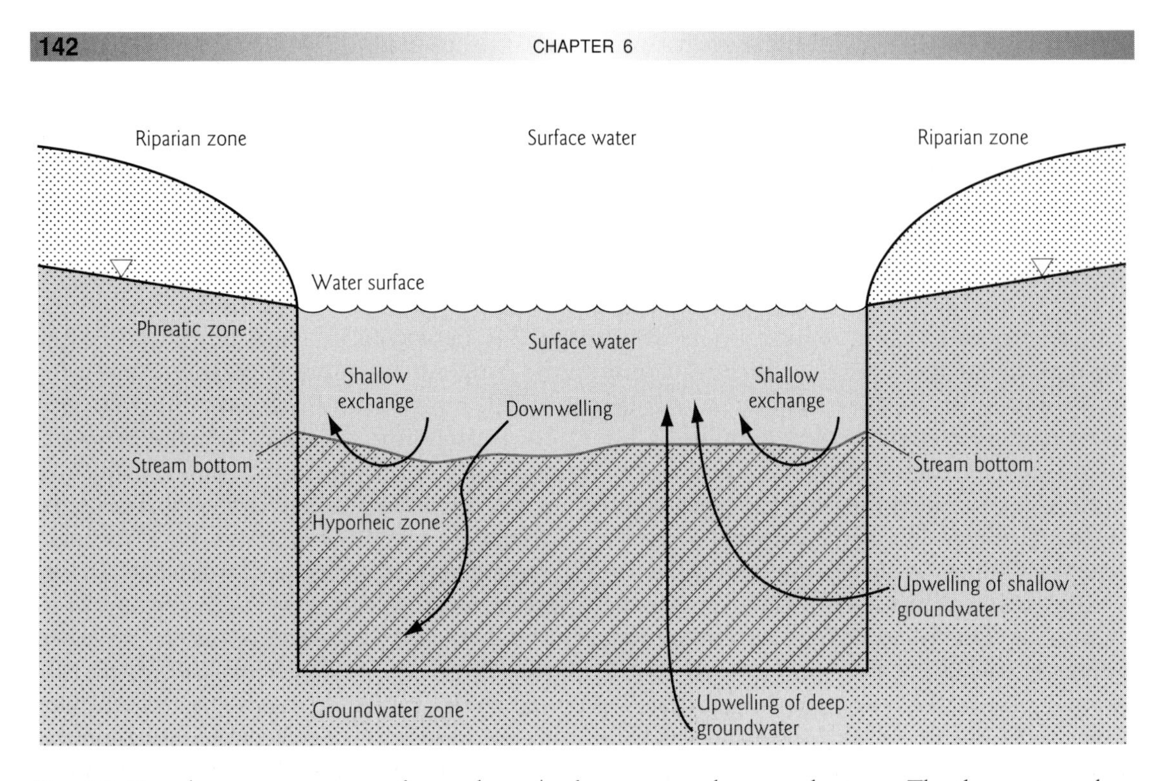

Fig. 6.2 Hyporheic zone processes of groundwater/surface water exchange and mixing. The three principal processes are: downwelling of surface water into the underlying aquifer; upwelling of deep groundwater into the stream; and shallow exchange of surface waters into the sediments immediately underlying the stream bottom. (Adapted after Kaplan and Newbold 2000.)

runoff. Contrasts in temperature are also usually evident between groundwaters and surface runoff. As was noted in Section 4.2.1, shallow groundwaters typically have a fairly constant temperature all year round, the absolute value of which usually approximates closely the local mean annual air temperature. Consequently, surface runoff tends to be warmer than groundwater in the summer months, but significantly cooler than it in the winter months. Furthermore, oxygen is more soluble in cold waters than in warm waters. This means that temperature and dissolved oxygen concentrations are inversely correlated.

Most freshwater life-forms are highly sensitive to fluctuations in temperature and to the availability of dissolved oxygen: stable temperatures and high dissolved oxygen concentrations are generally favorable for most creatures. In the hyporheic zone, stable temperatures are associated with a high proportion of upwelling groundwater, whereas high dissolved oxygen concentrations are associated with downwelling surface runoff. Hence the vigor with which upwelling groundwater mixes with shallow exchange waters can have considerable importance for the survival of organisms within the bed sediments. In the majority of cases, the deeper into the hyporheic zone one penetrates, the lower the dissolved oxygen content will be, but the more stable the water temperature will be. The "ideal" niche for many organisms will thus tend to be in the zone of approximately equal mixing of ground and surface waters, where dissolved oxygen levels are still high, but temperatures are stable throughout the year. Of course this generalization does not apply equally to all organisms: evolutionary adaptations which have bequeathed an ability to cope with wider variations in temperature and/or dissolved oxygen can bestow competitive advantages on different species in different hyporheic niches.

The supply of nutrients to hyporheic organisms is just as critical to their well-being as is the maintenance of favorable temperature and dissolved oxygen conditions. Downwelling surface waters can deliver nutrients originating from the breakdown of plant debris; these nutrients are directly usable by some of the smaller animals which inhabit the hyporheic zone, such as protozoans and annelid worms (Table 6.1). Quite large fragments of plant debris can be transported into porous gravels; however, in finer-grained streambed sediments only small (colloid-sized) fragments are likely to reach the deeper portions of the hyporheic zone. Far more important in terms of overall energy flows are dissolved microbial nutrients, which include not only dissolved organic carbon compounds, but also inorganic molecules containing oxygen (most notably nitrate and sulfate) which are important nutrients for specific groups of microbes (Findlay and Sobczak 2000). The growth of such microbes within the bed sediment represents an important source of in situ primary production, resulting in the accumulation of organic debris which is subsequently available for grazing by protozoans, rotifers, and worms (cf Table 6.1). These tiny animals, referred to collectively as the **meiofauna** (a term meaning "middle-sized animals"), are in turn predated by larger organisms (the **macrofauna**), especially crustaceans and insects (cf Table 6.1). Such predation hierarchies (also known as **trophic chains**) together form the threads that constitute the entire hyporheic food web.

The dependence of predators on prey ensures that the availability of dissolved nutrients (which cannot be used directly by crustaceans or insects) is reflected in the distribution of invertebrates within hyporheic zone sediments (see Boulton 2000). Indeed, the microscale distributions of temperature, dissolved oxygen, and nutrients exert fundamental controls on the numbers and diversity of the microbial, meiofaunal, and macrofaunal communities within the hyporheic zone. For the macrofauna, a further important control is exerted by the pore size distribution of the sediment: the larger the invertebrate, the larger the pores it requires for easy passage. It has been noted, for instance, that crustaceans are more common in gravels than in fine-grained sands (e.g. Hakenkamp and Palmer 2000). Figure 6.3 summarizes the overall controls on the well-being of the hyporheic macrofauna.

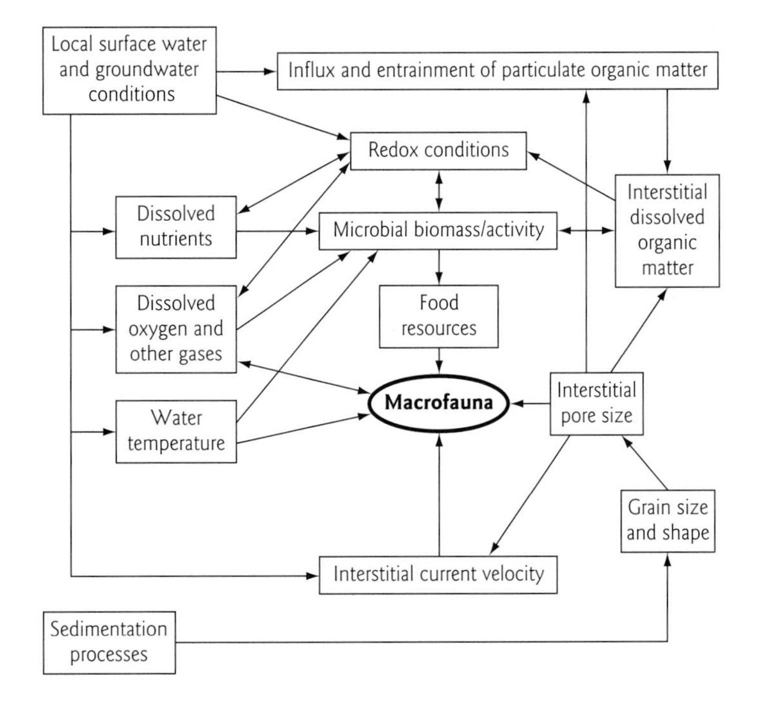

Fig. 6.3 The principal physical, chemical, and biological factors which govern the well-being of the macrofauna of the hyporheic zone. Groundwater upwelling primarily affects the temperature, dissolved oxygen content, and redox conditions, hence a shift in the relative proportions of downwelling and upwelling waters in the hyporheic zone can have major implications for the survival of macroinvertebrates, the viability of salmonid eggs buried in redds, etc. (Adapted after Boulton 2000.)

Most of the same considerations apply to the survival of fish eggs within the hyporheic zone. Migratory salmon and trout[ii] bury their eggs in hollows which they make with their tails in stream bed gravels. After the eggs are laid, the fish bury them in sediment. The resultant pockets of buried eggs, known as **redds**, are quintessential ecological features of the hyporheic zones of many of the world's finest fishing rivers. Studies of the impact of groundwater–surface water interactions on egg survival in redds were initiated in North America (Power et al. 1999). Those studies highlighted the beneficial effects of the sustained discharge of deep groundwaters during the harsh winters typical of the northern continental interior, when upwelling of relatively warm waters into otherwise ice-bound rivers ensures the persistence of nonfrozen refugia for fish. There is evidence that fish migrate long distances along rivers to take advantage of such zones of strong hyporheic upwelling (Power et al. 1999). At finer scales of resolution, it is possible to derive reasonably accurate correlations between hyporheic-zone water temperatures and the timing of key stages in the development of embryos within redds (Acornley 1999). More recent work in the highlands and eastern lowlands of Scotland (e.g. Malcolm et al. 2002, 2003, 2005) has demonstrated close correlations between egg survival and variations in the dissolved oxygen content of hyporheic zone pore-waters: where mean dissolved oxygen contents were less than 7.6 mg/L, almost all embryos died before hatching, whereas almost 100% survival rates were observed where dissolved oxygen exceeded 11.5 mg/L (Malcolm et al. 2003). Low dissolved oxygen contents are primarily associated with zones of groundwater upwelling (Malcolm et al. 2002, 2003), although infiltration of fine sediment (blocking pores and hindering through-flow of oxygenated water) was also associated with fatally low oxygen contents in some hyporheic zone waters (Soulsby et al. 2001). These potentially negative impacts of natural groundwater discharge are highly variable in both space and time. During periods of dry weather, the major foci of groundwater upwelling (and resultant redd damage) are fairly stable. However, during periods of storm runoff (which are very frequent in that part of the world), previous zones of upwelling can become zones of downwelling, with consequent rapid changes in dissolved oxygen availability in the hyporheic zone.

6.4 Groundwater ecology

6.4.1 The emerging paradigm of groundwater ecology

It is fair to say that the old proverb "out of sight, out of mind" applied to subsurface ecosystems prior to 1990. Given the fundamental importance of photosynthesis to life on Earth, biologists can be forgiven for having long dismissed the subsurface as a zone of low biotic diversity and productivity. Such dismissiveness was, however, predicated on an important oversight: it failed to recognize the potentially important role of groundwater as a transport medium for nutrients ultimately derived from photosynthesis. That groundwater systems are far from being utterly sterile began to become evident in the 1970s, when researchers began to realize that organic contaminants were undergoing biodegradation during their passage through aquifers. However, the microbes responsible for these biodegradational processes proved unresponsive to laboratory culturing techniques then available. The key breakthrough came in 1983, when US EPA researchers in Ada, Oklahoma, successfully pioneered culturing techniques appropriate to native subsurface bacteria (Wilson et al. 1983). Applying these new techniques, it soon emerged that aquifers, which had long been considered to be almost devoid of life, actually host microbial ecosystems that are comparable (in terms of species diversity and organism numbers) to nutrient-rich lakes in temperate regions. In the wake of this pioneering work, the study of microbial ecosystems within aquifers proceeded very rapidly, initially in relation to the characterization of pollution incidents and the development of bioremediation strategies, and subsequently as a pure science topic pursued for its intrinsic interest (Chapelle 2000; Griebler 2001).

In parallel with these developments, a number of pioneering researchers were also busy investigating the meio- and macro-fauna of a range of aquifer systems. The first substantial synthesis of this work appeared in the mid 1990s (Gibert et al. 1994), providing a firm foundation upon which "groundwater ecology" has begun to be erected as a new subdiscipline of ecology (see Wilkens et al. 2000; Griebler et al. 2001; Danielopol et al. 2003).

By the mid 1990s, the concept of "applied groundwater ecology" was beginning to emerge (Malard et al. 1996), with the proposal that evaluations of groundwater ecosystem status might be useful within wider groundwater quality monitoring programs (just as invertebrate surveys of streams are used to assess the overall quality of surface waters). As in surface water studies, the initial proposal of Malard et al. (1996) was to use invertebrates as "**biomonitors**" of aquifer quality. A number of pioneering investigations have been launched to investigate the feasibility of this proposition. One such study, within a karstic aquifer (cf. Section 6.4.4), found that the relative abundances of certain classes of invertebrates proved to be a reliable indicator of sporadic sewage pollution, which might have gone undetected using conventional water sampling techniques.

It is beyond the scope of this book to provide a thorough introduction to groundwater ecology, let alone its potential applications to aquifer monitoring and management. Rather, the following sections offer brief summaries of the key findings of studies of particular aspects of groundwater ecosystems. For more detailed insights into the techniques and findings of such studies, the references cited in the text should be consulted.

6.4.2 Natural microbial communities in aquifers

Our understanding of the microbial ecology of aquifers is an area of extremely active research (Chapelle 2000). The recent emergence of molecular probing tools is revolutionizing the ecological investigation of all types of microbial systems, including groundwater ecosystems. It is now clear that microbes are present at great depths in all types of aquifers. Great numbers of bacteria are typically present, usually numbering between 10^5 and 10^7 per gram of dry sediment. In silt and clay strata, most bacteria are "Gram-positive" species (i.e. they respond to dyeing with the so-called Gram dye), whereas in sands and gravels, Gram-negative bacteria predominate. In karst aquifers, bacteria normally associated with surface waters are commonly found at depth, undoubtedly reflecting wash-in by recharging waters. Moderate densities of protozoans are typical of many aquifers, principally represented by amoebae and cyst-forming flagellates. Fungi and algae are occasionally encountered, especially in karstic aquifers; as for surface water bacterial forms, the presence of fungi and algae indicates recent recharge/downwelling of surface waters.

6.4.3 Macrofaunal ecology of alluvial aquifers

As in all other subdisciplines of ecology, the burgeoning literature concerning invertebrate groundwater ecology is replete with specialist terminology. Animals which live in the subsurface (and therefore for the most part within groundwater systems) are referred to as **hypogean** fauna. The three most important types of hypogean organisms are (see Gibert et al. 1994):

- **Stygoxene:** an organism belonging to a species which is utterly alien to the subsurface environment. By definition, stygoxenes tend to be present in the subsurface only by accident (e.g. following burial beneath sediment during a storm or by wash-in via karst dolines etc).
- **Stygophile:** an organism belonging to a species which is adapted to living in the subsurface for at least part of its life-cycle, but which is also at home in surface environments.
- **Stygobite:** an organism belonging to a species which is only truly at home in the subsurface.

In any one alluvial aquifer, it is typical to find 30–40 species of stygophile/stygobite invertebrates. Most studies of alluvial aquifer systems in Europe and North America have found the hypogean invertebrate fauna to be dominated by crustaceans (especially amphipods; cf. Table 6.1). Next in abundance, in most cases, are nematodes and oligochaete worms, though in some cases stygobite gastropods are also abundant.

Detailed ecological studies of the alluvial aquifers of the Rhône and Danube valleys have been sustained without significant interruptions since the early 1970s (Gibert et al. 1994; Griebler 2001). Taken together with comparable data gathered in the USA, the results of these long-term studies have revealed a number of apparently general characteristics of groundwater ecosystems in alluvial aquifers. First, the difficulties experienced by groundwater invertebrates in attempting to migrate over large distances result in a very high incidence of endemism.[i] Second, strong vertical gradients in the proportions of the various types of hypogean organisms are typically detectable at any one study site: stygoxenes tend to become less abundant with increasing depth, whereas the relative (though usually not absolute) abundances of stygophiles and stygobites tend to increase with depth. Of course "depth" *per se* is not the controlling factor on the occurrence of hyopgean fauna; rather, depth is a collective, surrogate measure of water pressure, pore size, temperature variability, and the availability of dissolved oxygen and key nutrients.

6.4.4 Karst aquifer ecology

Karst terrains are landscapes dominated by the presence of caves, dolines, and other features commonly associated with the dissolution of permeable bedrock (see Ford and Williams 1989; Gillieson 1996; Moore and Sullivan 1997). As such, karst is principally found in areas underlain by limestone and/or gypsum. The allure of caves has captivated many people since the earliest days of humankind; indeed, humans have formed an integral part of many cave ecosystems since the emergence of our species. Our archetypal "caveman" ancestors are simply one more example of the many vertebrates (most famously bats) which shelter in caves, but hunt out in the open. The importance of this type of cave dweller to the overall ecology of karst terrains cannot be overstated, for the activities of cave-sheltering vertebrates effectively transfer nutrients from the photosynthetic world to the subsurface via the feeding–defecation cycle. A voluminous and steadily growing literature exists on the ecology of karst terrains, both above and below the water table (see Wilkens et al. 2000 for an extensive review). Here we are concerned solely with **groundwater ecology** in karst terrains.

As in other groundwater ecosystems (cf. Section 6.4.3), a specialist terminology has been developed to describe the subsurface affiliations of different organisms. Thus a "troglobite" is defined as an organism which can **only** live in caves. "Troglophiles" are organisms which prefer the cave environment, but are able to flourish elsewhere if necessary. Finally "trogloxenes" are essentially surface life-forms which use caves for shelter.

As in other subsurface ecosystems, the distributions of the various categories of organisms throughout karst terrains reflects the availability of various environmental resources. Take a trip into a cave (preferably one not perpetually lighted for the benefit of tourists) and you will quickly appreciate how soon daylight is left behind. Once you are past the reach of daylight, no more green plants will be present; the only "flora" will be the vari-colored microbial biofilms which are sporadically present on cave walls and ceilings. With the loss of green plants, there is an abrupt decline in the availability of detrital organic carbon, which is an important source of nutrition for the microfauna. In the furthest recesses of the cave, atmospheric oxygen is usually also in short supply, and carbon dioxide may reach dangerously high concentrations (promise me you'll turn back before you reach that zone . . .). The relative isolation of many cave atmospheres from the surface means that cave environments tend to have far more stable climates than the overlying surface environment

(Chapman 1993; Gillieson 1996). This is manifest particularly in the stable temperature and humidity regimes of most deep portions of caves.

The distribution of organisms within karst groundwater systems generally reflects the changing availability of environmental resources with increasing depth. As depth increases, the abundances of troglophile and troglobite species increase relative to those of trogloxene species. All cave fauna and a significant proportion of the cave flora (especially at shallow depth) are "heterotrophic," that is they feed on the remains of other organisms. For instance, the bacteria and algae involved in the decay of fecal material and dead animals lying in caves are heterotrophic; so are the protozoans that feed on the bacteria and algae, the invertebrates that feed on the protozoans, and the cave fish that eat the invertebrates. This is a typical example of a heterotrophic "food chain." At its base is photosynthesis, the form of "primary production" of nutrients which sustained the surface-dwelling/visiting animals upon whose waste and remains the bacteria and algae feed. The key point here is that, at shallow to medium depth in karst ecosystems, the food chain is still utterly dominated by the supply of nutrients from the surface environment. Even where carrion is not brought into the cave, the wash-in of nutrients in recharging groundwater perpetuates the supply of nutrients from the photosynthetic surface environment (Moore and Sullivan 1997).

On this basis, it might be assumed that, at great depth within karst ecosystems (beyond the typical penetration limits of bats, rats, and mice, and beyond the point at which the nutrients in incoming recharge waters have already been consumed by bacterial metabolism), the karst system would be utterly sterile. Such is not the case. This is because primary production is sustained even in the absence of inputs of nutrients from the surface by the activity of "autotrophic bacteria," which are capable of acquiring all of the nutrients they need from inorganic sources (such as the atmosphere and minerals). While the rates of primary production by autotrophic bacteria are modest in comparison to those levels observed in surface environments, they represent the difference between life and sterility in the depths of karst aquifers. Amongst the many autotrophic bacteria documented from cave systems, the example of an anaerobic bacterium *Perabacterium spelei* may be cited: this species can fix nitrogen from the cave atmosphere and obtain its supply of carbon (and energy) by the oxidative weathering of the iron carbonate mineral siderite. Primary production by *Perabacterium spelei* has been shown to be sufficient to support colonies of *Niphargus* (a well-known troglobite amphipod) in the absence of any inputs from heterotrophic primary production (Moore and Sullivan 1997).

Amphipods such as *Niphargus* are not the only invertebrates acting as primary consumers in the water-filled portions of cave systems: other crustacean species, flatworms, and gastropods are also prominent (Chapman 1993). The only vertebrate troglobites are salamanders and cave fish. The latter are particularly important in terms of groundwater ecology. Truly troglobitic cave fish (as opposed to surface forms which have accidentally entered cave systems) are usually blind and have pigment-less skin (variously described as white or translucent). What their skin lacks in color it makes up for in utility: most troglobite fish have very sensitive vibration receptors in their skin, which represent a wonderful adaptation to facilitate swimming and feeding in total darkness. Troglobite fish rarely exceed 10 cm length and are usually very thin: for instance, the Brazilian cave fish *Pluto infernalis* is about 50 times as long as it is wide! Such peculiar body shapes suit cave fish for migrating long distances through small conduits. Overall, about 60 species of troglobite fish have been recorded worldwide, though to date none have been recorded in Europe. In the USA, the 14 known species of troglobite fish are all found to the south of the late Quaternary glacial limit (Chapman 1993), perhaps suggesting that the former glaciation of much of Europe is responsible for the lack of troglobite fish there.

In addition to troglobite fish, many trogloxene fish are found in caves. The brown trout is

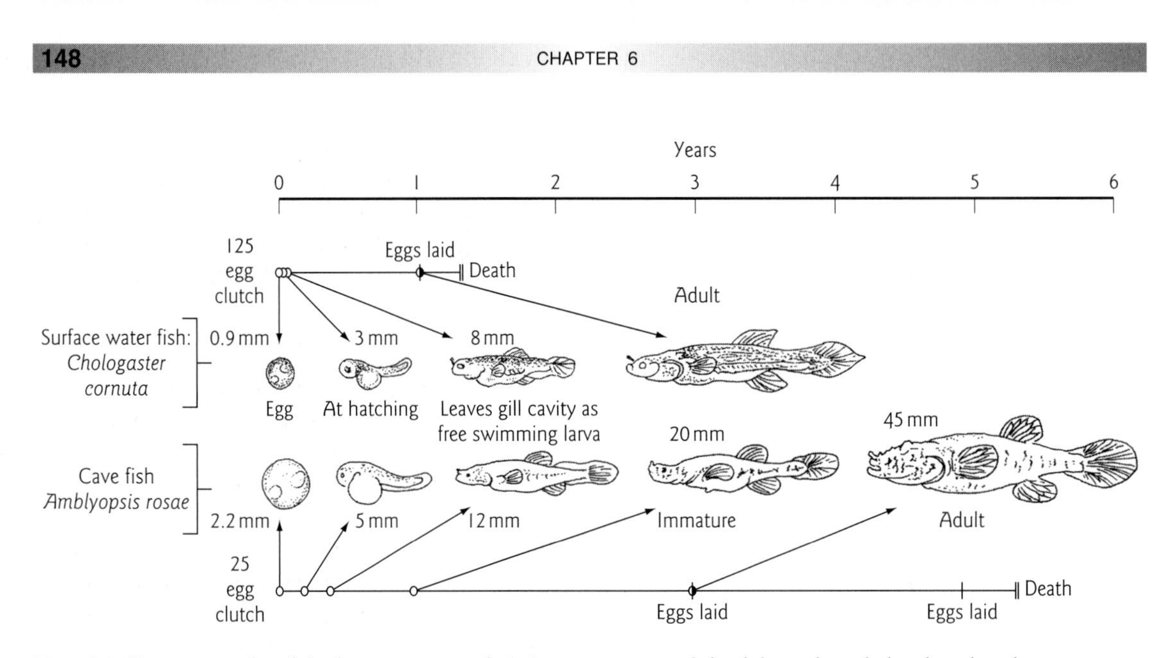

Fig. 6.4 For a stress-free life, live in an aquifer! A comparison of the life cycles of closely related species of fish, one of which is a conventional surface water inhabitant, the other a cave-dweller. The surface water fish has a shorter life-expectancy (around 16 months, compared with 5 years for the cave-dwelling form), spawns after only 1 year (compared to 3 years for the cave dweller), and never attains the same size as its subterranean counterpart. These marked differences show evolutionary adaptations to the marked lack of predators in the flooded cave system. (Adapted from Gillieson 1996, after Poulson and White 1969.)

abundant in many caves which have hydraulic connections with surface watercourses (Chapman 1993). Some brown trout in caves have been found to have lost their skin pigmentation, suggesting they have spent most (if not all) of their lives in the subsurface.

The relative stability of the subsurface environment relative to surface streams has already been noted. Besides the steady temperatures and oxygen contents of many karst aquifer waters, the lack of predators also makes cave systems rather benign environments for fish. This stability is reflected in the fact that the lives of cave fish are often much longer than those of relatives which dwell in surface waters (Moore and Sullivan 1997). Furthermore, the life-cycles of cave fish tend also to be more "attenuated" than those of surface water forms (Figure 6.4), with maturity, spawning, and signs of aging all appearing much later in the case of cave fish. Such are the blessings of an unhurried life, free from the stress of predators.

For all of their fascinating idiosyncracies and hidden beauty, karst aquifer ecosystems are extremely fragile. In part this is due to the extreme incidence of endemism[i] amongst troglobites: although a total of 50,000–100,000 troglobite species are estimated to exist worldwide, many of these are restricted to only one or two caves. Indeed, many troglobite species are known from only one or two specimens. The largest known population of any single troglobite species (the crayfish *Orconectes* within Pless Cave, Indiana, USA) numbers only 9090 individuals; this is a very low number when compared to surface water populations of closely related crayfish. More than 22 troglobite species are now on the "red list" of endangered species (Gillieson 1996), though this number almost certainly underestimates the true degree of "endangerment" of cave fauna in the face of growing disruption by human activities (e.g. waste dumping in dolines, mining, water well pumping, cave exploration, etc.).

Endnotes

i "Endemic" is an adjective, describing the distribution of given species; by extension, the noun **endemism** described the tendency of species to be restricted to small geographical areas.

ii Salmon and trout are often referred to collectively as "salmonids," a term which includes all members of the taxonomic family Salmonidae, which not only encompasses all species of salmon and trout, but also char, whitefish, and grayling.

7

Groundwater as a Resource

Saint Cuthbert said: "The place I have chosen [for my hermitage] is without a well. Pray with me, I beseech you, that He who 'turns the solid rock into a standing water and the flint into fountains of water' may open a spring of water for us on this rocky ground". They dug a pit and found it next morning full of water springing up from beneath . . . This water, strangely enough, kept its original level, never spilling over onto the ground nor sinking as it was drawn out. God in his bounty willed that there should never be any less, nor any more, than was needed.

(St Bede The Venerable, 673–735 AD, *Life of St Cuthbert*)

Key questions

- How is groundwater already used in different parts of the world?
- What are the physical limitations on groundwater availability?
- How does the quality of groundwater affect its utility?
- Does groundwater use have any negative side-effects?
- What do we need to do to abstract groundwater from springs and wells?
- Can groundwater and surface waters be used together?
- Can we make better use of aquifers for temporary water storage?
- How can groundwater be used as a source of energy?

7.1 Current resource utilization of groundwater

7.1.1 Groundwater vs. surface water as a source of supply

Groundwater accounts for a large proportion of the water used by humankind every day for a range of purposes from drinking to irrigation. But how large is this proportion? It is frequently claimed that groundwater accounts for between a quarter and a third of all water use worldwide. Other texts claim that "more than a quarter of people worldwide rely on groundwater for drinking" (e.g. Clarke and King 2004). However, it is difficult to interpret what this claim might mean in terms of the proportion of *total* water use that is satisfied by groundwater resources. This is because, out of a total daily need averaging 37.5 liters per person (Cole 1998), drinking water *per se* rarely accounts for more than 7% of the total water needs of an individual person; on average, 26% is used for preparing and cooking food, and 67% for personal hygiene and laundry.

Even when we turn to national statistics on water use, which might be expected to be more revealing, we run into problems with variant definitions of water sources. For instance, a report published in 1998 claimed that public water supplies in the UK comprise 28% groundwater and 72% surface water (Turton 1998). These figures no doubt derive from abstraction[i] licence records held by the Environment Agency. However, in many Environment Agency regions, abstractions from springs are classed administratively as "surface water" abstractions, thus masking the total contribution to water supply coming directly from subsurface sources. Many other countries explicitly differentiate in their records between spring abstractions, pumped groundwater abstractions (i.e. from wells/boreholes), and surface water intakes. However, even in these cases it is difficult to adequately assess the relative importance of groundwater to total supply due to complications arising from river–aquifer interactions. For instance, as we saw in Chapter 5, natural groundwater discharge sustains the flows in many rivers through dry periods, when water demand is often at its peak. On the other hand, many so-called "groundwater abstractions" are sufficiently close to major rivers that they induce a component of localized indirect recharge (cf. Chapter 2), so that the water pumped from a given well often includes a significant surface-derived component (which might itself have begun life as a natural groundwater discharge into the river upstream . . .).

For these reasons, it is rarely possible to be sure how best to interpret statistics concerning the groundwater component of water supplies. Furthermore, many underdeveloped countries in Africa, Asia, and Latin America do not collect statistics on water use.

Notwithstanding all of the foregoing caveats, Table 7.1 summarizes the proportion of water use considered to be supplied from groundwater sources in a selection of countries around the world. In many cases, the absolute values of groundwater abstraction and the rates normalized per capita (i.e. divided by the population of the country) are more revealing than the percentage figures. Even brief scrutiny of Table 7.1 leads to the alarming conclusion that the per capita rate of groundwater abstraction is highest in the hottest, driest countries (Saudi Arabia, Libya, and Iran) where recharge is naturally least scarce. Given that there's only so much water one person can drink, and that there are only so many hours in a day during which they can wash themselves or their clothes, it is intuitively obvious that the huge per capita demands in these countries reflect other water-use priorities (principally irrigated agriculture). In the following section we will consider the different uses to which abstracted groundwater is applied.

7.1.2 Categories of groundwater usage

Water use can be assigned to four principal categories:

A Agricultural (for irrigation, livestock watering, and cleaning purposes).
B Big industrial uses (including passive "use" of unwanted water intercepted by mines).
C Cooling for electricity generation plants.
D Domestic (and small-scale commercial uses, such as office blocks).

In any one country the breakdown between categories A through D will reflect the realities of climate and economic geography: in relatively dry countries with large arable agriculture sectors, "A" will dominate total use. In other situations, "B" may far exceed "A." There are few countries in which category "D" is predominant. Table 7.1 includes examples of groundwater usage for a range of countries in different climate zones. It is apparent that in every continent except Europe, agriculture is by far the greatest user of groundwater. In much of northern Europe, it is

Table 7.1 Statistics on direct use of groundwater for a range of countries worldwide. (Collated from Clarke and King (2004) plus World Resources Institute (www.earthtrends.wri.org). Most data relate to 2000/01. Countries not shown in the table apparently don't collect (or don't publish) the relevant statistics.)

Country	Total annual groundwater withdrawal (km^3)	Groundwater abstraction averaged *per capita* (m^3/year)	Groundwater abstraction as % of total water abstraction*	Groundwater use by category† of application (as % of total groundwater abstracted)		
				A	B‡	D
World§	600–700	106–124	25–30	65	15	20
Asia and Oceania						
Australia	2.2	143.2	11	75	10	15
Bangladesh	10.7	97.6	17	86	1	13
India	190	223.3	35	89	2	9
Japan	13.6	108.2	16	30	41	29
Malaysia	0.4	19	5	5	33	62
Thailand	0.7	15	0.01	14	26	60
Africa and the Middle East						
Algeria	2.9	117.1	58	49	5	46
Chad	0.1	15.7	52	71	0	29
Egypt	5.3	85.1	8	42	0	58
Israel	1.2	204.5	60	80	2	18
Jordan	0.5	100.7	49	66	4	30
Kuwait	0.3	142.7	61	99	0	1
Lebanon	0.4	153.2	39	78	9	13
Libya	3.7	734.9	81	87	4	9
Morocco	2.7	97.9	23	84	0	16
Niger	0.1	17.9	9	40	0	60
Saudi Arabia	14.1	899.3	56	90	0	10
Senegal	0.3	39.2	23	72	4	24
South Africa	1.8	64.9	18	84	6	11
Syria	1.8	133.5	11	83	4	13
Tunisia	1.6	181.8	63	86	4	10
The Americas						
Argentina	4.7	180.4	23	70	19	11
Brazil	8.0	57.0	16	38	25	37
Canada	1.0	37.3	2.5	55	11	34
Mexico	25.1	275.4	35	64	23	13
Peru	2.0	139.4	19	60	15	25
USA	109.8	432.3	25.5	68	8	24
Europe						
Austria	1.4	172.5	66	5	43	52
Denmark	0.9	169.8	71	24	11	65
France	6.0	103.8	15	17	27	56
Germany	7.1	89.4	15.6	4	47	48
Greece	2.0	195.7	27	58	5	37

Table 7.1 (*Continued*)

Country	Total annual groundwater withdrawal (km³)	Groundwater abstraction averaged *per capita* (m³/year)	Groundwater abstraction as % of total water abstraction*	Groundwater use by category† of application (as % of total groundwater abstracted)		
				A	B‡	D
Hungary	1.0	96.5	12.6	18	48	35
Ireland	0.2	62.3	21	29	38	35
Italy	13.9	243.2	31.5	58	4	39
Netherlands	1.0	70.2	14	23	45	32
Norway	0.4	97.5	20	0	73	27
Poland	2.0	51.5	12	0	30	70
Portugal	3.1	311	28	39	22	39
Spain	5.4	137.2	15	80	2	18
Sweden	0.6	72.8	2	0	8	92
Turkey	7.6	124	22	60	9	31
Ukraine	4.0	77.5	10	52	18	30
UK	2.5	42.4	26	2	47	51

* "Total water abstraction" in some cases includes not only surface water, but also any desalinated water, plus water abstracted in another country (from whatever source) and imported.

† Categories as defined in Section 7.1.2: A, Agriculture; B, Big industrial uses; D, Domestic/minor commercial.

‡ In this table, category B incorporates cooling water for power stations (C), for which groundwater is rarely used anywhere in the world.

§ Ranges denote uncertainties due to data scarcity and interannual variability.

possible to pursue arable agriculture using only natural rainfall. In southern Europe, however (such as in the dry areas of Spain, Italy, and Greece), irrigation is necessary to sustain significant arable production. Consequently, in the latter areas groundwater use patterns conform more closely to the global norm, in which agricultural use of groundwater typically exceeds domestic by a factor of two to four.

The *scale* of water demand varies considerably between the four use categories. Table 7.2 summarizes the typical demands associated with particular water uses within the four categories. Variation *within* each category is also considerable, depending on climate (for agricultural activities in particular), political priorities, and expectations. For instance, while most people in Europe would object if their personal access to water was restricted to less than about 40 liters per day, residents of a remote village in central Africa who had endured severe water scarcity would likely perceive 10 liters per person per day as an abundance of water.

In assessing the viability of meeting demands for certain uses, it is important to consider the **disposition** of the water immediately after use. In this context, "disposition" refers to whether abstracted water is used consumptively or else returned to the natural environment after use.

A **consumptive use** of water effectively removes water from the environment. Examples are the export of water in finished products (such as the water present in fruit and vegetables sent to market, the moisture present in quarry products and manufactured goods) and loss to the atmosphere (the typical fate of much of the

Table 7.2 Typical water demands for different use categories. (Adapted from information presented by Cole (1998), plus unpublished data held by author.)

Category	Purpose	Typical water requirement*
A: Agriculture	Livestock drinking water	20 L/day per head (cattle) 4 L/day per head (sheep and goats)
	Rice and cotton (tropical crops)	10 ML/day per hectare of crop
	Wheat and maize (warm temperate cereal crops)	1 ML/day per hectare of crop
B: Big industrial uses	Steel-making	0.1 ML per tonne of steel produced
	Petrochemicals	0.5 ML per tonne of product
	Other chemical industries	>5000 liters per tonne of product
	Breweries	2.5 liters per liter of beer
	Food processing	10,000 liters per tonne of product
C: Cooling water†	(i) Once-through cooling circuits	3.5–5 ML/day per megawatt of electricity generated
	(ii) Closed loop cooling circuits	0.15–0.25 ML/day per megawatt (MW) of electricity generated
D: Domestic and small-scale commercial uses	Drinking water	2.5 liters per person per day
	Preparing and cooking food	10 liters per person per day
	Hygiene (sanitation, bathing, laundry)	25 liters per person per day
	Target minimum daily water provision for people everywhere	20 liters per person per day
	Interim target: minimum safe water needed for drinking/hygiene in developing areas	10 liters per person per day
	Commercial laundries	500 liters per tonne of throughput
	Hotels	200 liters per resident
	Garages	500 liters per employee
	Retail premises	100 liters per employee
	Space heating (groundwater-fed heat pump)‡	1.5–4 L/day per watt (W) of thermal energy required§ (≤6 L/day per W if space cooling also required)

* Note on units: 1 ML (megalitre) = 1×10^6 liters, or 1000 m^3.
† Condensers for cooling steam turbine exhaust streams in thermoelectric turbine power plants.
‡ Generally a nonconsumptive use, as the entire flow is typically returned to the subsurface after heat extraction; in rare cases the cooled water is discharged to a surface watercourse and the use is then arguably consumptive.
§ Derived from figures quoted by Sachs (2002, p. 35).

water which is converted to steam in the cooling towers of thermoelectric plants). In many cases, **return flows** correspond to large percentages of the original abstraction. For instance, many big industrial users and electricity generation plants return more than 80% of the water they use to the natural environment (albeit it may have become warmer and/or chemically altered). Even agricultural activities typically result in return flows in excess of 50%.

The term "return flow" can sometimes be a little misleading when applied to groundwater abstractions. Although it is often technically feasible to re-inject used groundwaters back into the aquifers from whence they were originally abstracted (see Section 7.4.2), it is often much cheaper to discharge such waters to the nearest surface watercourse. By contrast, it is extremely rare for "spent" surface waters to be injected into an aquifer. The relative virtue of a mode of water use which results in a high-percentage return flow must be judged from an holistic environmental perspective, rather than being regarded as inherently virtuous without further scrutiny.

7.2 Constraints on groundwater utility

7.2.1 Groundwater, surface water, or both?

Supposing that all of the demands for various uses have been identified, it is possible to proceed to consider the feasibility of meeting these needs from the available natural water resources. In some cases, a three-way choice will exist between using surface water alone, using groundwater alone, and using a mix of groundwater and surface water sources. A straight choice between groundwater and surface water resources will often be resolved on the basis of water quality, as we shall see in Section 7.2.3. Occasionally, selection of a groundwater source may be counterindicated by the likelihood of undesirable side-effects, such as subsidence or salinization (Section 7.2.4). However, where neither water quality nor unwanted side-effects are an issue, choosing between groundwater and surface water (or a hybrid of the two) will generally be an economic issue. For instance, surface water courses cover a far smaller land area than is underlain by useable aquifers. This means that groundwaters are often more readily available locally than are surface waters. This is one of the two main reasons (the other being water quality – Section 7.2.3) why rural water supplies tend to draw predominantly on groundwaters, even in countries which have highly developed public water supply systems based on the capture of abundant surface waters.

7.2.2 Water quantity constraints

Let us suppose it has been decided that development of a groundwater resource is likely to provide the most effective means of meeting water demands in a given area. At this point, there are two questions to ask:

- How much water can the local aquifer(s) potentially yield?
- How much water can we feasibly deploy?

In underdeveloped areas, the second question may actually be irrelevant: if there is no existing water-supply infrastructure in a given area, then the only constraint on the quantity of water which can be deployed will be the cost of installing the infrastructure needed to deliver water to potential consumers at rates equaling their projected demands. These days, however, relatively few hydrogeologists enjoy the experience of developing an aquifer "from scratch," in the utter absence of pre-existing developments. More usually, incremental improvements in aquifer utilization are being sought, and these must inevitably fit in (as far as is reasonable) with existing water-supply infrastructure. Therefore the question "How much groundwater is feasibly deployable?" is very pertinent in most cases.

The starting point for most assessments of the physical constraints on groundwater utility requires the quantification, for each specific case, of the following two properties:

- **Potential yield.** This is defined as "the yield of a commissioned source or group of sources as constrained only by well and/or aquifer properties for specified conditions and demands" (Beeson et al. 1997). (In this context a "source" refers to an individual pumping well or spring.)
- **Deployable output.** This is defined as "the output of a commissioned source or group of sources or of bulk supply as constrained by: licence (if applicable); water quality; environmental issues; water treatment system capacity; the spare capacity of raw water mains and/or aqueducts; and limitations of pumping plant."

Of these two, potential yield represents the truly physical limitations on groundwater availability, whereas deployable output reflects wider technical and socioeconomic issues. Further discussion of deployable output is therefore reserved to Chapter 11, where management strategies are examined.

Potential yield is almost wholly determined by hydrogeological factors. Of prime importance are the transmissivity and storativity of the aquifer, which fundamentally determine the ability of the aquifer to yield a certain volume of water to one or more specific pumping wells (or springs) over (a) given period(s) of time. Rarely do groundwater resource managers experience fortune as blessed as that of St Cuthbert (whose tale of hydrogeological felicity opens this chapter). It is rare indeed that demand for water neatly coincides with the potential yield. Usually, it is very difficult to accurately define potential yield, for the yield which is potentially available in any one period of time may well differ from that which will be available in another period of similar duration.

Although groundwater flows slowly, the quantity of water stored in any aquifer waxes and wanes over time in response to seasonal and multiannual changes in rates of recharge and natural discharge.

In a classic paper published in 1940, Charles V Theis explained that there are ultimately only three possible sources of water available to wells:

- Removal of water from long-term storage in the aquifer (which is a "once-off" source of water).
- A decrease in natural discharge from the aquifer.
- An increase in recharge.

Figure 7.1 illustrates the key concepts. The development of a cone of depression around a pumping well was explained in Section 3.3.3. As the cone of depression spreads outwards from the well, it will inexorably remove groundwater from storage until such time as the rate of pumping is matched by an increase in the overall availability of water within the aquifer. The most common way for water to become available (as shown in Figure 7.1) is for the drawdown in the aquifer to result in a flattening of the water table in the vicinity of natural discharge zones (see Chapter 5), which leads to a decrease in the rate of natural outflow from the aquifer. The water "intercepted" in this manner thus becomes available to the pumping well. Another way in which water can become available is by a lowering of the water table in the vicinity of a river or wetland, such that surface water arriving at that point is induced to infiltrate into the subsurface as a new source of indirect recharge (see Chapter 2). The limiting factors on the ultimate availability of water to pumping wells is therefore *not* the total (predevelopment) recharge rate, as the proponents of the old "safe yield" concept would argue, rather, the true potential yield of pumping well (or wellfield) equals the sum of (i) the amount of storage which can be depleted over a given period of time without giving rise to undesirable consequences (Section 7.2.4) *plus* (ii) the magnitude of any decrease in natural discharge *and/or* (iii) the increase in natural recharge which the pumping of the well will induce (Theis 1940).

Despite the very clear exposition of these principles by Charles Theis back in 1940, many water resources managers have since adhered to an alternative and utterly erroneous conception of the limits on groundwater availability. This misconception equates the maximum amount of water available for abstraction from an aquifer with the long-term predevelopment recharge rate. If one could quantify the rate of recharge to the aquifer, then the mean annual recharge rate could simply be relabeled as the "safe yield" of the aquifer. Subsequently, this simplistic concept of "safe yield" was modified to take into account interannual variations in recharge and subsurface outflow from an aquifer to determine the "maximum perennial yield" (Todd 1980) or "reliable yield" (Khan and Mawdsley 1988). As late as 1982, the misconceived equation of "safe yield" with total natural recharge rate remained so prevalent that a team of leading US hydrogeologists felt it necessary to go on the offensive against it (Bredehoeft et al. 1982). Using both conceptual and mathematical arguments, these

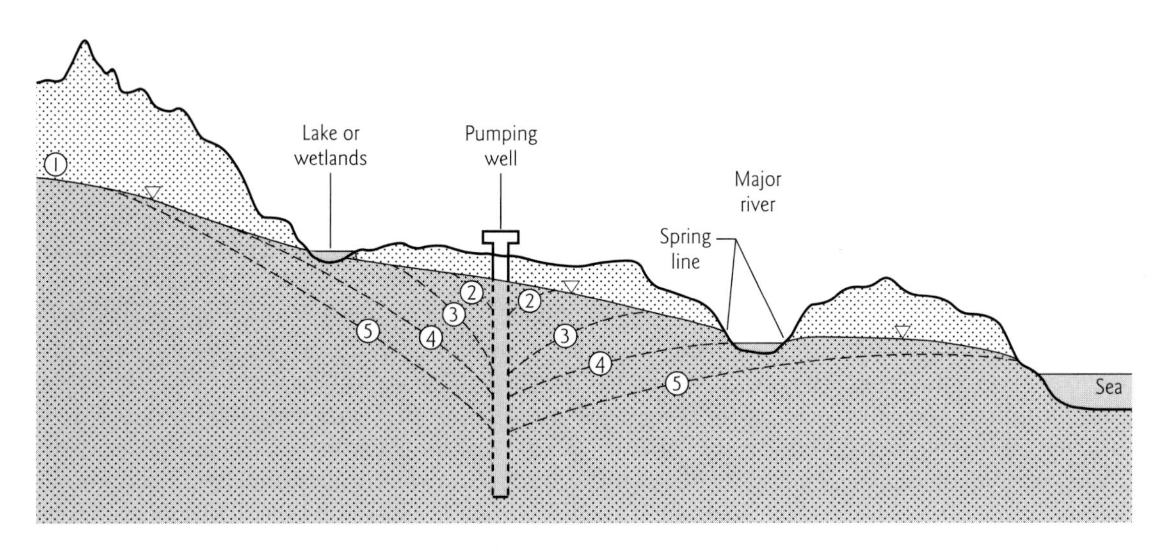

Fig. 7.1 The source of water derived from wells. The original pre-pumping water table (1) is disposed such that it feeds discharging groundwater to the lake or wetlands near the hills, to the spring lines and the major river, and finally to the sea. During the early stages of pumping (2, 3), all of the water arriving at the well is being removed from storage (the volume now occupied by the expanding cone of depression). By the time the cone of depression has reached position (4), the water table has fallen below the lake or wetlands (which will now leak further water into the aquifer, augmenting recharge), cut off discharge from the left-hand spring line above the river (a decrease in natural discharge), and reversed the water table gradient at the river itself (both a decrease in natural discharge and an induced increment in recharge). By the time the cone of depression has reached position (5), all inland freshwater bodies are hydraulically disconnected from the water table (cf. Figure 5.4c) and will be acting as recharge features only; furthermore, discharge to the sea is now being reduced, due to the lessening of the seaward hydraulic gradient, and if this continues, sea water intrusion to the aquifer can be anticipated (cf. Figures 3.6 and 7.2). (Inspired by Theis 1940.)

hydrogeologists comprehensively debunked the "safe yield" myth. Using several examples from around the USA, Bredehoeft et al. (1982) clearly demonstrated that the available resources in a wide range of aquifers bear no relation to the predevelopment recharge rate. Further examples were subsequently presented in even greater detail, both from the USA (Johnston 1989; 1997) and from the UK (Younger 1998). Nevertheless, the erroneous "safe yield" concept continues to be submitted in argument both by scientists (e.g. Das Gupta and Onta 1997) and by well-meaning advocates of sustainable water management (e.g. Clarke and King 2004). In reality, determination of potential yield is only one of many constraints that need to be considered when attempting to establish rules for aquifer management. We shall now consider some of the other key constraints, paving the way for a discussion of management options in Chapter 11.

7.2.3 Water quality constraints

Having established that there seems to be a sufficient quantity of groundwater available to meet a given demand, the next question is whether the water is of suitable quality. Although technologies now exist which can transform even the world's most polluted waters into potable waters, all but the simplest of water treatment technologies are very expensive. Traditionally, one of the key reasons for discarding a surface water source in favor of a groundwater source is that the latter is far less likely to be contaminated. The groundwaters pumped by most public supply wells are largely free from pathogens and need only be subjected to minimal treatment[ii] before being despatched for human consumption. On the other hand, because they are not typically equipped to provide treatment for other potential contaminants, many groundwater sources are very vulnerable if the quality of the raw water suddenly changes: often, well abandonment will be the only economic option.

The intended use of a given groundwater critically determines the limits within which specific quality parameters must lie. It is therefore worthwhile returning to the predominant groundwater use categories (A through D) (cf. Section 7.1.2).

Category A: Agriculture

For both livestock and arable farming, the principal issues are the salinity of the water (most conveniently measured by conductivity) and the toxicity of particular dissolved substances. Guidelines for livestock drinking water were developed in the early 1970s by the US National Academy of Sciences (see Soltanpour and Raley 1999). Waters with conductivities less than 1500 μS/cm are considered excellent livestock drinking waters. However, brackish waters are still useable in many cases: few large mammals will experience anything worse than temporary diarrhea if they drink waters with conductivities ranging up to 11,000 μS/cm. Poultry are much more sensitive, and are unlikely to flourish if obliged to drink water with a conductivity in excess of 5000 μS/cm. In terms of individual toxic substances, guidelines for livestock drinking waters generally resemble those for waters destined for human consumption, in terms of both the elements of concern and the maximum recommended concentrations (albeit these are generally more lax than those recommended for humans by a factor of between 2 and 5).

For arable agriculture, the most widely used water quality guidelines remain those originated by the United Nations Food and Agriculture Organization in 1978 (later revised; Ayers and Westcot 1985). Table 7.3 summarizes these guidelines. Salinity is an issue in arable agriculture for two reasons: it impedes the uptake of water by plants, and it can affect the soil structure such that infiltration becomes impeded. Sensitivity to salinity varies from crop to crop. In general terms, fruits, most root vegetables, sunflowers, and corn are most sensitive to salinity, ideally requiring water with a conductivity of less than 1200 μS/cm. Most common "greens," plus turnips and tomatoes, will tolerate conductivities up to about 2200 μS/cm. The majority of cereal crops can cope with conductivities up to 3500 μS/cm, but even the hardiest crops (oats, rye, barley, soya, and sugarbeet) will not flourish where conductivities significantly exceed 5000 μS/cm. While it is possible to use soil flushing techniques to successfully raise a range of crops using waters which would generally be regarded as too saline (Table 7.3), the design of soil water management strategies for such purposes is far from easy, and expert guidance must be sought. In terms of specific toxicity effects, boron is particularly problematical, as it is toxic to many plants at relatively low absolute concentrations (~3 mg/L), which are sometimes exceeded even in otherwise fresh groundwaters. Besides their role as components of salinity, sodium and chloride are directly toxic to many plants when present at high concentrations (Table 7.3). Most other toxicity problems are highly crop-specific; for detailed guidance the reader is referred to Ayers and Westcot (1985).

Category B: Big industrial uses of groundwater

Every industry has its own specific requirements. If the water is only required for swilling down work yards, then any old groundwater will do. However, in many cases the water quality needs are so exacting that it is unreasonable to expect any natural groundwater to meet the users' requirements; consequently, much pre-conditioning of feed water is undertaken by industry. In the most stringent sectors (e.g. some activities in the microelectronics industry), "preconditioning"

Table 7.3 Guidelines for quality of waters suitable for the irrigation of arable crops. (Adapted after Ayers and Westcot 1985.)

Potential irrigation problem	Degree of restriction on use		
	None	Slight to moderate	Severe
Salinity (affects water uptake by crops):			
Assess in terms of conductivity (μS/cm)	700	700–3000	>3000
Soil infiltration (affects infiltration rate of water into the soil; evaluate using conductivity and SAR* together)			
Conductivity (μS/cm)			
for SAR (meq/L) = 0–3	<2900	2900–5000	>5000
for SAR (meq/L) = 3–6	<2900	1300–2900	>2900
for SAR (meq/L) = 6–12	<1900	1900–5000	>5000
for SAR (meq/L) = 12–20	<1200	1200–3000	>3000
for SAR (meq/L) = 20–40	<700	700–2000	>2000
Specific ion toxicity (affects sensitive crops)			
Sodium (Na)			
Surface irrigation (meq Na/L)	<3	3–9	>9
Sprinkler irrigation (meq Na/L)	<3		>3
Chloride (Cl)			
Surface irrigation (meq Cl/L)	<4	4–10	>10
Sprinkler irrigation (meq Cl/L)	<3		>3
Boron (B) (mg B/L)	<0.7	0.7–3.0	>3.0
Miscellaneous effects (affects susceptible crops)			
Nitrogen (NO_3–N) (mg N/L)	<5	5–30	>30
Bicarbonate (HCO_3) (overhead sprinkling only) (meq/L)	<1.5	1.5–8.5	>8.5
pH		Normal range 6.5–8.4	

* Sodium Adsorption Ratio (SAR) = $[Na]/\sqrt{(([Ca] + [Mg])/2)}$, where the square brackets denote the concentration of the indicated metal in meq/L.

may actually require that all raw water be deionized (e.g. using reverse osmosis), and then selectively re-mineralized by adding trace quantities of desirable metals. (Given such a wide variation in requirements, sector-specific guidance is beyond the scope of this book.)

Category C: Cooling for power plants

This category is very seldom an important component of groundwater use, both because surface waters and marine waters can usually be deployed far more conveniently to satisfy the huge pumping rates typically required by thermoelectric plants, and also because many groundwaters are close to saturation with calcium carbonate, which can precipitate to form scale within condensers.

Category D: Domestic and small-scale commercial uses

For all such applications, the principal concern is to ensure suitability for human consumption: whether one is at home or at work, it is necessary to have access to safe drinking water.

Specifying drinking water guidelines for general observance is complicated by the fact that many countries have established their own drinking water guidelines. For instance, in the USA the federal Environmental Protection Agency (EPA) sets the standards which are then implemented by the individual states. Similarly, central standards have been established by the European Union, and these have then been translated into national regulations by individual EU Member States. Although minor differences can be detected between quality standards in different jurisdictions, most suites of guidelines have converged on very similar maximum admissible concentrations (MACs) for safe human consumption. In the vast majority of cases, these MACs closely match the "guideline values" published by the World Health Organization (WHO). The current WHO guidelines (WHO 2004) represent the closest approximation to truly worldwide standards for drinking water quality. The WHO guidelines cover a range of water-borne **pathogens** (i.e. disease-causing microbes) and some 128 chemicals which may be found in drinking waters. Table 7.4 summarizes some of the principal recommendations.

Fortunately, most groundwaters are relatively secure from gross contamination by pathogens. However, in shallow gravel aquifers and in karst terrains, it is always prudent to assume that pathogen contamination is a possibility. Also, where pumping wells or springs are inadequately engineered (Section 7.3) localized introduction of pathogens into the abstracted water is a frequent cause of apparent microbial infestation of groundwater. A vast range of microbes (viruses, bacteria, protozoa, and helminths) have the potential to cause gastroenteritis, diarrhea, and dysentery. In the surveillance of groundwater sources, microbial water quality is usually assessed simply by testing for the presence or absence of easily detected organisms which unequivocally indicate the presence of fecal contamination. In the vast majority of cases, the organism which is tested for is the bacterium *Escherichia coli* (usually written simply as *E. coli*), the presence of which provides conclusive evidence of recent fecal contamination. WHO guidelines simply stipulate that *E. coli* should not be detectable at all in water intended for human consumption. Of course it may well be tolerable that *E. coli* is present in a raw water source, provided that a treatment system is in place which will ensure that no pathogens persist in the final drinking water supplied to humans. It should be noted, however, that *E. coli* does have its limitations as an indicator species. This is because many viruses and protozoa are more resistant to disinfection than is *E. coli*, so that absence of the latter does not necessarily indicate the absence of all potential pathogens. Where there are high levels of potentially water-borne viral and parasitic diseases in the local population, it would be wise to test for the presence of more resistant microorganisms, such as bacteriophages and/or bacterial spores (WHO 2004).

In most groundwaters, chemical pollutants are commonly more problematical than pathogens (Fetter 1999). While it is certainly the case that the natural quality of some groundwaters renders them unsuitable for drinking water use, in many cases human activities necessitated the development of the guideline values shown in Table 7.4. In the case of manmade organic compounds (e.g. many chlorinated pesticides, chlorinated solvents, and refined hydrocarbons), the link to human activity is clear. However, even natural inorganic substances (such as nitrate and many toxic metals) are also released into solution due to human activities, ranging from agriculture through mining to various industrial processes. (Further discussion of the sources of common groundwater pollutants can be found in Chapter 9.)

Although domestic/small commercial uses only account for around 20% of groundwater abstractions globally (Table 7.1), when one looks at the overall amount of water used for drinking purposes in any one country, groundwater is often the predominant source. This is certainly the case in most of Europe, for instance (Hiscock et al. 2002). For this reason, there is a tendency to regard drinking water quality guidelines as general quality targets for groundwater *in*

Table 7.4 Drinking water quality guidelines: a simplified summary. (This summary is based on the guidelines of WHO (2004) and groundwater-specific information (Fetter 1999); reference to these sources is essential before embarking upon detailed planning for water quality management.)

Constituent	Guideline limit value	Comments on nature/sources
Pathogens		
Escherichia coli	Zero organisms per liter	Usual indicator for fecal contamination
Other pathogenic bacteria/ protozoa viruses/helminths	Zero organisms per liter	Fecal contaminants; only investigated in specific cases
Manmade organic compounds*		
Alachlor	20 μg/L	Herbicide
Aldicarb	10 μg/L	Insecticide
Aldrin and dieldrin	0.03 μg/L	Insecticides. (Sum of both)
Atrazine	2 μg/L	Herbicide
Benzene	10 μg/L	Hydrocarbon-based products
Bromodichloromethane	60 μg/L	Solvents, fire-extinguishers
Chlordane	0.2 μg/L	Pesticide
Chloroform	200 μg/L	Fumigants, propellants, biocides
Dichlorobenzene	300 μg/L	Solvents, insecticides
Dichloroalkanes	20 μg/L	Solvents, fumigants
Dichlorprop	100 μg/L	Pesticide
Endrin	0.6 μg/L	Pesticide
Ethylbenzene	300 μg/L	Solvents hydrocarbons coal tars
Lindane	2 μg/L	Pesticide
Mecoprop	10 μg/L	Pesticide
Pentachlorophenol	9 μg/L	Biocide
Simazine	2 μg/L	Herbicide
Styrene	20 μg/L	Plastics, resins, protective coatings
Tetra-(or Per-)chloroethene (PCE)	40 μg/L	Degreaser, dry-cleaning agent
Toluene	700 μg/L	Adhesives, solvents, many uses
Trichloroethene (TCE)	70 μg/L	Degreaser, dry-cleaning agent
Vinyl chloride	0.3 μg/L	Polymers, adhesives, organic synthesis
Xylenes	500 μg/L	Aviation fuel, coal tars, gasoline, solvents
Inorganic substances		
Antimony	20 μg/L	Common in alloys
Arsenic	10 μg/L	Natural, some mine waters
Barium	700 μg/L	Alloys and lubricants
Boron	500 μg/L	Natural, alloys, semiconductors
Cadmium	3 μg/L	Batteries, zinc ores
Chromium	50 μg/L	Alloys, paints, nuclear facilities
Copper	2 **mg/L**	Paints, wiring, ores
Cyanide	70 μg/L	Polymer production, gold mines
Fluoride	1.5 **mg/L**	Natural, aluminum smelting
Lead	10 μg/L	Paints, batteries, roofing, mine wastes
Mercury	1 μg/L	Electrical apparatus, pharmaceuticals
Manganese	400 μg/L	Natural, purifying agent
Molybdenum	70 μg/L	Pigments, lubricants, alloys
Nickel	20 μg/L	Ceramics, batteries, alloys, plating
Nitrate (as NO_3^-)	**50 mg/L**	Fertilizers, food preservatives
Nitrite (as NO_2^-)	**3 mg/L**	Fertilizers, food preservatives
Selenium	10 μg/L	Electronics, ceramics, catalysts
Uranium	15 μg/L	Nuclear industry, U mine wastes

* Only a representative selection listed here; for more complete listings see WHO (2004) and Fetter (1999).

situ. Where legal obligations are in force regarding the quality of groundwater within aquifers, the blanket application of drinking water quality guidelines is likely to be unrealistically rigorous in many cases, and a more nuanced approach which recognizes the likely uses of a given groundwater (e.g. irrigation, livestock watering) is likely to lead to less dissipation of limited managerial resources. This is a point to which we shall return in Chapter 11.

Finally, it is worth noting that there is little logic in supplying water suitable for drinking merely to serve purposes such as toilet flushing and pavement cleaning. It is therefore always worth asking whether the bulk of the water supply to a given building really need be drinking-quality water, or whether water of poorer quality can be used for most purposes, with high-quality water being reserved solely for drinking purposes.

7.2.4 Undesirable side-effects of groundwater usage

While there is no doubting the enormous benefits to humankind of groundwater use, it must also be acknowledged that, on some occasions, the abstraction of groundwater has negative side-effects. The good news is that these side-effects are by no means universal: many groundwater resources have been heavily developed for many decades (or in some cases centuries) without any negative side-effects becoming apparent.

Figure 7.1 clearly illustrates one category of side-effects associated with the development of wells and wellfields: diminution of surface water flows from springs and in rivers. Lowering of the water table can also lead to drops in water levels in wetlands and lakes. There is potential for all such changes to have detrimental effects on freshwater ecosystems (see Chapter 6), and indeed clear examples of ecosystem damage due to groundwater withdrawal can be cited (e.g. Box 6.1). However, in many cases the "springs" which cease to flow as a consequence of pumping a well in the adjoining aquifer amount to nothing more than a number of minor seepages, none of which support identifiable ecosystems. It is arguably more beneficial to "gather these waters together" by means of pumping a well, rather than have them flow away to the benefit of neither people nor ecosystems. Even where diminutions in surface flows are not so negligible, it is possible to site return flows of used water so that any such impacts are minimized. This is often done in the case of quarry dewatering, for instance, where there is little choice over the location of maximum water table lowering, but the clean pumped water can be disposed at a position which ensures maximum benefit to surface ecosystems.

Related to the diminution of surface flows is the problem of **saline intrusion**. We have already seen (Figure 7.1 and Section 7.2.2) that the lowering of the water table due to pumping of a well can induce inflows to the aquifer from adjoining surface water bodies. When the water body in question is the sea, the induced inflow will be saline. Consequently, continued pumping of wells located too close to the coast can result in saline waters intruding deep into previously fresh aquifers. The coastal plains of the world are littered with the remains of wells which have had to be abandoned because their operators failed to appreciate this phenomenon of saline intrusion. Before condemning their stupidity too vigorously, it is worth examining Figure 7.2. As the cross-sections there show, because sea water is more dense than fresh groundwater, there is a natural tendency for a "wedge" of saline water to intrude beneath the land surface, such that a relatively thin wedge of fresh water rests directly above the brine. If a well begins to pump in this fresh water wedge, the density of the sea water often proves particularly treacherous, as it causes the interface between saline and fresh waters tending to rise about forty times as fast as the water table falls. Viewed in this light, it is difficult not to feel some sympathy for the disenchanted coastal well owners of old.

Excessive irrigation with groundwaters rich in sodium and chloride can lead to **salinization**, in other words the accumulation of salt in the soil. This destroys soil fertility, rendering the land

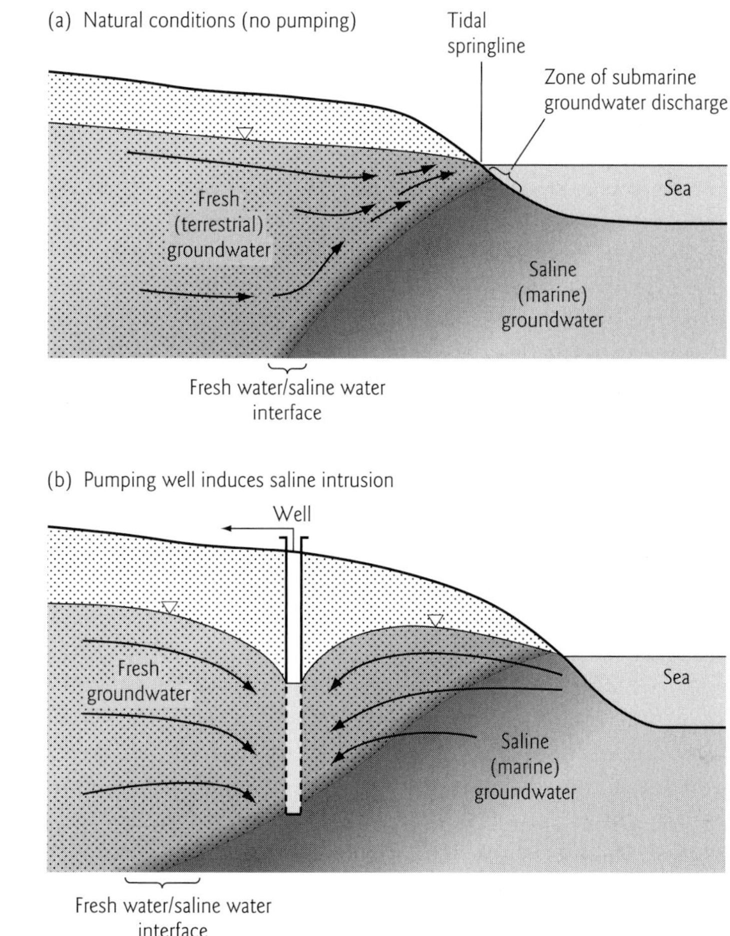

Fig. 7.2 Fresh/saline groundwater interactions in a typical coastal zone: sketch cross-section (cf. Figure 3.6). (a) Under natural conditions, the fresh/saline water interface slopes gently inland (the higher density of the saline water making it tend to sink beneath the fresh groundwater). (b) With excessive pumping, the fresh/saline water interface rises and intrudes further inland into the formerly fresh water aquifer. This is the process known as "saline intrusion." Eventually, the proportion of saline water entering the well may be sufficiently high that the well becomes unusable for water supply purposes. (Only 10% sea water will typically make a well unusable, due to elevated Na, Cl, and SO_4 concentrations.)

unproductive. The risks are particularly high in hot countries where evapotranspiration rates are high. Once salinized, it is virtually impossible to rehabilitate the affected soils at any reasonable cost and over any reasonable time-scale. Avoidance of salinization can be achieved by observance of the limitations on sodium adsorption ratio and conductivity laid out in Table 7.3. While it is possible to irrigate using moderately saline groundwaters, specialist guidance (e.g. Rhoades et al. 1992) must be followed to ensure attainment of the necessary levels of design and management.

One of the more dramatic potential side-effects of groundwater abstraction is land **subsidence**. Thankfully it is a rather rare phenomenon, effectively restricted to aquifers comprising either:

- Unconsolidated sands interbedded with soft mud layers.
- Certain types of karst terrain.

Chapter 8 discusses land subsidence due to groundwater abstraction in considerable detail, and hence it is discussed no further here.

7.3 Methods of groundwater abstraction

7.3.1 Spring sources

Springs have been used as water sources since time immemorial. In ancient times, the life-giving nature of spring waters was so revered that springs themselves came to be attributed with supernatural significance (Bord and Bord 1986). The cult of spring waters is the ultimate origin of the "wishing well," artificial versions of which may nowadays be found serving as collection boxes for charities in many shopping malls. Given how long springs have been cherished by humanity, it might seem that there is little that can be said about them. However, bitter experience has shown that few people seem to instinctively use springs in an altogether wholesome manner. Unimproved spring sources are often contaminated with pathogens, which are introduced to the water from the hands and feet of people and animals. Only if a spring orifice is protected from such inadvertent contamination can it truly be used as a safe source of drinking water.

Spring protection is achieved as shown in Figure 7.3. The first step is to install a floor slab immediately in front of the spring orifice. This is usually made of concrete, cast in situ. A permeable wall of large stones is then installed on the floor slab, effectively burying the natural spring orifice in a stack of large stones. This pile of stones is then sealed-in on all sides by impermeable walls, which are either made of bricks and mortar or else of concrete (cast in situ using form-work). A roof is added, typically incorporating a removable access plate, thus forming a completely enclosed chamber. Water can only exit this chamber by flowing through a pipe which is fitted through the front wall of the chamber. The entire chamber is disinfected using strong bleach before it is brought into service. Water flowing from the spring chamber outlet pipe can either be collected in containers brought by individual users (a common arrangement in developing countries), or else the pipe can be connected to a water distribution network, which will carry the water to individual taps elsewhere. It is often worth installing a few chambers to capture the flows of several small springs, which together give a more useful yield than any one spring could provide on its own. In some cases, the same effect has been achieved by driving tunnels into hillsides to intersect the fracture systems that feed a number of discrete springs.

In terms of resource assessment, springs offer us only "Hobson's Choice" (take it or leave it). A spring will only flow within a given discharge range, and there is really nothing we can do to increase the yield of a spring. To assess the resource available from a given spring, therefore, we need to measure its flow rate regularly over a period of at least one year (and preferably many years), and then construct a flow–duration curve[iii] (see Box 5.2). Once a flow–duration curve is available, we can use it to estimate percentile flows (as shown in Figure 5.9), with Q_{95} being a particularly popular choice for resource assessment purposes. Given Q_{95}, we can judge

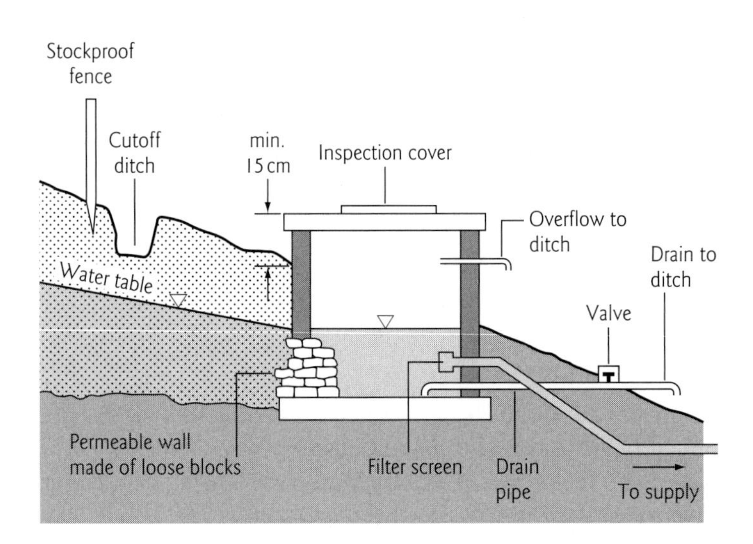

Fig. 7.3 Spring protection: a secure structure to maintain a spring discharge in a suitable condition for potable supply, by preventing access for humans and animals to the groundwater source. (Adapted after Brassington 1998.)

what level of demand can be reliably met by the spring (bearing in mind the possible need to leave residual surface outflows to sustain streams or ponds of amenity or ecological value).

7.3.2 Boreholes, wells, and wellfields

To the novice hydrogeologist, the plethora of names which are applied to vertical holes accessing the water table can be bewildering. Albeit English has now become the global lingua franca of science and engineering, there is still little consistency in such terminology from one place to another. Table 7.5 presents a concise summary of the terms most likely to be encountered in practice.

The drilling, completion, and development of water wells has already been described in Section 3.3.1. Figure 3.8 illustrates the typical arrangements within a modern pumping well. We now need to consider the capability of adequately equipped water wells to meet demands for water from the agricultural, industrial, and domestic sectors. The assessment of the potential yield obtainable from the aquifer entails the quantification of aquifer transmissivity (T) and storativity (S) (see Section 3.4), and the assessment of the possible hydrological consequences of pumping from one or more wells (as discussed in Sections 7.2.1 and 7.2.4). For any one well, we can use T and S to directly calculate the anticipated drawdown for a given pumping rate (sustained over a specified period of time). In reality these predictions always yield smaller estimates of drawdown than are experienced in reality. In other words, our wells are never 100% efficient: i.e. the well yield for a given drawdown never equals 100% of the rate we would predict using the T and S values for the aquifer at that point. In fact, a properly designed and commissioned well can only be expected to have an efficiency of the order of 60–70%. The difference between predicted and observed water levels is attributable to two factors: inadequate well engineering and turbulent upflow head losses. Examples of poor engineering would include screening the well in an inappropriate interval, fitting the well with an inefficient filter pack, or inadequate development, which fails to counteract clogging that may have been introduced during drilling. Turbulent upflow losses arise from the incompatibility of turbulent flows with all common drawdown prediction techniques for aquifers which assume all flow to be laminar, whereas within the well casing (at least) the flow tends to be turbulent. The consequences are a greater drawdown per unit pumping rate (cf. Section 3.2.4), when compared with predictions that assume only laminar flow. The difference between the real water level observed during pumping and that which one would anticipate taking only aquifer T and S into account is termed the **well loss**.

If one could find a way of measuring the drawdown in the aquifer immediately outside of the well casing, then one could subtract this from the drawdown measured within the well to directly determine well loss. In reality, it is not usual to install a monitoring well immediately adjoining the casing of a pumping well. In practice, therefore, well losses are usually estimated from test-pumping data. Most estimations of well loss are based upon analysis of a kind of pumping test known as a **step-drawdown test** (or "**step-test**"). A step-test is performed by pumping the well in a series of "steps" of incrementally greater pumping rate. During each step the pumping rate will be held constant while the drawdown is monitored until it seems to stabilize. The pumping rate is then abruptly increased, thus initiating the next step. Typically, step-tests include four or more steps, each of which will typically last for at least 1 hour (see Clark 1977; Brassington 1998). A thorough discussion of step-test interpretation is beyond the scope of this text; interested readers are referred to a recent radical reformulation of the requisite methodology (Karami and Younger 2002).

Besides determining well losses, step-tests are also useful in basic decision-making over the maximum yield which any individual well (as an engineered structure) can sustain. To apply this method, all that is required is that the pumping rate and drawdown in each of the steps are plotted against one another on ordinary graph paper

Table 7.5 Types of borehole and wells – a summary of terminology.*

Term	Definition	Sometimes also known as . . .
Borehole	A hole, usually vertical or near-vertical, created by drilling	Drillhole, well
Well	A general term encompassing all the specific types of boreholes and wells	(*All other terms in this table . . .*)
Water well	A borehole drilled for the principal purpose of obtaining a supply of water	Abstraction hole/borehole/well; pumping borehole/well; production hole/borehole/well; tubewell[†]
Hand-dug well	A large-diameter, shallow water well constructed by manual digging	Open well; Dug well; Well
Test well	A borehole drilled to test an aquifer by means of pumping tests	Test hole/borehole; pilot hole/borehole/well
Exploration borehole	A borehole drilled to obtain information on the geology/groundwater conditions in a specific place. (In some cases these are retained and equipped as monitoring wells)	Exploration well; investigation hole/borehole/well
Monitoring well	A borehole constructed to allow collection of long-term data on variations in groundwater levels or quality. (See also *piezometer*)	Observation well/borehole
Piezometer	A small-diameter *monitoring well* specially constructed to measure hydraulic head at a specific depth within a groundwater system. (The screened section in a piezometer is very short compared to that in an ordinary monitoring well)	Piezo. (Misnomers ("monitoring well," "observation borehole") are common, due to mistaken identification/ignorance on the part of the person describing the feature)
Well-point	A small-diameter, shallow (typically < 10 m) well, which is suction-pumped in consort with many similar well-points nearby to achieve lowering of the water table in superficial deposits. (Usual applications on construction sites)	Jetted well/jet well (only if the well-point was installed by jetting)
Driven well	A small-diameter, shallow well installed by hammering a complete steel tube (fitted with a piezometer-type tip) into the ground	Well-point (if used for that purpose); piezometer (if used for that purpose)
Jetted well	A small-diameter, shallow well installed by displacing the soil with a high-pressure jet of water	Well-point (if used for that purpose); piezometer (if used for that purpose)
Injection well	A well which is used for purposes of artificial recharge, i.e. to inject water into an aquifer	Injection borehole; artificial recharge borehole/well; recharge well

* The definitions used here generally follow those of Clark (1988), with the addition of a few extra terms.
† This term is peculiar to Bangladesh, Pakistan and adjacent parts of other countries.

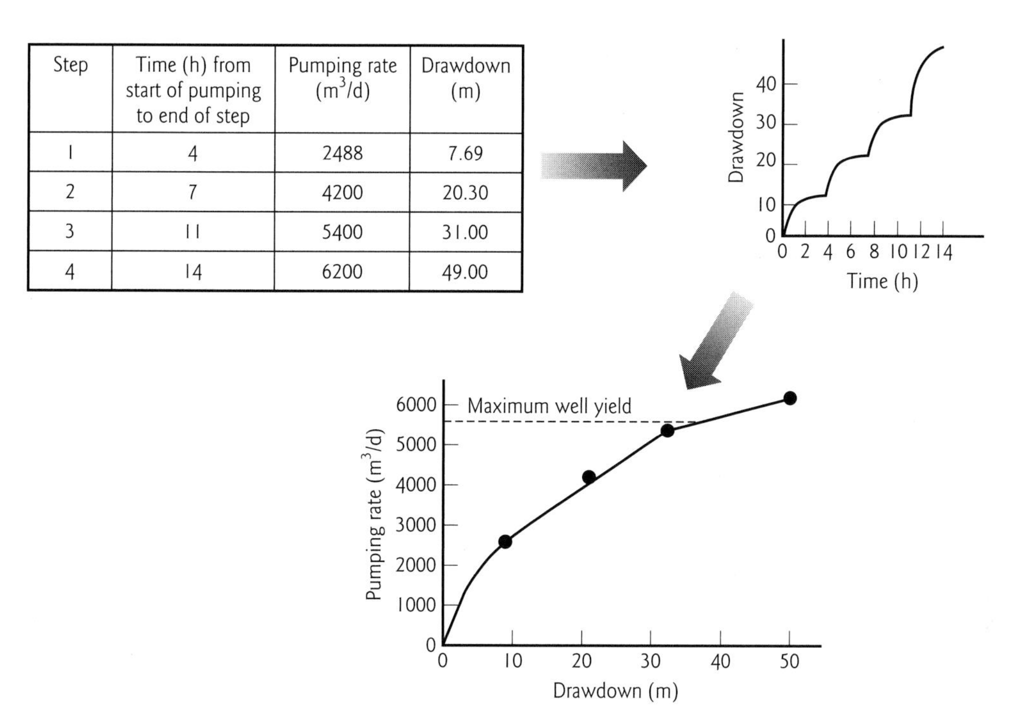

Step	Time (h) from start of pumping to end of step	Pumping rate (m^3/d)	Drawdown (m)
1	4	2488	7.69
2	7	4200	20.30
3	11	5400	31.00
4	14	6200	49.00

Fig. 7.4 Step-test: data collection and use. The raw data shown in the table produce the typically stepped profile when plotted on the raw data graph (top left) of time versus drawdown. A plot of Q versus drawdown (bottom plot) will yield a linear relationship as long as the well is operating within its mechanical limitations. The pumping rate at which this plot departs from linearity defines the maximum yield for this particular well.

(Figure 7.4). In theory, the ratio of pumping rate to drawdown should be constant as long as flow remains predominantly laminar (cf. Section 3.2.4). Sure enough, on Figure 7.4 the first few points on the graph fall roughly on a straight line. However, at higher abstraction rates, as the proportion of turbulent flow increases, even small increments in pumping rate result in large increases in drawdown. The last point on Figure 7.4 shows this, with the plot deviating from the earlier straight line. The maximum yield of the well is determined by the point on the plot where deviation from straight-line behavior sets in. For the case shown in Figure 7.4, this suggests a maximum well yield of the order of 5500 m^3/day.

Many texts concerning wells (e.g. Driscoll 1986; Clark 1988) present their material as if individual, isolated water wells were the norm. For many applications (ranging up to total yields of around 10,000 m^3/day) single-well installations may indeed suffice. However, where the aquifer is to be called upon for substantially greater yields (e.g. to feed a large industrial plant, or to supply water to a large town), it is more common to use a group of wells located in relatively close proximity to each other. Such groups of wells are termed **wellfields.**[iv] The key concept behind the design of wellfields is that if a number of wells each pump at modest rates, the resultant drawdown will be far more evenly spread through a given area of aquifer than would be the case were we to obtain the same total flow rate by overpumping a single well. As noted above, the more heavily we pump any one well, the greater will be its well loss. Thus even if the aquifer is more than capable of supplying the demand which we wish to place on it, there will come a point at which two or more wells will be able to extract the water much more efficiently (i.e.

with lower well losses) than a single well. In practice, the construction costs of additional wells always exert a restraining influence on the hydrogeologist's zeal for optimum wellfield efficiency.

7.4 Conjunctive use of groundwaters with surface waters

7.4.1 Basic principles of conjunctive use

If the impression has been given that the development of groundwater resources is a strict alternative to the use of surface waters, banish that thought. Not only is it possible to develop groundwaters in conjunction with surface waters; there are often very sound economic and operational reasons for doing so. Possible motivations include:

- Taking advantage of the fact that groundwaters often remain available at times of year when surface waters are scarce. (Due to the slow nature of seasonal recharge processes, many groundwater systems do not reach maximum water levels until the spring, when surface runoff may already be in decline.)
- Taking advantage of the vast storage capacity of many aquifers, thus lessening the need to develop costly and controversial surface reservoirs.
- Blending mineralized groundwaters with less-mineralized surface waters to obtain a final product that complies with drinking-water standards or (if that was never in issue) is simply more palatable for consumers.

It is actually the issue of palatability that tends to mean that simply switching supply from a groundwater source to a surface water source for a given season of the year (which would be the most obvious form of conjunctive use) is rarely used in practice. Although many people would regard water as tasteless, in fact we are quite sensitive to sudden changes in the mineral content of our drinking water. For instance, in the city next to mine, around two thirds of the population receives groundwater year-round, while the other third generally receives surface water. However, at times of high demand, surface water may have to be diverted to supply other towns which (because of the layout of distribution mains) simply cannot receive groundwater. When this happens, a considerable number of properties which normally receive surface water are suddenly switched onto the groundwater source. This always results in a large number of complaints from the households affected by the change, who claim to find the taste of the "new" source of water unpleasant. In areas where people receive the same groundwater without interruption, on the other hand, the residents consider their tapwater to be the finest in the city, and they turn their noses up at the (surface) water they are offered in other neighborhoods. The moral of this story is: conjunctive use of groundwaters and surface waters needs careful planning and management. Three of the more successful strategies for conjunctive use are outlined below.

7.4.2 Riverside wells, induced recharge, and bank filtration

It is axiomatic that flow through porous media tends to result in purification of waters. This is precisely why many water treatment plants incorporate sand filters as essential "unit processes" in the conditioning of surface waters for potable use (see Binnie et al. 2002). Recognition that subsurface flow physically filters out many pathogens, and also facilitates many biogeochemical processes which serve to strip pollutants out of infiltrating waters, is behind the logic of choosing a groundwater source in preference to a surface water where both are potentially available (see Sections 7.2.3 and 9.3.2). In some cases, large rivers might provide more reliable *quantities* of water than could be supplied by local aquifers, but the quality of these surface waters may be such that they could not be used for public supply without extensive filtration. Where permeable sands and gravels occur in close proximity to such rivers (see Box 5.1), it is often possible to draw upon the river water by pumping wells drilled into the sands and gravels not far from the river banks. The resulting influx of water from the river to the aquifer is termed **induced recharge**,[v] and in most cases it will be accompanied by an improvement in water quality, which is usually referred to as "**bank filtration**" (e.g. Hiscock and Grischek 2002). In addition to pathogen removal, marked reductions in dissolved concentrations of nitrate, phosphate, sulfate, natural organic matter, and manmade organic substances are commonly observed during bank filtration (e.g. Grischek et al. 1998; Hiscock and Grischek 2002). Detailed investigations over the years have demonstrated that some of the most profound improvements in the quality of induced recharge occur very close to the surface of the streambed (e.g. Younger et al. 1993), where silty sediments rich in organic matter are often found (see Box 5.1). So profound is the localization of water quality improvement in the bed sediment that, in some instances, dredging of the river for navigational purposes has resulted in rapid contamination of riverside wells which had previously been supplying good quality water for many years (e.g. Younger et al. 1993). Major bank filtration wellfields are well documented throughout Europe; indeed larger proportions of the total public water supply come from bank filtration sources in some European countries (e.g. Hungary, 40%; Finland, 48%; France, 50%; Switzerland, 80%) (Tufenkji et al. 2002). Although less common than in Europe, the USA also boasts some substantial bank filtration sources, e.g. on the Ohio River (at Louisville, KY), the Rio Grande (at El Paso, TX), and the Platte River (at Lincoln, NE).

7.4.3 Groundwater-based river augmentation systems

As we saw in Chapter 5, natural groundwater discharge is a major contributor to the sustenance of flows in rivers during dry periods. Where these rivers are relied upon for major municipal/industrial water abstractions, the natural rate of sustenance in the very driest periods might not be sufficient to maintain healthy aquatic ecosystem conditions *and* continue to meet abstraction. In such cases, the obvious solution is to obtain water from elsewhere to sustain flows in the river. Intercatchment transfers of surface water may be an option in some case, though there are increasing ecological objections to this practice, as (*inter alia*) it has a tendency to propagate diseases from one fish population to another. Another option might be to construct a surface reservoir in the affected catchment, so that winter runoff can be retained for slow release during the summer. However, the damming of headwater catchments is an increasingly unpopular option in many parts of the world. One option that avoids the objections associated with these surface water-based approaches is to seasonally pump groundwater into the river, using wells constructed for the purpose in nearby aquifers. This approach, which is normally referred to as "**river augmentation**," was widely implemented in the UK during the 1970s and 1980s (Downing et al. 1974, 1981; Owen et al. 1991; Headworth 2004).

As may be imagined, the relatively straightforward principle of river augmentation using pumped groundwater often becomes clouded by complications in practice. The complications principally arise from the natural interconnectivity of aquifers and rivers (Downing et al. 1981). Given the high costs associated with laying pipelines from boreholes to rivers, there is a strong economic incentive to pump those aquifers which lie close to the rivers in question. However, the closer the aquifer lies to the river, the more likely it is that pumping wells in that aquifer will decrease the rate of natural groundwater outflow to the river. Moreover, where drawdowns spread as far as the river, water may cause induced recharge. There is thus potential for a "law of diminishing returns" to beset the implementation of river augmentation wellfields. The **net gain** of a river-augmentation wellfield is defined as the ratio of the volume of pumped groundwater to the sum of the volumes of decreased natural outflow plus induced recharge from the river (Downing et al. 1981). River augmentation schemes are unlikely to prove economic where the net gain drops much below 60%; the normal aspiration is to achieve net gains in the range 70–80%.

In most river augmentation systems currently in use, it is normal practice to pump groundwater from storage during a really dry summer, then simply leave the groundwater system unpumped for one or more winter seasons while the depleted storage naturally recovers. In principle, the whole process could be speeded up, and thus made available for annual use, if surplus surface waters were actively injected into the subsurface during the winter. Implementation of this option would be an example of aquifer storage and recovery.

7.4.4 Aquifer storage and recovery

The deliberate introduction of water into the subsurface is known as **artificial recharge**. This practice dates back many decades (see Todd 1980). There are a number of reasons for practicing artificial recharge (European Commission 2001), most notably:

- To take advantage of natural aquifer processes (especially filtration and geochemical reactions) in order to improve the quality of the injected waters.
- To compensate for the depression of the water table caused by historical groundwater abstraction, or ongoing artificial dewatering operations (for mining or construction; e.g. Wardrop et al. 2001).
- To create hydraulic barriers within aquifers, in the form of high "ridges" on the water table, to prevent the migration of polluted waters or the intrusion of saline waters.
- To create a subsurface reserve of freshwater which may be used in another season (when surface water is naturally scarce) or as an emergency reserve when surface water intakes are unusable for whatever reason.

The last of these motivations provides the first step in a chain of operations which has now come to be termed "**aquifer storage and recovery**" (**ASR**) (e.g. Pyne 1995). The basic concept of ASR is to use artificial recharge to build up a large volume of fresh water within an aquifer, from which it is later pumped during periods of high demand. In many cases, an individual cycle of injection/recovery will extend over several years.

There are two basic techniques of artificial recharge:

- Spreading basins, which are open pits excavated into the aquifer materials; clearly these are only applicable where the aquifer lies close to the surface. The water will later be pumped from the aquifer using water wells located beyond the margins of the spreading basin.
- Injection wells, which can be used to introduce water into aquifers at virtually any depth. In many cases, injection wells can also serve as pumping wells in the seasons of high demand.

The principal limitation on both techniques is clogging (European Commission 2001), which can be temporary (due to entrapped air) or permanent (due to accumulation of suspended solids and/or precipitation of minerals). Careful planning and management is required to minimize this problem.

Given that a source of water suitable for artificial recharge has been identified, the principal challenges in designing a successful ASR scheme relate to the properties of the intended host aquifer. Clearly there must be sufficient "freeboard" in the aquifer (i.e. scope to raise the water table without causing flooding above ground level) to enable it to receive a worthwhile quantity of water. To have a reasonable prospect of future recovery of injected water, it is necessary that the aquifer is not so permeable that it will quickly transmit the injected water away from the ASR wellfield before the next abstraction phase of the ASR cycle commences.

A number of major ASR systems are now in operation around the world. Box 7.1 summarizes three interesting examples.

7.5 Groundwater as a thermal resource

7.5.1 Categories of hydrogeothermal resources

In many parts of the world groundwater is spectacularly warm, and the notion of exploiting it as an energy source is obvious. The geysers of Iceland (e.g. *Geysir* itself, and Strokkur), the

Box 7.1 ASR in practice: from Key West to Kuwait.

ASR in the Comprehensive Everglades Restoration Plan (CERP) (USA)

The Florida Everglades have been heavily altered by human activities over the last century. With concerns over this invaluable ecosystem beginning to reach new heights in the mid 1990s, the US Congress authorized the CERP with the aim of improving the supply of fresh water to critical areas of the Everglades. ASR plays a central role in the CERP, with more than 300 injection/recovery wells in southern Florida which together handle in excess of 7000 ML/day in season. Much of the water injected is effectively storm runoff that previously flowed to the sea in winter. As using surface reservoirs was out of the question in this area, ASR provides the crucial means of storing this water for subsequent use in dry periods. Different parts of the CERP ASR system have different operating strategies. In the Lower East Coast area, ASR principally serves to provide dry season reserves in the Biscayne Aquifer (the main water supply aquifer), and it also helps alleviate flooding in urban areas. In the catchment of the Caloosahatchee River, ASR is the basis for a river augmentation system, allowing water supply abstractions to continue without leading to the problems of hypersalinity which formerly plagued the Caloosahatchee Estuary. Finally, in the Lake Okeechobee area, ASR is designed to improve the health of the lake, both by preventing the development of excessively low water levels in dry periods and by helping to abate extreme high-stands of water in exceptionally wet periods. In the process it maintains a strategic resource of fresh water which has improved the reliability of water supply for municipal and agricultural uses (CROGEE 2002).

Thames Water's North London ASR Scheme (UK)

The North London ASR Scheme was one of the first such systems in the world to become fully operational. By the early 1980s, centuries of heavy pumping from many wells in London had led to the development of a deep depression in the water table of the Chalk aquifer. Around the same time, the London Ring Main was being commissioned by Thames Water. This vast distribution main encircles London, facilitating the flexible delivery of water to different areas as demands wax and wane. During periods of low demand (especially at night), pressures in the north London section of the Ring Main are often very high – and high pressures in water mains mean high leakage rates. Rather than have water flowing to waste from multiple small leakages, a strategy was developed in which water would be taken from the ring main at times of high pressure and injected into the Chalk aquifer. In this way a strategic reserve has been built up, which is pumped back into supply during periods of drought (Owen et al. 1991).

Kuwait: balancing peaks in desalination with peaks in freshwater demand

Kuwait is one of the driest inhabited areas in the world. Being rich in oil, Kuwait obtains its energy from oil-fired power plants which are cooled by flash distillation units, which produce copious quantities of deionized water. However, the peak in electricity demand (in the winter months) is completely out of synchronization with the peak in water demand (in the height of summer). Furthermore, the deionized water produced by the power plants is *too* pure for direct human consumption, and must be partly re-mineralized (by mixing with brackish groundwaters) before it is suited for public supply. With a view to solving both of these problems, the government of Kuwait instigated a number of investigations into the feasibility of injecting deionized water into the local aquifers (which naturally contain brackish waters), creating large "lenses" of fresh water which can then be withdrawn in periods of high water demand. Detailed numerical modeling of the growth, lateral migration, and potential control (using smart "gradient-control well" strategies) of injected freshwater lenses (Al-Otaibi 1997) paved the way for full-scale field piloting of ASR in Kuwait, which is underway at the time of writing.

USA (e.g. Old Faithful at Yellowstone and The Geysers in California), and New Zealand (e.g. Pohutu at Rotorua) are legendary symbols of the power of superheated groundwater. All three of the countries named have impressive track records in the utilization of natural reservoirs of steam and very hot (>100°C) water (Dickson and Fanelli 2005). Although examples of this type can be cited from many countries, especially those around the Pacific rim, the incidence of superhot groundwater is rather sporadic in terms of proximity to urban (or even potential future urban) demand centres. Fortunately, lower-grade geothermal resources are far more widespread. In broad terms, natural hydrogeothermal resources can be categorized as follows:

1 **High-enthalpy resources** (i.e. steam and superheated water at temperatures > 150°C), which are essentially restricted to areas of current/recent volcanism.
2 **Intermediate-enthalpy resources** (100–150°C), usually representing the cooler portions of the same hydrothermal systems which give rise to high-enthalpy resources.
3 **Low-enthalpy resources:** aquifers at reasonably shallow depths (<3 km) containing groundwater at temperatures in the range 25–100°C. They are associated with areas of the Earth's crust in which there is an abnormally steep geothermal gradient (see Section 4.2.1). Many such areas are associated with high rates of heat production due to natural decay of radionuclides. Though far more widespread than categories **1** and **2**, low-enthalpy resources are by no means ubiquitous.
4 **Ground-source heat resources:** ubiquitous shallow groundwaters, with temperatures close to the local mean annual air temperature, which are potential sources of thermal energy only if processed using electrically or mechanically actuated heat pumps.

While high- and intermediate-enthalpy geothermal resources (**1, 2**) can be used to generate electricity, the temperatures of low-enthalpy and ground-source heat resources (**3, 4**) are usually too low so support significant electricity generation. Advances in heat pump technology over the last two decades have resulted in **3** and **4** being widely harnessed for direct space heating application. As **3** and **4** are now the most widely available hydrogeothermal resources, they are considered further in the following sections.

A further category of geothermal energy known as **Hot Dry Rock (HDR)** has potential in some parts of the world. HDR prospects can exist where natural warm groundwaters are absent: they are predicated on drilling to intersect fractures in warm rock at depth, opening these fractures further (using explosives or hydraulic fracturing techniques), and injecting cool water which is then heated by the rock (typically to temperatures of around 80°C), and pumped back to the surface for exploitation (e.g. Downing and Gray 1986). As they do not rely on natural groundwaters, HDR prospects cannot really be classified with the four categories of natural *hydro*geothermal energy listed above.

7.5.2 Low-enthalpy geothermal resources

The principal controls on groundwater temperatures were outlined in Section 4.2.1, where it was noted that the geothermal gradient tends to average about 2.0–2.5°C per 100 m depth. Low-enthalpy geothermal resources are located in those parts of the Earth's crust characterized by a coincidence of (Barker et al. 2000):

- An above-average geothermal gradient (generally ≥ 3°C/100 m), such that usefully high temperatures can be accessed by boreholes less than approximately 2.5 km in depth
- Aquifers which are sufficiently permeable at the target drilling depths to yield reasonable quantities of water to production wells without inducing excessive drawdowns.

In principle, these two conditions can be met in a wide range of geological conditions, and certainly in a wider range of localities than those in which low-enthalpy resources have so far been successfully harnessed. Typical production wells abstract groundwater with temperatures higher than 70°C from depths of 2 km or more. As these groundwaters are nearly always very saline, the heat that they contain is usually transferred into fresh water using a heat exchanger, so that it is noncorrosive, fresh hot water which is distributed (via insulated pipes) to meet the space-heating and hot water requirements of buildings. The cooled saline water leaving the heat exchanger tends to require careful handling to avoid saline contamination of streams or shallow aquifers. Where the geothermal development lies close to the sea, direct disposal to marine waters may be feasible. In many cases, however, the saline water is re-injected into its original aquifer using a second well.

Only a few countries can yet boast extensive exploitation of low-enthalpy aquifers. For instance, in France more than 200,000 homes in the Paris Basin are heated by low-enthalpy borehole sources. In the Pannonian Basin in Hungary, where crustal thinning associated with the formation of the Alps has resulted in geothermal gradients as high as 6°C/100 m, more than 500 ML/day of water with temperatures between 30 and 100°C are used for space heating, agricultural applications (heating greenhouses and soils), and spas (Korim 1994). While more widespread use of such resources is likely in the future, the risk expenditure associated with exploratory drilling to assess deep subsurface prospects is still something of a disincentive, given the current costs of established sources of energy.

7.5.3 The ground-source heat revolution

Only 20 years ago, it seemed that there was little prospect that shallow groundwaters with temperatures below 20°C would ever assume any importance as potential hydrogeothermal resources (e.g. Downing and Gray 1986). Subsequent developments both in heat-pump technology and, crucially, in the willingness of people to contemplate different ways of heating their properties are now transforming the face of the geothermal energy business worldwide (Sanner et al. 2003). After 20 years of exponential growth in uptake, around 500,000 ground-source heat pumps were estimated to be in use worldwide in 2004. These yield some 6.7 GW of energy, which compares closely with the total output of all wind turbines currently in use (7.2 GW). This form of geothermal energy utilization is now considered more likely than any other single technology to yield reductions in CO_2 emissions capable of exerting a significant mitigating effect on global warming (Banks et al. 2004).

In the early years of ground-source heat-pump installation, developments in the USA and Europe progressed in relative isolation from each other (Sanner et al. 2003). Although differences in the ultimate mode of heat distribution within buildings continue to divide the USA from Europe, on all of the essentials there is now a consensual trans-Atlantic accord on the concepts and practice of capturing heat from shallow subsurface sources. The heart of the process is the heat pump, a long-established component of many familiar appliances, such as refrigerators and window-fan space cooling systems. By cleverly exploiting the differences in boiling and condensation temperatures of different liquids, heat pumps can efficiently extract heat from a large-volume, low-temperature substance such as groundwater, and transfer it to a smaller volume of another fluid which can then be used for space heating. In this USA, the "fluid" is usually air, which is then blown through the building; in Europe, it is usually fresh water, which is circulated in a closed-circuit underfloor heating system. In order for this system to work, the heat pump requires inputs of external energy, which in most modern applications is electricity. Fortunately, the energy input is usually far less than the energy yielded: in typical European applications (in which groundwater at around 10°C is used to sustain temperatures in a heating system at about 50°C) the ratio of thermal power produced (as heat) to electrical power consumed is typically 3:1 to 4:1. The overall saving in heating costs compared to conventional space heating is typically in excess of 50%, and can exceed 60%.

There are two alternative approaches to extracting heat from the groundwater (Sachs 2002). It is often possible to simply pump water from a well in the normal manner (cf. Figure 3.7), and pass the pumped water through a heat pump. Typical water requirements for such applications are given in the final row of Table 7.2. In most cases, re-injection of the now-cooled water into another borehole would be recommended. For most large-scale applications, pumping groundwater will be by far the simplest and least expensive option. However, for smaller-scale applications (e.g. individual houses, small housing blocks, small workshops, etc.) it is often both feasible and cost-effective to tap the heat present in the ground water by dangling one end of the heat pump exchanger apparatus down one or more boreholes.[vi] The drawback of this approach is that the requisite number of static boreholes quickly multiplies with increasing heat demand, so that for large schemes the capital costs of drilling large numbers of boreholes soon overwhelms the projected savings in revenue costs during the operating lifetime of the heating system.

Whether cooled water is reinjected or groundwater is cooled in situ by contact with downhole heat exchangers, the economic sustainability of the operation will depend on the degree of interference between the cooled groundwater and the higher-temperature native groundwater. This problem can be addressed to some degree where the same equipment is used to provide cooling to the building during summer. This can be achieved simply by running the heat pump in reverse, so that heat is extracted from the building and passed to the groundwater in summer. Over successive seasons, the heat distribution in the groundwater can be expected to even itself out. However, there remains a need for further research on the long-term response of shallow groundwater systems to these types of perturbations.

Even wider applications of **underground thermal energy storage (UTES)** may well prove possible in coming years (Sanner et al. 2003). In principle, it is possible for UTES systems to use heat storage capacity within an aquifer; alternatively, wholly cased boreholes can be the sole receptacles for stored waters, which simply exchange heat with the surrounding groundwater (via thermally conductive casing grouts), instead of actively mixing native and injected waters. While storage of cold water has already proven feasible in several applications, high-temperature heat storage (>50°C) is currently still in the demonstration phase (Sanner et al. 2003).

Endnotes

i In the field of water resources, the term "abstraction" signifies an artificial withdrawal of water from the natural environment, such as the pumping of groundwater from an aquifer via a borehole.

ii Treatment of groundwaters for public supply typically involves nothing more than contact-tank chlorination, which adds the "residual" concentration of disinfectant needed to ensure that the water remains free of pathogens during its passage through the water distribution pipe network to the point of consumption.

iii In cases where a wait of one or more years is not feasible, it is sometimes possible to synthesize a flow–duration curve for a spring by correlating a few "spot" measurements of flow with those from another spring for which a flow–duration curve already exists. By identifying the percentile points corresponding to the spot measurements, and assuming both springs respond similarly to seasonal variations in recharge, a synthetic flow–duration curve for the ungaged spring can be constructed graphically.

iv Another common application of wellfields is in the realm of construction dewatering (see Section 11.3.4).

v A synonym for induced recharge is "induced infiltration."

vi Downhole heat exchangers can even be used above the water table, to extract heat from the soil atmosphere; indeed where land availability is not a problem, the expense of drilling boreholes can be avoided by laying heat-exchange pipes in trenches (≥1.5 m deep) which are then back-filled with soil.

8
Groundwater Geohazards

But the mountain falls and crumbles away, and the rock is removed from its place; the waters wear away the stones; the torrents wash away the soil of the earth; so you destroy the hope of mortals.

(Book of Job 14:18–19)

Key questions

- What role does groundwater play in landslides?
- Where and why does quicksand occur?
- Are land subsidence problems ever attributable to groundwater processes?
- Isn't a "groundwater flood" a contradiction in terms?
- Are there any naturally poisonous groundwaters?
- What groundwater geohazards can beset building sites and mines?
- What happens when groundwater levels rise in urban areas?

8.1 Geohazards and hydro-geohazards

8.1.1 Geohazards

Geohazards are threats to life and property in which geological processes are the principal causative factor. The concept of "geohazards" burst into public consciousness worldwide following the tragic events of December 26, 2004, when an earthquake measuring 9 on the Richter Scale occurred above the Sunda Trench subduction zone near the Indonesian island of Sumatra, triggering a tsunami which killed more than 156,000 people. Large-scale tectonic events such as this are equalled only by volcanic eruptions as natural agents of destruction. Such "mega-geohazards" represent the extreme upper end of a continuum, varying in scale and intensity down to the mere risk of tripping posed by minor undulations on the ground surface. In between are a wide range of potentially destructive phenomena such as avalanches, landslides, and collapses in the ground surface (cf. Koch 1994). Where groundwater is the principal agent of destruction we may legitimately talk of hydro-geohazards.

8.1.2 Hydro-geohazards: introductory summary

Groundwater is the principal causative agent of a wide range of geohazards, which are experienced across a very broad range of scales (physical and temporal). At the largest of scales, sudden groundwater ingress to magma chambers is responsible for the most destructive form of volcanic eruptions, so-called "phreatic"[i] eruptions (e.g. Mastin 1997). **Phreatic eruptions** are extremely explosive, due to the violence with which cool groundwater is instantaneously vaporized. During both volcanic eruptions and some earthquakes, shallow groundwater can be forced to the surface, leading to **liquefaction** of loose sediments, which contributes to the generation of devastating mudflows. High groundwater heads are also responsible for the triggering of many **landslides** (Box 8.1; Sections 8.2.1 and 8.3.1).

At more modest scales, groundwater is implicated in many forms of **subsidence** (Section 8.2.3), i.e. the localized lowering of the ground surface, forming closed depressions. Such subsidence depressions can be hazardous if they develop beneath buildings or roads, or in other locations where humans or animals might be injured by falling into them. Groundwater is implicated in triggering subsidence in a range of hydrogeological environments (Section 8.2.2); some of the most dramatic examples have been induced by human activities (Sections 8.3.1 and 8.3.2).

A hydro-geohazard well known to all *aficionados* of Hollywood movies (especially those of the "western" or "jungle adventure" genres) is **quicksand** (Section 8.2.2): to judge from its frequency of deployment as a dramatic device, one could be forgiven for thinking it a widespread phenomenon. Thankfully it is relatively uncommon in nature, though it is quite easy to induce quicksand behavior by careless management of groundwater during excavations (Section 8.3.3).

All of the hydro-geohazards we have listed so far are physical in nature; however, groundwaters containing **toxins** are also hazardous. Although most shallow groundwaters are bacteriologically pure and free from natural toxins, some groundwaters do contain dissolved elements that are potentially harmful to human health (Section 8.2.5). Add to these natural phenomena the widespread introduction of toxins and carcinogens to groundwaters by human activities over the centuries, and the scale of the hazards posed by contaminated groundwaters increases dramatically.

Fortunately the story is not all doom and gloom. Policies and technologies exist to prevent many of these hydro-geohazards from occurring in the first place. Remedial technologies also exist for most of these hydro-geohazards, although few are claimed to be able to completely restore damaged aquifers and land areas to pre-disruption conditions. Management strategies for hydrogeohazards are outlined in Section 11.3, in the final chapter of this book. In the following sections we will explore the role of groundwater in the most common geohazards (excluding those exclusively associated with active volcanoes and earthquakes), examining natural and human-induced phenomena in turn.

8.2 Natural hydro-geohazards

8.2.1 Landslides

"A landslide is the movement of a mass of rock, earth or debris down a slope" (Cruden 1991). As such, the term "landslide" embraces a very wide range of phenomena, from innocuous slippages of top-soil to the catastrophic collapse of entire mountain sides. Landslides can be extraordinarily destructive; for instance, a landslide in February 2006 killed more than 1800 people in the village of Guinsaugon in the Philippines. Many large landslides destroy sections of roads, topple pylons, breach underground cables, and burst water pipes and sewers; the water released by the latter can add to the fluidity of the landslide mass, expediting its propagation and intensifying its destructive power (see Box 8.1).

What role does groundwater play in landsliding? Clearly a prerequisite for a landslide is the presence of steep ground underlain by soils or rocks which are susceptible to movement. With the exception of landslides triggered by volcanic or seismic activity, virtually all landslides are triggered by water. Given that most major landslides occur during (or immediately after) periods of intense rainfall (or, at high altitudes/latitudes, snowmelt) it might be assumed that hydrological triggering of landslides must be a strictly surface water phenomenon. However, groundwater plays a significant role in the initiation of many landslides. Most experts now agree that the immediate cause of most land-sliding is the build-up of excessive pore water pressures within the slope materials. Where the water table is already close to surface, excessive pore water pressures develop more readily than where the water table lies at great depth. In permeable ground, steep slopes will seldom retain a water table close to the surface unless the soils/rocks are of low permeability. This is why landslides occur more commonly on mud-rich soils than on sands and gravels. "Once a slope in a sensitive soil has been over-steepened by erosion at the toe or by excavation work and the groundwater table is high, the stage is set for a landslide to occur" (Eden 1971). Under such conditions, intense rainfall lasting only a few hours may be sufficient to trigger a substantial landslide.[ii]

Whether the combination of a given surface slope and the water table which it hosts is sufficient to give rise to a landslide can depend on the wider hydrological setting. For instance, where a steep slope adjoins a large body of surface water, the water table at the toe of the slope will correspond to the lake water level, and the heads within the higher parts of the slope will mimic the surface slope. Abrupt lowering of the surface water level removes a significant weight which was previously supporting the toe of the slope. Furthermore, as the water table within the slope cannot drop as quickly as the surface water level (it takes a considerable time to down-drain a significant saturated thickness in mud-rich soils), the pore water pressures can suddenly become "excessive" within the local slope setting, so that the slope becomes prone to landsliding. The triggering of a landslide by sudden drops in water level at the toes of steep slopes is fairly common along many coastlines, lake shores, and river banks, especially where sustained periods of elevated water levels are suddenly ended. For instance, in rivers this can occur when the river stage drops rapidly at the end of a period of spate flows associated with the spring snowmelt period; in the ocean, a rapid drop in sea level often follows a storm tide.

Major landslides can result in the deposition of so much sediment in valley floors that they clog streams, reducing channel capacities such that frequency of overbank flooding increases dramatically. In some cases, landslides dam the principal stream channels in valleys, creating new lakes. Natural groundwater/surface water interactions can be significantly altered in such circumstances.

8.2.2 Quicksand

Quicksand is a body of saturated sand in which the movement of water lifts the grains away from one another, resulting in the sediment having far less load-bearing capacity than would normally be expected. In ordinary sands (wet or dry) the grains are all in constant contact with one another. If you step onto an ordinary body of sand, the grains beneath your feet will be pushed more tightly together until the mass of your body is carried by the sand. When you step onto quicksand, the grains will move aside to allow your foot to enter the sand/water mixture: you will sink rapidly into the quicksand. This treacherous behavior has led people to view quicksand with fear and suspicion. The very name "quicksand" is derived from an archaic meaning of the adjective "quick," i.e. "living." A truly "living" sand would indeed be a rather terrifying phenomenon, a point which has not been lost on the screenplay writers of the world.

What causes quicksand? It occurs where the hydraulic gradient acting across a body of sand exceeds a value of about 1. (This surprisingly "round" number arises from considerations of the porosity and relative density of most natural sands and silts; see Capper et al. 1995.) Such steep hydraulic gradients are not very common in natural aquifers; they are largely restricted to groundwater discharge zones, where groundwater is flowing upwards to discharge via springs, streambeds, or similar features (cf. Chapter 5). For this reason, quicksand is most commonly found along the banks of rivers crossing major aquifers, and also in some coastal areas where groundwater is upwelling to the sea. Indeed minor manifestations of quicksand, only a centimeter or two in depth, occur on almost all sandy beaches during low tide. Persistent quicksand locations tend to be well known locally. Indeed, a particular manifestation of the quicksand phenomenon known as "boiling springs," in which sand grains can be seen gyrating in a basin of upwelling water, have long been celebrated as fascinating attractions. Several towns in the USA take their name from such springs, and a particularly fine example forms the centerpiece of Boiling Springs State Park in northwestern Oklahoma: in this location, the quicksand phenomenon is associated with groundwater discharge converging on the North Canadian River from a regional aquifer comprising Permian sandstones.

Although the literature universally refers to quick*sand*, in many cases the sediment behaving in the "quick" manner is actually silt. Some muds also exhibit quick properties when subjected to high hydraulic gradients, although the powerful intergranular attractions typical of many clays tend to prevent sustained grain separation. Peats are particularly prone to quick behavior, and given the low relative density of organic materials, they can do so at hydraulic gradients considerably lower than 1.

Just how hazardous are quicksands? Contrary to the impression conveyed by numerous movies, most quicksands are relatively shallow and are unlikely to engulf the unsuspecting pedestrian above knee-depth. Deeper quicksands can be more hazardous, and people have certainly drowned in some of these. The problem is that, although quicksands are more buoyant than open water, they are also more viscous. This has the unfortunate consequence that instinctive efforts to "tread water" (as one would in a swimming pool) can actually lead to deeper immersion, as the viscous sediment/water mixture offers frictional resistance to the rapid raising of the hands towards the surface. By filling your lungs with air and resisting the temptation to move arms and legs, it is possible to float to the surface of the quicksand, whence rescue ought to be possible. In the case of most coastal quicksands, the real danger arises from the hindrance to rapid walking, thus entrapping struggling travelers in the path of a rising tide.

8.2.3 Subsidence

Natural subsidence is a widespread consequence of the collapse of subsurface voids, which in turn owe their existence to the erosive action of groundwaters.[iii] The largest-scale and most widespread instances of this genre of subsidence are found in karst terrains, where groundwaters have dissolved limestone, gypsum and other soluble rocks. Closed depressions in the ground surface are abundant in most karst terrains; indeed they are so widespread that they have been proposed as *the* diagnostic feature of karst terrains (Ford and Williams 1989). These closed depressions are known by a variety of names, including "sinkholes,"[iv] "potholes," and "shakeholes." The preferred term in karst science circles is "**doline**." While all dolines share in common the basic form of a surface depression, this uniformity belies a wide range of origins (Ford and Williams 1989):

- **Solution dolines** are formed by direct dissolution of the bedrock surface (either exposed or beneath a thin soil cover) by recharge waters.
- **Suffosion dolines** are craters left behind as loose sediments fall/are washed into underlying caves in karstified bedrock.

Fig. 8.1 A collapse doline formed in 1997 by failure of the roof of a flooded, natural cave in a bed of gypsum. The collapsed building was a double garage, adjoining a recently constructed detached house on Ure Bank Terrace, Ripon (North Yorkshire, UK). The gypsum bed containing the cave lies within a sequence of dolostone aquifers of Permian age, which are in turn overlain by a major sandstone aquifer of Permian age and permeable river gravels in hydraulic continuity with the River Ure. Vigorous cross-stratal groundwater flow is the root-cause of this dramatic subsidence feature. (Photograph courtesy of Dr John Lamont-Black.)

- **Subsidence dolines** are broad, shallow depressions formed due to down-warping of overlying rocks as an underlying soluble rock layer is gradually dissolved.
- **Collapse dolines** are formed due to the collapse of caves, as the void "migrates" through successive collapse towards the ground surface.

The first three of these types of doline tends to form rather gradually. They tend to be landscape features of long standing, which can be easily avoided by house builders, and can be made safe for pedestrians by the use of fencing and warning signs. Some subsidence dolines are so large that their presence is not obvious; over periods of several decades, ongoing dissolution of the deep soluble layer can lead to extensional deformation around the margins of the feature, though this rarely results in more than superficial damage to buildings.

By far the most hazardous category are collapse dolines: this is because they can form in a matter of seconds, replacing a previously level area of ground with a deep gaping hole (Figure 8.1). The dangers associated with the development of a collapse doline such as that shown in Figure 8.1 require no commentary. Groundwater is implicit in two key processes which give rise to collapse doline hazards. First, groundwater is the key agent of **speleogenesis** (see Klimchouk et al. 2000), i.e. the combination of processes responsible for the formation of caves in the first place. As long as caves are filled with groundwater, they will experience significant **buoyant support** from the water itself. However, natural lowering of the water table (seasonal or long-term) can remove this buoyant support, increasing the probability of void roof collapse and thus the initiation of collapse doline formation.

It is possible for collapse dolines to develop in areas other than those underlain by soluble rocks. Physical scouring by rapidly flowing groundwaters can create collapsible voids in unconsolidated clastic sediments, and in some siliceous rocks (including sandstones, quartzites and even some granites). Such physical erosion by groundwater occurs along distinct flow paths, normally in close proximity to springs or other discharge features through which entrained sediment can be expelled from the aquifer. Indeed, initial enlargement of flow paths by physical erosion usually occurs right at the point of surface discharge by **seepage erosion**, i.e. the entrainment of grains by discharging groundwater. As with the quicksand effect, seepage erosion is most likely to initiate where hydraulic gradients are in excess of 1. If sustained over a long period of time, the eroded face will retreat back into the aquifer materials, giving rise to a conduit which may eventually extend over distances of tens (and exceptionally hundreds) of meters. Within such a conduit, flowing water shears the bed and walls just as in any surface stream, adding significantly to the mobilization of grains due to seepage erosion. This combination of conduit enlargement processes is termed **piping** (Higgins 1984). It has been reported from groundwater discharge zones in sands, peats, mudstones, sandstones, and quartzites (Higgins and Coates 1990; Younger and Stunell 1995). Over long periods of time, piping can give rise to dendritic cave systems very similar to those found in true (solution-dominated) karst terrains. The collapse of caves formed by piping also results in doline formation, with equally devastating impacts at the surface. Thankfully, collapse dolines arising from piping are rarer, and often smaller in diameter, than those found in carbonate and evaporite karst terrains. Indeed, collapse dolines are by no means formed in all rocks or sediments subject to piping: the retreat of the eroding face in a seepage erosion zone is often accompanied by simultaneous collapse of any incipient roof, so that instead of forming a conduit, sustained seepage erosion gives rise to a valley. Development of valleys in this manner is known as **sapping** (Higgins 1984); the characteristically steep-headed "theatre-headed valleys" and "light bulb-shaped valleys" to which it gives rise (see Section 3.3.1 and Box 3.2) are in themselves dangerously prone to landsliding and toppling.

8.2.4 Groundwater-fed surface flooding

Floods are usually considered to result primarily from surface runoff. However, as we saw in Section 5.2.2, groundwater is intimately involved in the generation of surface runoff in some areas. At one extreme, where baseflow indices (cf. Table 5.1) are very high, aquifers can sometimes be the principal sources of floods. Where this does happen, the consequences can be rather different from those associated with more common surface water flooding. For instance, most surface water floods peak on time-scales of hours to days, and generally decline again over similar periods of time. When an aquifer is the source of a flood, high water can persist for periods of weeks to months.

Realization that groundwater can be the principal agent of flooding is a fairly recent development. Indeed it was only in the wake of the major floods which affected northern France, Belgium, and southeast England in the winter of 2000/01 that the term "groundwater flooding" was coined. The areas that were flooded in 2000/01 all overlie (or adjoin) the outcrop of the Chalk, a limestone of Cretaceous age, which is one of the most extensive and transmissive aquifers in northwestern Europe (Downing et al. 1993). That it was Chalk groundwater which was responsible for the floods was immediately apparent to residents in many of the normally dry valleys in the vicinity of Brighton (England). Dry valleys are a hallmark of the Chalk landscape; they are believed to have developed under periglacial conditions many millennia ago. Many valleys in which surface runoff had never been recorded were extensively flooded in the winter of 2000/01. Unlike in surface water floods elsewhere, in which the runoff is invariably turbid, it was crystal clear water that flowed out of the ground to flood the valleys around Brighton; the flood water was typical, high-quality Chalk groundwater.

Meanwhile, in nearby northern France and Belgium, groundwater monitoring stations recorded a sudden rise in the water table just before the floods struck. The valley of the River Somme was particularly affected by the floods. Subsequent analysis of the flood hydrograph for the River Somme in the vicinity of Abbeville revealed that the volume of flood water conveyed by the channel far exceeded the total volume of rainfall in the preceding weeks (Mul et al. 2003). Subsequent hydrological analysis resulted in the following explanation: Some of the water which recharges the Chalk aquifer is "trapped" above the regional water table forming lenses of "perched" groundwater. Under normal conditions, these lenses drain very slowly. Intense rainfall can trigger a sudden acceleration of the drainage of these lenses. As the water table rises, the lenses are progressively submerged, and the volumes of water that they contain become part of the main saturated zone, flowing towards the discharge zones as part of the three-dimensional flowfield. As Mul et al. (2003) conclude: "After submersion of these [lenses] in the groundwater, the combined system drains as one linear reservoir, causing a massive groundwater-induced flood."

Similar instances of groundwater flooding have been reported from limestone aquifers elsewhere. For instance, in southern Spain, groundwater floods in 1996–97 caused extensive and costly damage to prime agricultural land and rural residences overlying the Sierra Gorda karst aquifer (López Chicano et al. 2002). After a prolonged drought which lasted around 5 years, precipitation during the winter of 1995–96 was more than 60% greater than average. The resultant rise in the water table within the Sierra Gorda aquifer was as great as 175 m in places. Although the rains of the following winter were not as intense, because the water table was already very high and there was no moisture deficit in the unsaturated zone, the further rise in the water table led to flooding of large areas on the floor of a large-scale doline feature. Flooding initially took the form of two separate lakes which later coalesced.

Although evidence is at present limited to a few case studies, these few relate to limestone aquifers, which are typified by a combination of high transmissivity and low specific yield (see Younger et al. 2002b). It seems likely that large-scale groundwater flooding will be restricted to such aquifers, particularly those adjoining lowland areas which are sufficiently flat-lying to prevent discharging groundwaters from draining away rapidly. Hazard maps of areas possessing these characteristics, and which may thus be prone to groundwater flooding, are now beginning to be produced (e.g. Jackson 2004, pp. 70–71).

8.2.5 Naturally toxic groundwaters

It is part of Nature's bounty that most shallow groundwaters are potable; few groundwaters naturally contain toxins. Sadly, in the few cases where toxins are naturally present, they often go unsuspected until damage to health has occurred. Such was the case with what is now deemed to be the "greatest mass poisoning in history" (Smith et al. 2000). In what was originally a laudable effort to save people in Bangladesh from lethal diarrheal diseases associated with the use of untreated surface waters, a large number of deep water wells (known locally as "tubewells") were drilled in the 1970s. The groundwater yielded by these wells was crystal clear and tasted great; routine analyses confirmed that it contained normal quantities of the major cations and anions, and was free from bacterial contamination. However, arsenic was not then a "routine" analyte: at the time it was simply never suspected that arsenic might be present in these groundwaters in dangerous concentrations. Chronic arsenic poisoning is an ailment that manifests itself in visible signs only slowly. Consequently, people had been drinking these waters for many years before it was realized that many of them were showing symptoms of severe arsenic poisoning. The long-term health effects of arsenic poisoning include: skin lesions (painful black welts, especially on hands and feet); cancers (of the skin, bladder, kidney, and lung); neurological disorders; hypertension; cardiovascular disease; pulmonary disease; peripheral vascular disease; and diabetes mellitus (Smith et al. 2000). It is estimated that between

30% and 60% of the 125 million inhabitants of Bangladesh are now at risk due to drinking groundwater contaminated with arsenic.

Although arsenic is a common artificial contaminant (for instance in mine waters; LeBlanc et al. 2002; Loredo et al. 2002), the source of the arsenic found in these Bangladeshi groundwaters is natural (Burgess et al. 2002). Sedimentary iron oxyhydroxides rich in adsorbed arsenic are present at a distinct sedimentary horizon within the Holocene alluvial sediments which comprise the shallowest few tens of meters of the local sand-and-gravel aquifer. These iron oxyhydroxides were formed when the groundwater within them was oxidized; now that reducing conditions have been established in much of the aquifer, the oxyhydroxides are gradually dissolving in groundwater, releasing arsenic to solution in the process.

Arsenic is not the only naturally occurring element to be present at toxic concentrations in certain groundwaters. Fluoride is beneficial to the dental health of humans where it is present in drinking waters at concentrations close to 1 mg/L; at much lower levels, dental caries tends to be more prevalent in the population. Fortunately, most shallow groundwaters contain appreciable quantities of calcium, which reacts with dissolved fluoride to precipitate CaF_2 (the mineral fluorite), which is at equilibrium with about 1 mg/L of dissolved fluoride. However, where calcium is scarce (for instance, in groundwaters which have sodium as the dominant cation) fluoride may be present at far higher concentrations, sometimes as high as 15 mg/L. At such concentrations, sustained consumption of the groundwater will hinder the healthy development of bones, leading to severely debilitating conditions such as bowed legs and other limb deformities.

Besides fluoride and arsenic, very few pristine groundwaters contain toxic elements at dangerous concentrations. As the cases of fluoride and arsenic clearly show, however, it is always prudent to conduct thorough analyses of waters intended for human or animal consumption on a regular (if not frequent) basis, and to evaluate the results against regulatory limits such as those listed in Table 7.4.

8.3 Hydro-geohazards induced by human activities

8.3.1 Introduction: making matters worse

The physical processes which give rise to natural hydro-geohazards are also implicated in the triggering of analogous hazards by human activities (Cripps et al. 1986; Maund and Eddleston 1998). All of the hazards discussed below can be viewed simply in terms of human activities unwittingly exacerbating the destructive potential of natural processes. To that extent, we are still dealing with "natural disasters," for which the immediate "triggering" is the (usually inadvertent) introduction of imbalances in natural hydrogeological systems by careless excavation or water management. Mining and some of the larger civil engineering projects occur at a sufficiently large scale to physically disrupt groundwater systems. Degradation of groundwater quality arises from a wide range of agricultural and industrial activities.

8.3.2 Triggering landslides by careless groundwater management

Failure to take account of the hydrogeological conditions relating to manmade embankments (formed either by excavation or construction) can lead to catastrophic failures. As we saw in Section 8.2.1, the immediate cause of landsliding is the build-up of excessive pore water pressures within the slope materials. Where embankments are formed by excavation, if the natural water table lies above the base of the embankment, then excessive pore water pressures are likely to develop in at least part of the excavated slope. The risk of landsliding can be reduced by digging the slopes back to a shallower angle, though often this cannot be done due to limitations of space. Alternatively, groundwater heads behind the slope can be artificially lowered by installing low-angle boreholes ("relief wells") to intersect the saturated zone; these will drain water to surface by gravity alone (Section 11.3.2). Where embankments are *constructed* (by loose-tipping or more orderly piling of soils), two aspects of groundwater behavior need to be taken into account. First, it is important that the base of the embankment is not emplaced directly over natural springs or seepage areas, as continued inflow of groundwater from such features can significantly weaken the base of the embankment. Second, introduction of moisture from above needs careful management to avoid the build up of pore pressures within the emplaced soils. Some soils may themselves be wet at the time of emplacement; installation of drainage layers within the embankment is recommended in such cases. Interceptor drains (basically channels back-filled with cobbles) should also be installed above the crest of the embankment and on its surface to prevent infiltration of surface runoff. Failure to observe such precautions can have catastrophic consequences (Box 8.1).

Box 8.1 Black death: the devastating 1966 colliery spoil landslide at Aberfan, South Wales.

On Friday October 21, 1966, a substantial part of a very large, steep-sided colliery spoil heap suddenly began to landslide. The black muds and coal wastes formed a dense debris flow which slid rapidly down the mountainside below and into the village of Aberfan. Several homes were engulfed, together with part of Pantglas Junior School (see figure below). A total of 144 people

The Aberfan disaster of October 21, 1966 (South Wales, UK). The text in this box describes the geotechnical and hydrogeological triggers of this tragic debris flow from a colliery spoil heap, which claimed the lives of 144 people. (Source: Merthyr Tydfil Local Archives.)

were killed, 116 of whom were school children, buried alive in their classrooms along with their teachers. This was by far the worst peace-time disaster in the British Isles in the twentieth century (McLean and Johnes 2000). In the subsequent inquiry into this horrific incident, it emerged that the spoil heap (and several neighboring heaps) had been constructed directly over a natural spring line. Heavy rains in the autumn of 1966 appear to have resulted in very high flows from these springs into the overlying mass of spoil. Combined with introduction of water into the heap by tipping of wet spoil, and failure to include adequate drainage features to prevent infiltration of rainwater and surface runoff, the scene was set for the build up of excessive pore water pressures in the heaps which then led to collapse. As the debris flow crossed a road above the village, it breached a water distribution pipe, which added large quantities of water to the sliding mass of rubble, further lubricating its downhill surge. The lessons learned from the Aberfan disaster bore fruit in some of the strictest legislation in the world for the future management of similar mine waste piles in the UK. Elsewhere, however, the carnage continued. Many of the most devastating instances of landsliding caused by inadequate management of groundwater in manmade structures relate to tailings dams, in which fine-grained mineral processing wastes are typically disposed by means of settlement from aqueous suspension. Most tailings dams are operated securely, and never give rise to environmental or human safety problems. However, severe problems have been found to develop in circumstances where tailings dams are incrementally constructed over many years without reference to the original design criteria. In such circumstances, sudden catastrophic failure of the dam can occur, leading to major environmental pollution problems and/or human fatalities, as summarized below (see Younger et al. 2002a, for source references and further details):

Location	Year	Direct fatalities	Environmental impact
El Cobre (Chile)	1965	~200	Not recorded
Bafokeng (South Africa)	1974	12	Polluted a 45 km reach of the Kwa-Leragane River and the Vaalkop reservoir
Stava (Italy)	1985	268	Extensive impacts on river and floodplain ecology
Merriespruit (South Africa)	1994	17	Limited pollution of lake sediments
Omai (Guyana)	1995	0	Minor fish kill in Omai River
Río Porco (Bolivia)	1996	0	Fish kills recorded through 300 km of river
Aznalcóllar (Spain)	1998	0	Severe contamination of 40 km reach of a river of high conservation value
Baia Mare (Romania)	2000	0	Severe contamination of 2000 km of the Danube and its tributaries, resulting in massive fish-kills (thousands of tonnes)
Sebastião das Águas Claras (Brasil)	2001	5	A 6 km reach of the Córrego Taquaras (a tributary of the Río das Velhas) was buried up to 15 m deep in a torrent of red mud, which engulfed and uprooted trees. Imperiled 70% of the water supply of the city of Belo Horizonte

8.3.3 Subsidence due to groundwater abstraction

Abstraction of groundwater has the potential to cause land subsidence in two principal hydrogeological settings:

- In shallow karst aquifers (limestone or gypsum).
- In unconsolidated sand/gravel aquifers, especially those which are interbedded with beds of soft mud.

Inducement of karst subsidence

By far the most important karst subsidence process attributable to human activities occurs when lowering of the water table removes the buoyant support from the roofs of caves, which then collapse with subsequent void migration propagating upwards to form collapse dolines (e.g. Figure 8.1). Widespread triggering of collapse doline formation has been associated with the pumping of public water supply wellfields. A notable example is the State of Florida, where major abstractions from the limestones of the Floridan Aquifer have been expanding since the 1930s. Over the years, it was realized that the resultant lowering of the water table was triggering the formation of many large collapse dolines (Sinclair 1982). As doline formation can be exacerbated by loading of the ground surface with buildings, and by the focusing of surface water drainage into small areas, doline collapse is further promoted by urban development. The value of properties damaged each year by doline collapse in Florida presently exceeds $30M. Given that the Floridan Aquifer supplies more than 95% of the water demand in this fast-growing state, cessation of groundwater abstraction is not a feasible response to this problem (except in very localized circumstances). Rather, Florida State law simply requires all house owners to have sufficient insurance to cover their potential losses from subsidence.

Large-scale pumping of aquifers to facilitate mining has been known to cause similar problems. The most classic case of this type occurred in South Africa in the 1970s in the aftermath of a major inrush in 1968 at West Driefontein Gold Mine. To ensure the safety of future underground mining, a major project was initiated to entirely dewater the dolomite aquifer overlying the mined strata (Forth 1994). This was probably the largest mine dewatering project ever implemented, involving peak pumping rates as high as 340 ML/day from several wellfields. Hundreds of collapse dolines subsequently developed, ranging from small depressions a few meters across to vast craters many tens of meters in diameter and depth. So numerous and dangerous were these dolines that the dewatered area eventually had to be almost wholly evacuated. Mining continued of course, but even the mine sites were not immune: one particularly large doline suddenly engulfed an entire mineral beneficiation plant, killing most of the people working there.

Two other forms of karst subsidence can also be induced by groundwater abstractions:

- The development of suffosion dolines can be fostered by a lowering of the water table leading to the establishment of rapid drainage pathways from drift deposits into underlying karstified bedrock. Although such suffosion (or "cover collapse") dolines often form relatively slowly, they often have very large radii.
- The general change in groundwater flow patterns and velocities caused by groundwater abstraction can lead to the ingress of more aggressive groundwaters into parts of karst aquifers where the native groundwater was formerly at equilibrium with the surrounding soluble rock. This leads to a considerable acceleration of dissolution processes which, if sustained over many years, might eventually lead to the development (or extension) of large-diameter solution dolines (see Lamont-Black et al. 2002).

Inducement of subsidence in unconsolidated deposits

Withdrawal of unconfined groundwater from coarse clastic deposits *per se* does not give rise to measurable subsidence. While the withdrawal of buoyant support from sand or gravel clasts might be expected to lead to *some* settlement, in all but quicksands this turns out to be so small that is has no measurable effect on the elevation of the ground surface overlying the aquifer. Poorly designed abstraction wells in unconfined aquifers *can* cause localized subsidence if they withdraw so much sand from the aquifer that voids are created around the well. This form of subsidence is often referred to as being due to **migration of fines** or **sand pumping**. It occurs by the same processes of seepage erosion and piping which were described in Sections 8.2.3. As in natural manifestations of these processes, significant entrainment of sand in pumped groundwater can only occur where the hydraulic gradient exceeds 1. In practical terms this means that this type of subsidence is strictly limited to the innermost portions of the cone of depression developed around any one pumping well. However, because sand pumping damages expensive pumping machinery and clogs pipes and reservoirs, groundwater engineers go to considerable lengths to prevent it from occurring. Using industry-standard water well designs (with screens and gravel packs where necessary; see Figure 3.7), and proper well development immediately after drilling (see Section 3.3.1), most water wells yield no appreciable sand during long-term operation.

Where sand/gravel aquifers are confined by mudstones, or where thick beds of mud occur within the sand/gravel aquifer, major land subsidence can result from sustained groundwater abstraction. To understand why this occurs it is necessary to recall the concept of elastic storage which was introduced in Section 1.4. Groundwater is often under great pressure within confined aquifers. Unlike in unconfined aquifers, where drainage of pores leads to only negligible settlement, a significant drop in total head within a confined aquifer can lead to significantly tighter packing of the grains, and thus to measurable settlement. If the head is raised again, the sand/gravel clasts can be forced apart once more, so that the settlement is largely reversible. However, where beds of mud occur within and above the aquifer, irreversible settlement can result from the squeezing of the pores which occurs when the head drops. There are two reasons for the irreversibility of this form of subsidence: first, because the permeability of compacted clays is so low, re-entry of water is inhibited; second, electrostatic attractions between plate-like clay minerals packed tightly together will strongly resist later parting, even under high hydraulic heads. For most practical purposes, if we calculate the amount of settlement due to compaction of the mud beds, this will closely approximate the total long-term settlement due to the pumping of the aquifer as a whole. Techniques for predicting the amount of subsidence likely to result from groundwater abstraction in sequences of muds and sands/gravels are well established and reliable (e.g. Domenico and Schwartz 1997), although details are beyond the scope of this book. Large-scale examples of this form of subsidence may be cited from many of the world's major cities, including Shanghai (China), Venice and Ravenna (Italy), London (England), Tokyo and Osaka (Japan), Houston (Texas), and several areas in California (USA) (Holzer 1984; Poland 1984). Two of the most renowned and extreme examples of the genre are highlighted in Box 8.2.

One of the most perplexing aspects of this form of land subsidence is that suspension of pumping does not lead to an immediate cessation of subsidence. The reason lies once more in the low permeabilities of the mud beds: once internal hydraulic gradients are established within them to reflect the lowered heads in the adjoining aquifer layers, movement of moisture from the muds commences accordingly. Even after the head increases again in the surrounding aquifers, the low permeability of the mud layers means that a long period of time will elapse (months to decades) before the interior portions of the mud beds receive the changed hydraulic signal. Consequently, expulsion of moisture from the central portions of the mud beds will continue long after a rise in head in the adjoining aquifer layers.

In closed sedimentary basins, in which unconsolidated aquifer sediments are entirely surrounded by hard rocks of low permeability, regional-scale land subsidence leads to significant shrinkage in the volume occupied by the aquifer.

Box 8.2 That sinking feeling: subsidence due to groundwater withdrawal in two mega-cities.

Mexico City is one of the world's largest cities (population estimated to be 22M in 2005), and is still growing very rapidly. To meet the water needs of its booming population, the municipality has long resorted to pumping of groundwater from the extensive sandy aquifers which underlie the valley occupied by the city. Unfortunately these sandy aquifers are intimately interbedded with muds, the compaction of which has given rise to major subsidence (Figueroa Vega 1984), locally exceeding 15 m in total depth (measured relative to the original starting elevations). Historical records show that subsidence was certainly underway by the late nineteenth century, and although active subsidence was demonstrated by city engineers in the 1920s, it was not until 1948 that groundwater abstraction was realized to be the cause. Intensive studies in the 1970s showed that about 75% of the total subsidence was due to consolidation of the muddy horizons in the basin, with 25% due to compaction of the sandy aquifer horizons themselves. Although loading by buildings can locally exacerbate total subsidence, this has been shown to account for no more than 10–15% of total subsidence in this case. In the 1950s the rate of subsidence in the city center averaged around 0.5 m per annum. Following relocation of wellfields to peripheral areas of the city, the subsidence rate declined in the city center (stabilizing at around 6 cm per annum in the 1990s), but increased in the suburban areas (Rudolph 2001). The damage to buildings, roads, water mains, and sewers has been very costly. Where the city sewers used to drain by gravity, they now must be pumped continuously to avoid flooding in the city center. Whimsical instances of subsidence are popular with informed visitors, such as the well top which now stands 15 m above street level, and the 23 steps which have been added to maintain access to the city's Independence Monument (constructed 1910), which, thanks to deep-piled foundations, has remained at its original level while the land all around it has sunk.

Bangkok is Thailand's premier city and one of the fastest-growing mega-cities in southeast Asia (population 9.75M in 2005). Originally the city obtained its water supply from the Chao Phraya. By about 1950, this had become so polluted that drinking water began to be obtained by pumping the sand aquifers (unconsolidated and interbedded with beds of mud) that underlie the city. Pumping increased sharply during the second half of the twentieth century as Bangkok's population soared. As heads in the aquifer dropped by many tens of meters, subsidence reached rates as high as 10 cm per annum in the southern and eastern parts of the city. From 1987 to 2003 the maximum subsidence had reached 1 meter. While still small in comparison to Mexico City's subsidence, the impacts have been large because Bangkok is at almost sea level: Subsidence has therefore led to flooding, both from the tide in low-lying coastal areas of the city, and from rainfall: since subsidence set in, sewage and storm runoff have had to be pumped from much of the city center, and the pumps have sometimes been overwhelmed in the rainy season, leading to major floods in 1983, 1995, and 1996 which caused millions of dollars worth of damage. Subsidence has also adversely affected bridges, roads, rail tracks, and many private and public buildings. The availability of agricultural land near the city has also been affected by the subsidence problems, as wealthy Bangkok residents have bought soil to shore up their subsiding properties. By 1983, the municipal authorities had become aware of the link between subsidence and groundwater abstraction, and began to take steps to combat the problem. In 1994, the charges levied on groundwater abstractors were sharply increased, in an attempt to discourage groundwater use. Some major abstractions were discontinued, and water imported from wellfields located outside of Bangkok. These measures have met with some success: by 2003, subsidence rates in Bangkok had declined to between 1.5 and 5 cm per annum (a considerable improvement on the 5–10 cm per annum rates of the late 1980s). However, intensified groundwater abstractions outside the city limits have accelerated subsidence in those areas, prompting the public authorities to extend subsidence controls over an even wider area (Source: Bangkok State of the Environment Reports 2001 and 2003; see www.rrcap.unep.org).

Fig. 8.2 Earth fissure due to groundwater withdrawal. This major tear in the ground (note the snapped telegraph pole for scale!) is attributed to shrinkage of the sedimentary fill of an intermontane basin near Tucson, Arizona (USA) as a consequence of heavy and sustained pumping of groundwater. (Photograph courtesy of John Callahan, United States Geological Survey.)

As the sediment pile shrinks, it tears at the edges: the tapered edges of the sediment pile near the basin margins are subject to extensional deformation. In some cases, this deformation is so substantial that the desiccated sediment is ripped apart, leaving large open fissures at the ground surface (Figure 8.2). This process has been repeatedly documented from enclosed basins in the semi-arid southwest of the USA (Pewe 1990), with particularly dramatic examples occurring around Las Vegas (Nevada) and in Arizona, near Phoenix and Tucson. In the vicinity of the latter, the subsidence fissure shown in Figure 8.2 has now been mapped over a total length of around 16 km.

8.3.4 Groundwater hazards on construction sites

Almost all construction projects involve excavations, if only to install shallow foundations. Where excavations intersect the water table, a number of geohazards immediately begin to threaten the progress of construction (Rowe 1986; Preene et al. 2000). As we saw in Section 8.3.2, excavated slopes are far less stable when saturated than when drained. The sidewalls of excavations are therefore prone to collapse if continued below the water table without artificial support. Inter-locking steel sheet piles are often driven into

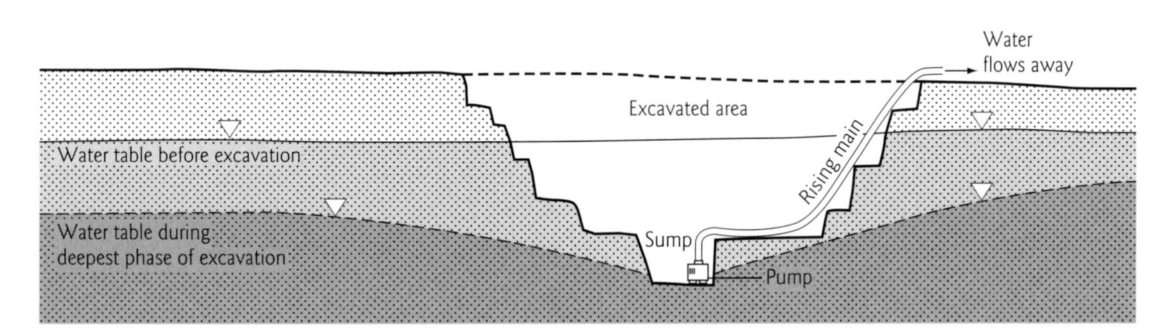

Fig. 8.3 Schematic cross-section showing the steepening of the hydraulic gradient (through drawdown of the water table) around an excavation, caused by pumping of groundwater from a sump.

the ground to provide such support,[v] prior to excavation of the soil inside the enclosed area. Such sheet piling not only provides support to the excavation walls, it also blocks possible groundwater flow paths, in many cases providing a partial contribution to the overall control of groundwater for a given site (see Section 11.3.4).

As it is usually necessary for workers and machinery to be deployed in the base of excavations, it is normal practice to pump all incoming groundwater out of an excavation. In many cases, however, simply pumping the water from the excavation itself will lead to a steepening of the hydraulic gradient in the surrounding soil (Figure 8.3). Where the hydraulic gradient at the base of the excavation exceeds 1, quicksand conditions are likely to occur, endangering the lives of workers as well as swamping mobile machinery. If steps are not taken to prevent the onset of quick conditions, sustained pumping from the excavation itself will simply lead to removal of large quantities of silt/sand. As more sand and silt is washed into the floor of the excavation by inflowing groundwater, seepage erosion and/or piping will occur, potentially leading to localized subsidence beyond the excavation boundaries.

Excess pore water pressures can also cause "floor heave," in which low-permeability beds which are confining groundwater below the base of the excavation are forced upwards under the force of excess head in the underlying aquifer layers. To inexperienced workers, the bulge in the floor of the excavation can be puzzling, and the temptation is simply to trim the excess material so that the floor is once more at the intended elevation. This can be disastrous, as it removes even more of the mass of material which was resisting the uplift, hastening the day on which the low-permeability beds will rupture, allowing the excavation to flood very rapidly. Fortunately, a number of groundwater control measures are available to prevent the establishment of quick conditions or floor heave (see Section 11.3.4).

8.3.5 Groundwater hazards during and after mining

Underground mines

Almost all underground mines receive inflows of groundwater. For the most part, groundwater is a nuisance rather than a hazard: it can make for uncomfortable working conditions; it can soften floor and roof strata and thus increase the burden of maintenance needed to keep the mine in production; it can accelerate the corrosion of mining equipment; it may necessitate the use of expensive, waterproof explosives; and its very presence in the material removed from the mine adds to the weight of material transported, without adding to its value (Younger et al. 2002a). All of these problems can be overcome to some degree by careful planning and management. For instance, during the sinking of mine shafts tight seals should be installed wherever aquifers are encountered. During mining, it is possible to manage extraction to avoid creating fracture connections from the active workings to overlying aquifers (e.g. Orchard 1975), with pumps and related drainage infrastructure being used to dispose of any residual water ingress (see Section 11.3.5).

Far more difficult to manage, and truly hazardous, are sudden unexpected inrushes of large quantities of groundwater, which have claimed the lives of thousands of miners worldwide. Death not only occurs by drowning, but also by entrapment of miners in isolated mine voids above the water line, where the end comes more slowly by asphyxiation or starvation. In the overall scheme of things, the fatality rate from water inrushes (about 1.5% of all deaths in mines) is relatively modest in comparison to other causes (Younger 2004b). It is, nevertheless, of the same order of magnitude as the fatality rate from explosions of methane and/or coal dust (3%), which generally receive far more attention in discussions of the hazards of mining.

There is no global register of inrushes to underground mines. Probably the best national register of data comes from the UK coal mining industry, which between the earliest record (1648) and 2002 recorded 67 major inrushes. An analysis of all of these inrush records leads to the conclusion that there are two principal sources of inrushes (Younger 2004b):

- Flooded old mineworkings, into which the modern workings accidentally stray.
- Natural bodies of water (aquifers or surface water bodies), to which highly permeable connections are accidentally made by mining-induced fracturing.

Of the two, inrushes from old workings are by far the more common. In the UK case, it is apparent that the risk from old workings was particularly acute prior to 1872, when a legal requirement to deposit mine plans for public reference was introduced. To this day, regulations are not sufficiently stringent in all of the world's mining districts, and inrushes still claim far too many lives.

Long after deep mines close, they can still give rise to geohazards (Younger 2002). These range from polluted discharges from the flooded workings to the reactivation of mining subsidence triggered by changes in groundwater levels. These generally arise due to physical destabilization of open mine voids by water, either through: (i) physical/chemical changes which weaken the floor, wall or roof strata; or (ii) downcutting of floors and undercutting of walls by rapidly flowing waters. Void migration triggered by such processes can result in surface collapse features which closely resemble natural dolines. In rare cases, flooding of extensive underground workings has been implicated in the reactivation of slippage on geological fault planes, which had previously been considered to be relict features incapable of further movement. Fault reactivation by rising mine waters can give rise to minor seismic tremors and the opening of fissures at the ground surface (e.g. Young and Culshaw 2001).

Surface mines

Because surface mines inevitably intersect all surrounding strata as they are sunk deeper, they often have much higher pumping requirements than underground mines in the same geological sequence. However, because they are open to the air, drowning hazards are far less marked in active surface mines than in underground mines. Hence it is generally possible to adopt a more relaxed attitude to problems of water ingress in surface mines than in deep mines. Water in a surface mine is still an inconvenience and undesirable expense, of course. However, the only real issues requiring hazard management relate to the destabilization of pit walls and floors by excessive pore water pressures (and, to a lesser extent, erosion by runoff originating as groundwater ingress) (Younger et al. 2002a).

Slope stability is crucial to effective surface mine operations. As we saw in Section 8.3.3, high pore water pressures dramatically reduce the stability of excavated slopes. Besides the obvious dangers posed to workers by landslides, and the loss of production associated with slide damage to mine roadways, if slopes have to be cut back to gentle angles to ensure stability, then both the area occupied by the mine and the amount of overburden requiring digging will increase dramatically, with negative implications for the economic viability of the entire operation. While competent rocks with high shear strengths may be cut into benches with faces as steep as 90 degrees, even if they contain groundwater, for most ordinary rocks (and all soils) it will be difficult to sustain face angles in excess of 45 degrees unless they are well drained.

As in shallower excavations associated with construction projects (Section 8.3.4), excess water pressures can give rise to "quick" conditions and "floor heave" in surface mine floors. Section 11.3.5 considers some of the common strategies used to combat such problems.

After surface mines are abandoned, geohazards associated with groundwater can continue to arise. In open pits which are not re-filled, accumulating groundwater often leads to the formation of pit lakes. Many pit lakes have poor water quality, and thus pose a hazard to wildlife or casual human visitors (Bowell 2002). If pit lakes decant to the surrounding environment, pollution of surrounding streams can ensue. Even where a surface mine has been entirely back-filled (as is common practice with opencast coal mines, for instance), the gradual rise in groundwater levels which often follows the completion of working can trigger settlement of the fill materials (Younger et al. 2002a). Above the water table, erosion by recharge waters flowing preferentially in rubbly zones can give rise to karst-like features (Groenewold and Rehm 1982), such as sinking streams and "pipes" within the back-fill (formed by winnowing out of fine-grained waste rock from between bouldery zones), which can give rise to surface hollows when they collapse.

8.3.6 Urban flooding due to rising groundwater levels

Geohazards associated with rising groundwater levels in abandoned mines have been noted above; however, problems arising from a suspension of groundwater pumping are also common in many urban areas. These problems are particularly acute in long-established cities, such as London and New York, where groundwater was formerly pumped heavily in what are now city center areas. Over the last century, many wells in city center areas have been switched off, as industries have re-located to less congested areas on the urban fringes with cheaper land prices, and as public water supply operators have sought to replace increasingly polluted urban groundwater with purer water from beyond the municipal limits. The result is that groundwater levels have been rising in many cities for several decades now (e.g. Morris et al. 2003), a process commonly referred to as **groundwater rebound**. In some cities, groundwater rebound has led to the flooding of basements, subways, and underground railways, and even to the destabilization of foundations for some tall buildings (which were designed on the assumption that the surrounding ground would always be unsaturated). If these undesirable outcomes are to be avoided, it is generally necessary to begin pumping again, so that drawdowns hold the water table below the level at which it becomes problematical. In some cases, the excess water is simply pumped to waste; however, in London, for instance, alternative uses for some of the pumped waters have been found, helping to defray the costs of keeping the Tube (underground railway) above water.

Groundwater rebound is not the only cause of rising groundwater levels in urban areas. In many cases, the water table is rising due to leakage of water from distribution pipes and/or sewers (e.g. Lerner 1986), and in some cases due to excessive irrigation of parks and gardens (Morris et al. 2003). Leakage from water pipes is not always bad news from a groundwater perspective: in Lima (Peru), for instance, leakage from pipes is a major component of groundwater recharge, making groundwater available for the use of poor communities in an area which is otherwise virtually devoid of natural recharge (Lerner 1986). In many cities, however, the rise in water table due to leakage and garden watering is giving rise to geohazards comparable to those associated with groundwater rebound elsewhere. Cities in the Arabian Peninsula are particularly badly affected, with problems reported from Riyadh, Jeddah, Kuwait, Doha, and several other municipalities. Detailed studies have revealed that about half of the extra recharge comes from amenity irrigation, a third from leaking water supply pipes, and the remainder from infiltrating wastewaters (Morris et al. 2003). Given the arid climate in this region, direct evaporation from the water table is intense wherever it lies within a meter or so of ground level. This has the effect of increasing the dissolved solids content in the water, one consequence of which is that sulfate concentrations can become so high that the water becomes capable of rapidly dissolving structures made from ordinary lime-based concretes, further exacerbating the potential weakening of foundations by the physical presence of the groundwater.

Endnotes

i "Phreatic" is an old synonym for "saturated." Formerly, unconfined aquifers were sometimes termed "phreatic aquifers." Though the term "phreatic" is now obsolete in hydrogeological literature, it persists in volcanology.

ii It should be noted that an elegant and detailed analytical framework for the relationship between pore water pressure and soil behavior exists (see for example Capper et al. 1995; Powrie 2004), and is used routinely by geotechnical engineers. Of particular relevance to landslide hazards is the "liquid limit," at which soil behavior begins to approximate that of a liquid (rather than the more familiar "plastic" behavior of damp soils).

iii It should be noted that subsidence due to dewatering of pores is rarely encountered in natural settings, being almost exclusively associated with artificial lowering of the water table by pumping (Section 8.3.3).

iv The term "sinkhole" is very popular in speleological circles in the USA. It is also widely used in the nonspecialist literature elsewhere. On the whole it is not the preferred term amongst speleologists internationally, partly because of the ambiguity which attends it: the "sink" element is sometimes thought to allude to the fact that streams sink into many such depressions. However, many sinkholes do not function as "sinks" for streams at all. The term "doline" has more neutral connotations and is therefore preferred (Ford and Williams 1989).

v Other techniques achieving the same purpose include "cut-off walls," which are concrete-filled slits in the ground, constructed either as trenches or as lines of adjoining boreholes.

9
Groundwater Under Threat

A nation that fails to plan intelligently for the development and protection of its precious waters will be condemned to wither because of its short-sightedness. The hard lessons of history are clear, written on the deserted sands and ruins of once proud civilizations.

(US President Lyndon B Johnson, 1908–1973)

Key questions

- How are aquifers under threat?
- Is physical destruction of aquifers possible?
- What are the consequences of overpumping aquifers?
- What impacts can be expected from climate change?
- How vulnerable are aquifers to pollution?
- What's the difference between point and diffuse pollution sources?
- How can soakaways, landfills, and leaking fuel tanks affect groundwater?
- Does intensive arable agriculture adversely affect groundwater quality?
- Can pumping operations themselves damage water quality?

9.1 Threats to groundwater systems

9.1.1 Types of threats to groundwater systems

"Out of sight, out of mind": it's an old proverb, but it unhappily has a great ring of modernity about it where human stewardship of groundwater resources is concerned. All over the world, our activities are either already damaging aquifers or posing the serious risk of doing so in the future.

What are the threats to which we are subjecting so many aquifers? There are three main categories:

- Physical destruction of aquifers.
- Depletion of the *quantity* of available groundwater.
- Degradation of groundwater *quality*.

Physical destruction of aquifers is wrought by large-scale mining activities. Although mining has wholly obliterated many minor sand and gravel aquifers, even the largest mining operations are generally too limited in areal extent to remove more than small portions of regional-scale aquifers. However, many surface mines *do* remove enough of the outcrop areas of major aquifers that natural recharge dynamics are dramatically altered; such impacts are especially marked in relation to major limestone quarries (e.g. Hobbs and Gunn 1998; Younger et al. 2002a). On the whole, physical destruction of aquifers is a geographically restricted phenomenon, and we will not dwell on it further here. The second and third of the threats identified above are very widespread, and they are therefore discussed in considerable detail in this chapter.

9.1.2 What is at stake?

The depletion and degradation of groundwaters negatively affect both the resource potential of aquifers and the viability of groundwater-dependent ecosystems. The most common financial impact of depletion of groundwater quantity relates to the increase in pumping costs necessitated by a decline in the elevation of the water table. Similarly, modest degradation of water quality does not generally render groundwater unusable; rather, it increases the cost of treatment needed to restore the water to usable quality. Ultimately, depths to water and/or groundwater quality may decline to such an extent that further use of the aquifer becomes prohibitively expensive. Resource-centred groundwater management strategies are essentially an exercise in avoiding this eventuality.

Increasingly, eco-centred groundwater management strategies are also being developed and applied. Common examples of the latter include specifying maximum drawdown thresholds for given points in aquifers in order to ensure healthy water levels in adjoining wetlands or rivers (e.g. Burgess 2002). At present, more sophisticated eco-centric groundwater conservation strategies are restricted to the drawing-board. For instance, the concept that conservation of groundwater ecosystems *per se* might be adopted as a goal for aquifer management has scarcely been considered in most jurisdictions: the Government of Western Australia is one of the first regulatory bodies to formally publish guidelines on how to take account of subterranean fauna in groundwater and caves during the preparation of Environmental Impact Assessments (Environmental Protection Authority 2003).

Notwithstanding the scarcity of such guidance, there are grounds for concern that depletion of aquifers may already be seriously degrading as-yet undocumented ground water ecosystems. As noted in Section 6.4.3, although stygophile and stygobite invertebrates are now known from the full range of aquifer types, the distribution of species is highly endemic. The relative hydrological stability of many aquifers under purely natural conditions may mean that ground water ecosystems are especially sensitive to artificially induced changes in water table elevation. However, there is some evidence to suggest that degradation of groundwater *quality* is not necessarily bad news for groundwater ecosystems: given that primary production in many aquifers is limited by the lack of nutrients, the artificial introduction of carbon- and nitrogen-rich organic pollutants can give rise to a boom in microbial populations, with benefits further up the food chain. A number of studies have shown that the numbers and diversity of stygobites are greater in the vicinity of polluted groundwaters than in pristine aquifer waters.

Alternative concepts of groundwater management are discussed in greater detail in Chapter 11.

9.2 Depletion of groundwater quantity

9.2.1 Causes of groundwater depletion

There are two principal ways in which human activities threaten the quantity of groundwater. First and foremost is excessive abstraction, which is widely termed **aquifer overexploitation**. Second, and far more subtly, insofar as climate change induced by greenhouse gas emissions alters recharge rates to aquifers, it is capable of leading to depletion of groundwater resources.

9.2.2 Aquifer overexploitation

If humankind restricted itself to using only water which flows naturally from springs, overexploitation of aquifers would essentially never occur: overexploitation is basically the pumping of excessive quantities of groundwater from an aquifer. But what constitutes "excessive"? Various attempts to define "overexploitation" in formal, quantitative terms have been made in the past (e.g. Simmers et al. 1992). The current consensus amongst hydrogeologists is that "overexploitation" is probably not amenable to a single, precise definition (Adams and MacDonald 1998; Custodio 2002). For instance, earlier formulations which compared the total groundwater pumping rate in an aquifer to the pre-existing natural recharge rate are too simplistic, for they often fail to take account of the important changes in rates of groundwater recharge which pumping itself often induces (Andreu et al. 2001; Custodio 2002). As we have already seen (see Section 7.2.2), to equate the "safe yield" of an aquifer with its predevelopment recharge rate is always fallacious and often dangerously misleading (Theis 1940; Bredehoeft et al. 1982; Johnston 1989, 1997). For these reasons, Custodio (2002) explained that "an aquifer is often considered [to be] overexploited when some persistent negative results of aquifer development are [experienced] or perceived, such as continuous water-level drawdown,[i] progressive water-quality deterioration, increase in abstraction cost, or ecological damage. [However], negative results do not necessarily imply that abstraction is greater than recharge: they may be simply due to well interferences and the long transient period that follow[s] changes in the aquifer water balance." It follows from this that the degree to which an aquifer is regarded as being overexploited is very much in the eye of the beholder: negative results "will be perceived differently by the exploiter, an affected third party, the licensing authority and environmentalists. Thus overexploitation is a relative concept dependent upon the criteria used to define it: qualitative, economic, social, ecological, etc." (Adams and MacDonald 1998). Box 9.1 presents some contrasting examples of recognized cases of aquifer overexploitation. In every case there are extenuating socioeconomic circumstances: the

Box 9.1 Overexploited aquifers: good or bad? – You decide . . .

Spain – heavy pumping of aquifers to support irrigation

Spain is the most arid country in Europe. Beginning around 1930, Spanish farmers have considerably increased the use of groundwater for irrigation. The value of the crops and the employment generated by the use of groundwater irrigation is higher than that from surface water irrigation (Llamas 2003). Most of this agricultural development has been made with scarce planning and very limited control by public authorities: although Spanish water law can boast one of the only codes anywhere to explicitly legislate against aquifer overexploitation (Royal Decree 849/1986), two decades after the law was enacted as many as 89 distinct aquifer units in Spain are now displaying symptoms of overexploitation (Andreu et al. 2001). Nearly all of these are in the dry southeast of Spain, where heavy pumping to support the production of water-intensive succulent crops (tomatoes, fruits, vines) has more recently been compounded by irrigation of golf courses and supply of swimming pools in some of Europe's most popular holiday destinations. The major drought of summer 2005 has placed the overexploited aquifers in this area firmly under the spotlight. All of the common symptoms of overexploitation are manifest here (Andreu et al. 2001): plummeting water tables (total drawdowns to date of between 50 and 270 m in the various aquifers), diminution of flows in the Vinalopó River, drying up of springs (including those of Cabezón del Oro, which once fed a popular spa, now abandoned), dramatic shrinking of groundwater-fed wetlands (at Salinas and Villena), substantial increases in the costs of continued pumping as pumps have to be suspended at ever-greater depths (e.g. 600–700 m now in the Crevillente Aquifer), and, in places, groundwater quality degradation (e.g. conductivity increasing from 2000 to 8000 μS/cm in the Cid Aquifer, due to upcoming of deep saline waters). How long can all of this be allowed to go on before politicians (in this area of powerful agricultural interests) will be bold enough to pursue a public debate on the optimal use of finite groundwater resources?

The High Plains Aquifer – watering the breadbasket of the USA

Often erroneously referred to by the name of one of its constituent hydrostratigraphic units (the Ogallala Aquifer), the High Plains Aquifer is a contiguous assemblage of sandy alluvial aquifer units which together underlie some 480 km^2 of eight states (Wyoming, South Dakota, Nebraska, Colorado, Kansas, Oklahoma, New Mexico, and Texas). By nature a semi-arid area (with total annual rainfall typically <500 mm), since about 1930, exploitation of the High Plains Aquifer for irrigation supplies has turned this region into one of the world's most productive agricultural areas. By 1980, total pumping rates exceeded the available direct recharge by a factor of about ten; not surprisingly, dramatic water table declines resulted, reaching 40 m in Texas, Oklahoma, and Kansas between 1950 and 1980. Subsequently, some changes in rates of water table decline became evident: in Nebraska, the Platte River was found to be feeding the aquifer with indirect recharge, locally leading to a *rise* in the water table; in the more southerly areas, increases in the costs of pumping spurred interest in more efficient irrigation technologies which reduced water demand sufficiently that water table decline slowed down (typically only 10 m further drawdown from 1980 to 2005) (McGuire 2003). Dire predictions heard on all sides 20 years ago have, thankfully, not nearly been borne out. Nevertheless, abstraction rates remain heavily out of balance in the southern areas of the aquifer. Given the contribution that High Plains farming has made to the growth of the US economy, is this really a case of intolerable "groundwater mining"[i] of waters recharged in pluvial times, or simply realization of a long-term national annuity?

The "Great Man-Made River Project", Libya

The most brazen example of nonrenewable groundwater pumping in the world! In this major development project, "fossil" groundwaters (i.e. groundwaters which were recharged during a pluvial period many millennia ago) are being pumped from beneath the deserts of southern Libya, and transported hundreds of kilometers northwards by pipeline, to be used for irrigation and industry in the urbanized coastal region around Tripoli (Salem 1992). Although the source aquifer will eventually be exhausted, the negative side-effects are considered to be minimal: it's a virtually uninhabited area, with no population to care about subsidence, and no surface ecosystems connected to the aquifer. The Great Man-Made River Project is thus seen simply as a major "once-off" development opportunity for Libya. Technically, it's over-exploitation; but is it a reprehensible or a responsible development?

acceptability of these instances of overuse thus becomes a question which must be answered by society as a whole.

As the data presented in Table 7.1 make clear, the predominant use of pumped groundwater worldwide is for agriculture, principally irrigation. The same can be said of most cases of aquifer overexploitation. In fact the advent of overexploited aquifers entirely mirrored the spread of large-scale irrigation during the second half of the twentieth century. As Llamas (2003) explains: "Intensive groundwater development is a [comparatively] recent development in most arid and semiarid countries. Usually, it is less than 30–40 years old. Three technological advances have facilitated this: (1) [the invention of deep-well] turbine pumps; (2) [the development of] cheap and efficient drilling methods, and (3) [numerous advances in] hydrogeology. [The] full costs [capital and revenue] of groundwater abstraction are usually low in comparison to the direct benefits obtained". The first two cases in Box 9.1 fully bear this out: irrigation is the principal reason for overabstraction, but the decision to irrigate has deep socioeconomic roots.

9.2.3 Climate change

Changes in weather and climate: an eternal reality

Climate change is a topic that has generated "as much heat as light" in many debates over the last two decades. All geologists are well aware of the falsity of the layperson's assumption that our planet has ever had a stable climate: if anything, the period of geological time in which *Homo sapiens* finally emerged (the Quaternary) has been characterized by more rapid and extreme fluctuations in climate than most previous eras. Nevertheless, the notion that human activities are now instrumental in promoting rapid climate change in ways that are damaging to ecosystems and human wellbeing is deeply disturbing.

Even so, consensus has not been attained on many key points concerning the processes, degree, and likely consequences of human-induced climate change. Part of the confusion arises from the failure of many people to appreciate the difference between climate and weather. **Weather** describes whatever is happening in a given place at a given time with regard to precipitation, temperature, wind conditions, and barometric pressure. As such, weather can change a lot within a very short time, which is why it is the subject of frequent reporting in the mass media. By contrast, **climate** describes the totality of all types of weather experienced in a given place over a period of years. (How long a period is a matter of debate.) Thus a description of the climate of a specific locality summarizes average weather conditions, regular weather sequences (such as those relating to the annual passing of the seasons), and special weather events (such as hurricanes, tornadoes, and extreme precipitation events). Failure to discriminate between weather and climate is at the root of the tendency for every extreme weather event to be greeted by the broadcast media with cries of "climate change." As long as the confusion of weather with climate persists, it will never be possible to move forwards to a more informed debate on climate change and its impacts. After all, every one us experiences weather; as long as we cling to the misapprehension that weather is synonymous with climate, we can all feel justified in regarding ourselves as self-appointed experts on climate change!

Impacts of ancient climate change on aquifers

Now, there is no doubt that climatic change can deplete aquifers. For instance, during the last glacial period, global sea levels were about 100 m lower than at present, and the water table in many regional aquifers was consequently depressed many tens of meters below present levels. Not only that, at latitudes above about 40 degrees, much of the groundwater in those aquifers froze to become permafrost (e.g. Younger 1989; Hiscock and Lloyd 1992). As in modern-day permafrozen aquifers (Williams 1970; Sloan and van Everdingen 1988), some unfrozen groundwater remained, much of it trapped below the base of the permafrost zone (Younger 1989). This water was often fairly saline, and much of it remains trapped at depth in modern aquifers to this day (e.g. Hiscock and Lloyd 1992; Elliot et al. 2001). Only with the warming of the climate at the end of the last glacial period did our modern-day aquifers begin to assume their present piezometric conditions, though this process was not completed until around 5600 years ago (e.g. Younger and McHugh 1995), when sea levels had finally reached their (approximately) modern level. The moral of the story is clear: dramatic climate changes, and dramatic responses on the part of aquifers, have long been the rule, many millennia before humans began systematically burning fossil fuels. What is also clear is that major changes in the base level of drainage (sea level in the above instance) can have more profound affects on groundwater levels than changes in recharge rates alone.

Carbon dioxide emissions and the greenhouse effect

There is no doubt that atmospheric concentrations of carbon dioxide (CO_2) have increased dramatically since the onset of large-scale burning of fossil fuels in the industrial revolution. Nevertheless, absolute atmospheric concentrations of CO_2 (~0.0375%) remain tiny compared with those of oxygen (~21%), nitrogen (~78%), and water vapor (≤4%, depending on circumstances). Perhaps more surprisingly, the CO_2 complement is even dwarfed by that of argon (0.93%).

Greenhouse gases are those which tend to increase the ability of the atmosphere to retain a blanket of heat close to the Earth's surface: without them, the Earth would be as frigid as the Moon, and there would be no higher life forms on our planet. The most powerful greenhouse gas is water vapor, followed by methane (typical atmospheric concentration 0.00017%), with CO_2 in third place. The hydrological cycle ensures that the water vapor content of the atmosphere has remained in dynamic equilibrium over recent millennia; methane concentrations are also controlled by a range of fairly brisk biogeochemical processes. However, possible natural controls on the upper limit of atmospheric CO_2 are sluggish in comparison, which explains the steady increase in its atmospheric concentration due to fossil fuel use over the last two centuries. Given its known "greenhouse" behavior, it is logical to suppose that this rise in CO_2 concentrations will have resulted in an increase in average air temperatures. Collations of data from around the globe do indeed show an increase of about 0.6°C (±0.2°C) over the last century, with a rise of between 0.2 and 0.3°C over the past 25 years (the period for which data are most reliable).[ii] The general increase in temperature has occurred despite local anomalies of cooling over the same periods of time.[iii] Further complications in the interpretation of global warming arise from the so-called **urban heat island** effect: because of the disposal of waste heat from multiple sources, urban areas tend to be rather warmer than rural areas in the same climatic zone. Although various means have been devised to adjust for this source of bias in long instrumental records of air temperature from weather stations in areas which have gradually urbanized over the years, the veracity of the adjustments remains a matter of scientific debate (McKendry 2003). Notwithstanding these complications and localized exceptions, the international scientific consensus is that artificially induced climate change is now upon us (Joint Science Academies 2005).

Hydrological consequences of human-induced climate change

Global warming can be confidently expected to give rise to changes in the atmospheric moisture regime, in terms of both evapotranspiration and precipitation (Allen and Ingram 2002). Given that these two variables are the principal controls on recharge (see Chapter 2), it is obvious that climate change is likely to result in changes in available water resources (e.g. Bouraoui et al. 1999). Assuming that the latest prognoses of forthcoming changes in precipitation and evapotranspiration are correct (Allen and Ingram 2002), then serious consideration will need to be given to ways of meeting the overall water supply–demand balance in future. While improvements in water conservation and demand management practices will have a considerable contribution to make, provision of further storage is virtually certain to become increasingly important in decades to come. With surface reservoirs becoming increasingly contentious in many parts of the world, groundwater storage is likely to assume ever greater importance in overall water resources management strategies (e.g. Price, 1998). Notwithstanding this likelihood, very few assessments of the water resources implications of climate change have yet considered groundwater explicitly.

Predicting climate change: General Circulation Models (GCMs)

The starting point for predicting the possible response of groundwater systems to climatic changes is to define future climate attributes, in terms of temperature, precipitation, and other variables. In the very earliest days of climate change impact assessment, it used to be crudely assumed that we could simply take present-day weather patterns and transpose them a certain number of degrees of latitude. This approach ignores crucial factors such as the decrease in both day length and the angle of incidence of solar radiation with increasing latitude. The effects of oceanic circulation patterns are also neglected if one, say, assumes the present-day climate of Marseille (on the relatively motionless Mediterranean Sea) is a good guide to the future climate of London (which receives much of its weather from the broad and dynamic Atlantic). In view of these difficulties, it is now accepted that the only defensible approach to climate change impact assessment is to derive future climate attributes for a given location from mathematical models which simulate global atmospheric and oceanic circulation. These so-called AOGCMs (Atmosphere–Ocean General Circulation Models) are a rapidly evolving suite of sophisticated computer codes which represent the entire global atmosphere as a patchwork (grid) of interconnected "cells" between which masses of air (and their contained heat and moisture) migrate in accordance with the laws of physics. The typical grid spacings (i.e. cell sizes) used in AOGCMs are typically squares of about 1° × 1°, amounting to about 250 km × 250 km near the equator, with occasional finer-scale grids using cells down to 40 km × 40 km. While the first few generations of General Circulation Models (GCMs) either ignored exchanges of heat and moisture with the oceans or represented them very simplistically, the present generation of AOGCMs handle these dynamics much more realistically (Allen and Ingram 2002). Because AOGCMs are being continually refined, climate change impact studies based upon them tend to have rather short shelf-lives. Application of AOGCM output to drive groundwater models is bedevilled by the difference in scale between AOGCM cells and the much smaller cells (rarely larger than 1 km × 1 km, and often as small as 100 m × 100 m) which are typically used in groundwater models (see Chapter 10). Full details of how the necessary transfer of AOGCM output to become groundwater model input is made are beyond the scope of this text; for an introductory review, the reader is referred to Younger et al. (2002b).

Implications of AOGCM predictions for groundwater systems

Mirroring the gradual evolution of GCMs and AOGCMs, investigations of the impacts of predicted climate change on aquifers have gradually proceeded from coarse-scale to fine-scale. For instance, an early attempt was made to predict possible global-scale changes in groundwater dynamics in response to global warming (Zekster and Loaiciga 1993); this scale is of no use to managers of real wellfields, of course. Subsequently, a number of studies treat entire aquifers as single "buckets," so that changes in recharge derived from GCM output could be considered the only process affecting changes in aquifer water levels and outflows. The inner workings of these simple input–output models vary from explicit calculations of water balances (e.g. Vaccaro 1992; Cole et al. 1994; Sandstrom 1995; Bouraoui et al.

1999) to statistically based regression models, in which correlations between *observed* monthly rainfall totals and minimum annual groundwater levels for a given period of time are used to develop simple predictive tools, which allow future annual groundwater-level minima to be modeled when they are fed with synthetic rainfall data based on climate-change scenarios (Bloomfield et al. 2003).

Such simple predictive approaches might suffice as long as the investigator is only concerned with long-term average volumes of groundwater discharge or simple low-stands of the water table; this might be so, for instance, in a study of the overall contribution of groundwater discharge to total surface runoff in a catchment. However, where interannual or seasonal variations in groundwater levels and discharge rates are of interest (which they nearly always are in water resource evaluations), the internal dynamics of aquifers must be taken into account with a greater degree of realism. Essentially, groundwater flow and storage processes effectively "smear" discrete, incoming parcels of infiltration to produce a relatively smooth, continuous aquifer discharge. Groundwater storage also results in aquifers having much longer residence times than river systems. In some cases the impacts of a given period of extreme rainfall (low or high) on aquifer responses may persist for several years; the groundwater floods of 2000/01 in Belgium, France, and England are a case in point (see Section 8.2.4). Hence aquifer responses to climate change can be expected to show a considerable temporal lag, which will not be picked up by simply equating total recharge to total discharge. Representing the internal dynamics of aquifers requires the use of numerical models, in which the entire aquifer is subdivided into hundreds of small cells (just as AOGCMs represent the world's atmosphere), each of which may be assigned different values of transmissivity, storativity and other parameters, if so desired. Chapter 10 explains how such models are developed and applied.

In broad terms, the concerns over groundwater susceptibility to global warming impacts revolve around resolving whether: (i) rainfall will increase or decrease; and (ii) evapotranspiration (ET) will increase. A number of scenarios can be easily identified in theory. For instance, rainfall might increase, but a concomitant increase in ET might compensate for the increment in rainfall, so that there is no net increase in recharge. Alternatively, rainfall might decrease as ET increases – certainly a "nightmare" scenario for managers of heavily used aquifers. On the bright side, evidence has begun to emerge that the increase in ET which is anticipated as a result of warmer air temperatures may be completely counteracted by changes in plant physiology likely to be induced by the increase in atmospheric CO_2, which will actually reduce the rate of release of water from leaf stomata (Eckhardt and Ulbrich 2003). Given such complications and the inherent uncertainties in the coupling between carbon emissions and atmospheric responses, the development of credible prognoses for real aquifer systems is a daunting challenge. Ideally, such prognoses should be formulated strictly in terms of probabilities (Allen and Ingram 2002). Unfortunately, thorough probabilistic models of coupled atmospheric/aquifer systems are both difficult to formulate and extremely demanding in terms of computing power. Not surprisingly, therefore, the few instances of the application of numerical groundwater models in climate change impact assessments to date have adopted methodologies which fall somewhat short of the fully probabilistic ideal.

Table 9.1 summarizes some published examples of climate change impact predictions for aquifers which have been derived using numerical models. These give some flavor of the range of predictions emerging from these types of investigations. Given that precipitation is (currently) predicted to increase in some high-latitude regions while it decreases in mid-latitude regions, the outcomes of the various predictive exercises are rather variable. In northern England, for instance, an *increase* in groundwater resources is anticipated. However, the predicted decreases in available groundwater resources in Mediterranean countries and in the southern USA (Table 9.1) are particularly alarming, in view of the present

Table 9.1 Numerical modeling predictions of the potential impacts of climate change on selected aquifers in the first half of the twenty-first century.

Country	Location	Type of aquifer	Predicted impacts	Source
India	Laccadive Islands	Recent coral limestones	Decline in available freshwater lens from 25 m to only 10 m due to sea level rise exacerbating saline intrusion	Bobba et al. 2000
Spain	Anoia (Catalunya)	Fractured limestones (little karstification)	Although average reduction in recharge not great (≤8%), greater interannual variability expected, leading to increased frequency of low flows from major springs	Younger et al. 2002b
Egypt	Nile Delta	Sand and gravel (multilayer)	Anticipated increase in groundwater pumping (0.5 m extra drawdown) could cause saline interface to advance ≤ 11.5 km further inland; exacerbated by sea level rise	Sherif and Singh 1999
Canada	Grand Forks (British Columbia)	Floodplain sand and gravel aquifer	Maximal and minimal variations in recharge will have little direct effect on water levels (fluctuations ≤ 0.05 m above or below present levels); increased flood levels in adjoining river will affect water table more profoundly (fluctuations ≤ 3.45 m above or below present levels)	Allen et al. 2004
England	East Yorkshire	Chalk*	Year-round *increases* in flow are likely (e.g. +9% change in total annual average flow)	Younger, et al. 2002b
Belgium	Geer Valley	Chalk*	No real change from present seasonal groundwater levels/flow rates	Brouyere et al. 2004
Spain	Sa Costera (Island of Mallorca)	Karstified limestones	Decline in recharge and average discharge of main spring ≤ 16%, with worsening trend to 2045 AD. Greater interannual variability	Younger et al. 2002b
India	Channai (Madras)	Sand and gravel (multilayer)	50 cm rise in sea level, coupled with heavier groundwater pumping with reduced recharge, will cause saline interface to intrude 0.4 km further inland	Sherif and Singh 1999
USA	Texas (Edwards Aquifer)	Karstified limestones	Without controls on future abstractions, aquifer likely to display symptoms of overexploitation under warmer climate conditions	Loaiciga et al. 2000
England	East Anglia	Chalk*	17–35% reduction in recharge rates expected to have little impact on groundwater levels in summer (≤2% decrease), but substantial decrease (≤14%) in autumn baseflows in connected rivers	Cooper et al. 1995; Yusoff et al. 2002

Table 9.1 (*Continued*)

Country	Location	Type of aquifer	Predicted impacts	Source
Germany	Swabian Alb	Karstified limestones	Increases in January spring flows, but decreases in all other seasons; interannual variability very high (exceeds trend in average recharge decline to 2045 AD)	Younger et al. 2002b
Scotland	St Fergus	Aeolian sands	No real change in hydrodynamics of the wetland fed by the aquifer	Malcolm and Soulsby 2000
England	Midlands	Indurated sandstones	Only modest declines in baseflow to rivers likely due to high aquifer specific yield	Wilkinson and Cooper 1993; Cooper et al. 1995

* Chalk is a fractured limestone with a highly porous rock mass; however, the pores in the blocks between fractures are very small and do not drain readily.

incidence of overexploitation in those regions (Box 9.1; Zeckster et al. 2005).

9.2.4 Aquifer depletion/overexploitation: symptoms and susceptibility

As we saw in Section 7.2.4, pumping of all but the most minimal quantities of water from an aquifer will always have *some* potentially undesirable side-effects. The full range of symptoms is:

1 Declining water levels, which in turn imply:
   (a) decreasing natural outflows to wetlands or rivers (see Chapter 5), which may well have ecological implications (see Chapter 6);
   (b) an increased risk of land subsidence (see Section 8.3.3);
   (c) an increase in the costs of both infrastructure (i.e. deeper wells with more powerful pumps) and revenue (electricity charges) to sustain the same rate of pumping.
2 Degradation in water quality, due to sea water intrusion, or to induced "up-coning" of poor quality waters from depth (Section 9.3.4).

Where aquifers are overexploited, or their available resources become depleted as a result of climate change effects, all of these symptoms may occur (e.g. Adams and MacDonald 1998; Andreu et al. 2001; Custodio 2002). Box 9.1 illustrates the manifestation of these various symptoms in a number of aquifers in Spain and the USA; several other case studies from the southwestern USA are presented by Zeckster et al. (2005).

Fortunately, the full range of symptoms are rarely manifest in any one aquifer. Rather, the particular combination of symptoms which can be expected in any one aquifer depends critically on its hydrogeological characteristics. As an aid to decision-making, Adams and MacDonald (1998) proposed a methodology for identifying the relative susceptibility of aquifers to displaying the symptoms listed above. **Aquifer susceptibility**[iv] can be defined as the likelihood that a given aquifer will develop declining water levels, ecologically damaging decreases in outflows, land subsidence and/or groundwater quality degradation as a consequence of specified levels of abstraction. In a pilot implementation of this concept, Adams and MacDonald (1998) erected a provisional scoring system for aquifers in terms of their susceptibility to water level decline, saline intrusion, and subsidence. The absolute values of scores yielded by this system are subjective, insofar as they do not correlate directly with measurable physical properties. However, they do provide a means of summarizing large volumes of both quantitative and qualitative information, thus facilitating comparisons between different aquifers. Thus aquifers with little susceptibility to the side-effects of exploitation will score less than 10, aquifers with intermediate susceptibility will score 11–20, and highly susceptible aquifers will score in excess of 20. For full details of this aquifer susceptibility index, the reader is referred to Adams and MacDonald (1998).

9.3 Degradation of groundwater quality

9.3.1 Causes of groundwater quality degradation

A bewildering array of human activities can degrade the quality of natural groundwaters (Figure 9.1). It is possible to categorize groundwater pollution sources in a number of ways. For instance the US Congress' Office of Technology Assessment developed a complex classification system for groundwater pollution sources, based on categorization of the original intended purposes of the various types of structures and activities which later caused groundwater pollution (Fetter 1999). While a focus on "intent" may seem logical enough to the legal mind, the resulting classification is unwieldy in practice. After all, an aquifer neither knows nor cares about the motives of the humans who polluted it. Furthermore, engineers designing a clean-up operation are more likely to be successful if they focus on the processes by which the pollutants passed through the soil and into the aquifer, rather than on the motives of the former land-user. For these reasons, a much simpler classification of groundwater pollution sources will be adopted here. We will classify all sources of pollution into two categories:

- **Point sources of pollution:** these are pollutant sources which are potentially identifiable as individual locations on the Earth's surface, such as the area onto which a toxic liquid was spilled, a leaking chemical storage tank, or a similar feature with a modest footprint (Figure 9.1). Further consideration of point sources follows in Section 9.3.3.
- **Diffuse sources of pollution:** these are typified as extensive areas within which a given type of pollutant solute might have entered the ground

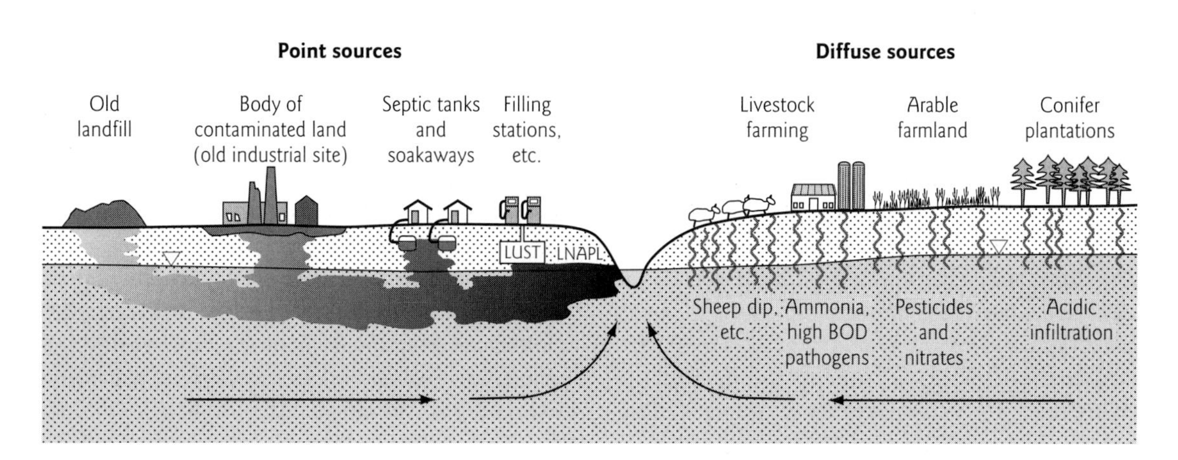

Fig. 9.1 Schematic cross-section through a particularly ill-fortuned (thankfully fictitious) aquifer, showing many of the major point and diffuse sources of groundwater pollution. (LUST stands for "leaking underground storage tanks," and LNAPL refers to "light non-aqueous phase liquids," which effectively float on the water table (see Section 9.3.3). BOD is "biochemical oxygen demand," a measure of the concentration of pollutant compounds in a given water which are biodegradable under aerobic conditions.)

surface at any one of millions of points. The classic examples of this genre are agricultural contaminants, such as pesticides and fertilizers (Figure 9.1), which are usually spread widely over the fields of many adjoining farms. Further consideration of diffuse sources follows in Section 9.3.4.

A third source of groundwater quality degradation exists, which does not fit neatly into a classification of "pollutant sources": these are natural bodies of poor quality water which are induced by heavy pumping to enter aquifers which previously contained good quality groundwater. Examples include sea water (see Figure 7.2, Section 7.2.4), deep saline groundwaters, and naturally toxic groundwaters (see Section 8.2.5). These and other sources of poor-quality groundwater whose mobility is aggravated by aquifer exploitation will be considered in some detail in Section 9.3.5. Before we go on to consider particular sources of groundwater pollution, it is necessary to consider the hydrogeological factors which determine the ease or difficulty with which any pollutant might enter a given aquifer.

9.3.2 Aquifer vulnerability

Groundwaters tend to be inherently less prone to contamination than surface waters, thanks to the protection afforded by filtration and biogeochemical reactions which occur during subsurface flow (see Section 7.2.3). Recognition of the potency of these natural purification processes also provides the rationale for many "bank filtration" wellfields currently operative worldwide (Section 7.4.2; Tufenkji et al. 2002). For all the admirable purifying powers of natural rocks, there are limits to the patience of Mother Earth: expose the ground to a sufficient quantity of pollutants for a sufficiently long period of time, and you will eventually overwhelm the assimilative capacity of the soil, subsoil, and aquifers. Pollution will ensue, with all of its attendant impacts on the economics of water supply and the health of ecosystems which receive discharging groundwaters.

The readiness with which a given aquifer (or portion thereof) is likely to succumb to pollution is termed its **vulnerability**. The term "aquifer vulnerability" was originally coined in France in the 1960s, and has since been taken up worldwide, with its meaning gradually evolving as it has passed from one group of workers to another (Vrba and Zaporozec 1994). Figure 9.2 graphically illustrates the difference between low- and high-vulnerability aquifer settings. The following characteristics contribute to the low vulnerability of the aquifer shown in Figure 9.2a:

- The water table lies at great depth (so that filtration and aerobic biochemical processes which occur in the unsaturated zone have plenty of scope to work before infiltrating water enters the saturated zone).
- A thick layer of clay-rich materials lies between the aquifer and the ground surface (providing a barrier both to water penetration and to pollutant entry).
- The aquifer is granular in nature, which means that it is lacking in highly permeable fissures (which can provide pollutants with short-circuit pathways to the saturated zone) and that it has a high specific yield (so that, for every cubic meter of aquifer, there is a large volume of groundwater available to dilute incoming pollutants).

By contrast, the highly vulnerable aquifer (Figure 9.2b) has a shallow water table, has no clay-rich protecting layer, and is a fractured hard rock aquifer, typified by a combination of high permeability and very low specific yield.

The sorts of features illustrated in Figure 9.2 will affect the vulnerability of the aquifer to *any* type of groundwater pollutant. For this reason, vulnerability designated on these strictly hydrogeological grounds, without reference to the properties of individual contaminants, is sometimes referred to as **intrinsic vulnerability** (Vrba and Zaporozec 1994). In some circumstances (e.g. in areas with intensive arable agriculture) it may be appropriate to consider specific geochemical properties of pollutants of major concern, such as chlorinated pesticides, as well as soil/aquifer properties which specifically affect the mobility

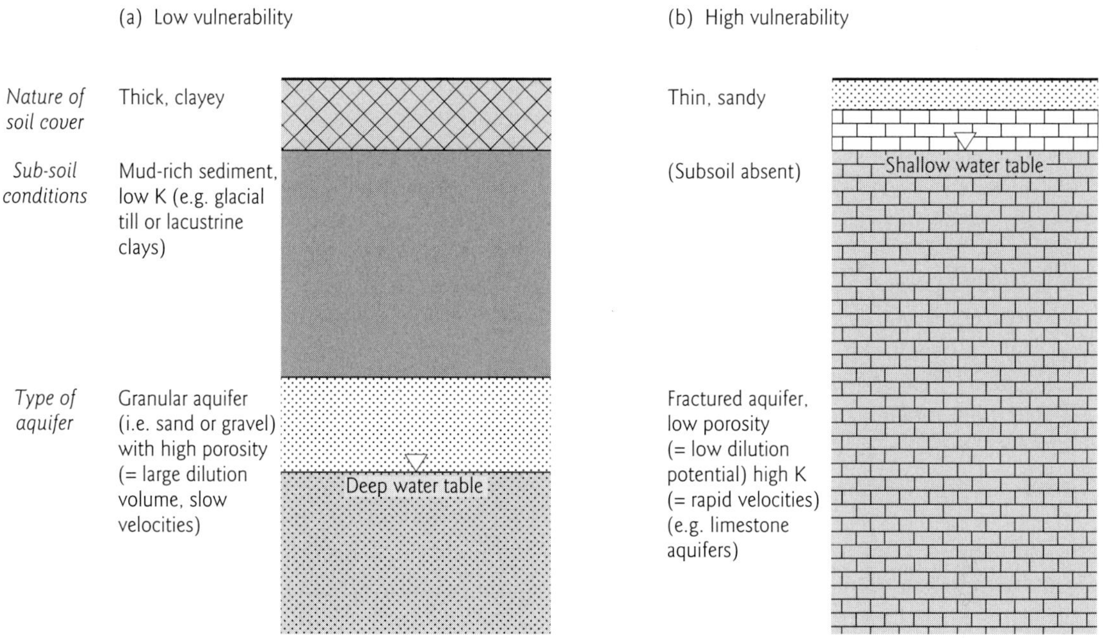

(a) Low vulnerability

Nature of soil cover Thick, clayey

Sub-soil conditions Mud-rich sediment, low K (e.g. glacial till or lacustrine clays)

Type of aquifer Granular aquifer (i.e. sand or gravel) with high porosity (= large dilution volume, slow velocities)

Deep water table

(b) High vulnerability

Thin, sandy

(Subsoil absent)

Shallow water table

Fractured aquifer, low porosity (= low dilution potential) high K (= rapid velocities) (e.g. limestone aquifers)

Fig. 9.2 Conceptual sketches illustrating the concept of aquifer vulnerability. In the low vulnerability setting (a), the soil cover is thick and clayey, which impedes infiltration of polluted waters. Those waters which do infiltrate are subject to long retention times in the thick unsaturated zone, in which aerobic degradation can destroy many pollutants. Finally, when it reaches the water table, it is in a high porosity aquifer offering a large dilution ratio per unit volume of rock. By contrast, in the high vulnerability setting (b), pollutants can infiltrate easily through the thin, sandy topsoil, there is no subsoil to hinder downward flow (hence pollutants quickly reach the water table) and the low porosity of the aquifer offers little dilution per unit volume of rock. (Adapted after NRA 1992.)

of these pollutants (e.g. sedimentary organic matter content, sorption to which can powerfully affect the mobility of pesticides). Vulnerability designated with respect to particular pollutants is termed **specific vulnerability**.

The determination of vulnerability has in practice been approached differently in many different countries (Vrba and Zaporozec 1994). One school of thought argues for the use of formal "point-scoring" systems, where, for every square kilometer (say) of an area underlain by an aquifer, factors such as "depth to water table" and "thickness of clay above aquifer" are assigned numerical rankings, with the sum total of rankings for all factors being used to index overall vulnerability. This is the logic behind two vulnerability assessment methodologies with particularly resonant acronyms:

GOD (**G**roundwater occurrence, **O**verall lithology of aquifer and **D**epth to water; Foster 1998), and

DRASTIC (**D**epth to groundwater, **R**echarge rates, **A**quifer media, **S**oil media, **T**opography, **I**mpact of unsaturated zone, and **C**onductivity (hydraulic) of the saturated zone; Aller et al. 1987).

A key criticism of both GOD and DRASTIC is that very different hydrogeological settings can yield very similar index values. In the case of DRASTIC, at least, this is arguably due (in part) to the way in which the methodology separately accounts for certain factors which are usually correlated. There is also a risk that assigning simple numerical values to a given area can lead to the unhelpful hiding of the basic hydrogeological data upon which vulnerability assessments ultimately depend (Foster 1998). For this reason, many of the more recent vulnerability assessment methodologies have focused on the thickness and lithology of the strata between the water table (or the upper surface of a confined aquifer) and the ground surface, with especial care being taken to account for the likelihood of intense fracturing of the strata (e.g. Misstear and Daly 2000; Lewis et al. 2000).

It should be noted that aquifer vulnerability assessment tends not to be carried out for its own sake, but rather as an important step in the development of groundwater protection strategies (see Section 11.4).

9.3.3 Point sources of aquifer pollution

The hallmark of point sources is that they give rise to more or less distinct "plumes" of contaminated groundwater within aquifers, which spread outwards from the point of origin in accordance with groundwater flow patterns (Figure 9.3a). There are two main kinds of point sources of groundwater pollution: those that are engineered specifically to discharge water (which may be polluted) into the subsurface, and those that "accidentally"[v] release pollutants to the subsurface.

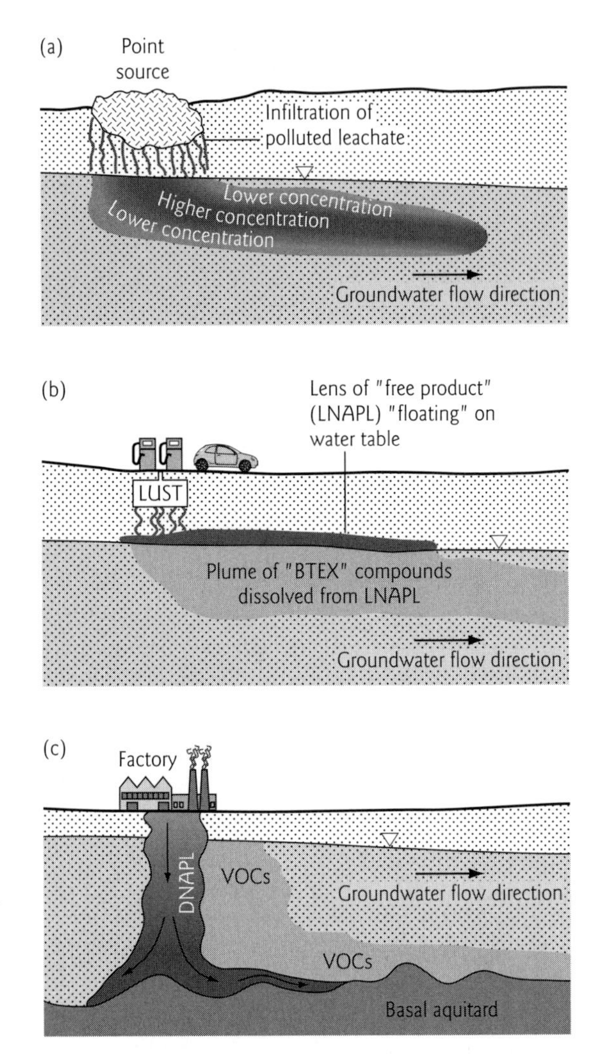

Fig. 9.3 Patterns of groundwater pollution associated with various point sources. (a) A typical plume, such as from downstream of soakaways and landfills. (b) LNAPL pollution downstream of a LUST, with "floating" free product on the water table and dissolution of BTEX compounds into the underlying saturated zone. (c) DNAPL sinking through an aquifer to form "pools" on top of the basal aquitard, whence VOCs dissolve into the water column. (Abbreviations are all defined in Section 9.3.3 of the text.)

Soakaways are the most widespread type of structure deliberately intended to discharge water into the subsurface. They typically take the form of 1–3 m deep chambers, usually wide enough to allow occasional access for maintenance. Modern soakaways tend to be empty chambers held open by perforated linings made of plastic or pre-cast concrete. Many older soakaways are not fitted with linings, but rather are back-filled with bricks and lumps of concrete, which provide support for the sides of the excavation while retaining lots of open pore space. Disposal of roof drainage to soakaways is rarely problematical from a groundwater quality perspective. It is no longer acceptable practice to dispose of road/parking lot runoff to soakaways unless the drainage line includes (as a minimum) an oil–water separator tank. However, it is still common in areas beyond the reach of municipal sewerage networks for soakaways to receive septic tank effluents, which are certainly charged with potential pollutants. Provided the density of septic tank soakaways is not very high (for instance, fewer than 10 single-property units per km^2), dilution in natural groundwater will usually be sufficient to ensure that gross pollution problems do not develop. Occasionally, however, a single-property septic tank will cause direct pollution of a nearby well; with adequate foresight at the planning stage, such problems can almost always be avoided.

Landfills are, essentially, holes in the ground filled with wastes. Over time, water that was in the waste at the time of burial, plus any infiltrating rainwater, together form "leachate," which typically carries high loadings of carbon, nitrogen, and sulfur pollutant compounds (sometimes augmented by metallic or organic contaminants depending on the type of waste present). Leachate naturally seeps downwards under gravity, and can leave the base of the landfill unless it is prevented from doing so by an impermeable liner. As such, it is not surprising that landfills have long been notorious as point sources of groundwater pollution. Fifty years ago, landfills all over the world were constructed without any impermeable liner, on the "**dilute-and-disperse**" principle: this held that the dilution capacity of natural aquifers was likely sufficient to disperse polluted leachates emanating from landfills without any problems. "Dilute-and-disperse" became a discredited philosophy during the 1970s and 1980s, following a string of infamous cases of groundwater pollution caused by leachates emanating from unlined landfills. A substantial legacy of existing groundwater pollution associated with long-closed unlined landfills is keeping contaminant hydrogeologists busy worldwide.

While unlined landfills are still in use in many less-developed countries, most recently developed landfills in North America and Europe are now fully lined, with elaborate leachate containment, collection, and treatment systems. These major improvements in engineering practice have certainly reduced the risk of groundwater pollution from landfills. For instance, the Wyoming Department of Environmental Quality recently reported that while almost 40% of landfills in the state were causing detectable groundwater pollution, not one of the polluting landfills had been constructed with engineered leachate containment systems. Elsewhere, however, fully lined landfills have been found to be leaking polluted leachate to aquifers. This has led regulators to conclude that the only way to ensure that newly constructed landfills will not eventually cause groundwater pollution is to ensure they are sited wisely, in locations where the risk posed to aquifers in the event of a leak would in any case be low. Building on the concept of aquifer vulnerability (Section 9.3.2), formal risk assessment techniques have been developed to assist in decision-making over the siting of future landfills (e.g. Leeson et al. 2003).

The arresting acronym **LUST** stands for **Leaking Underground Storage Tanks**, which rose to prominence as point sources of pollution in the 1980s, when many of the filling station petroleum-product tanks installed in the 1950s and 1960s began to reach the end of their design lives, finally succumbing to corrosion and beginning to leak. By the end of the twentieth century, some 370,000 LUST sites were being monitored in the USA alone, and the US EPA was committing some $70M per annum to clean-up of some of the most badly affected sites. It is well known that oil floats on water; in fact the beautiful display of Newton's Colours as a thin film of oil spreads out on the surface of a rain puddle can enliven an otherwise dreary urban day! In the subsurface, gasoline, oil, and similar **light non-aqueous phase liquids (LNAPLs)** will also "float" above the water table (Figure 9.3b) as a layer of **free product**. At the free product/water interface, various organic compounds dissolve from the LNAPL, migrating onwards with the bulk flow of groundwater (Fetter 1999). Because of these modes of pollutant migration, a very small leak of LNAPL can pollute a large amount of groundwater. Besides the fire hazards and oppressive heavy vapors associated with free product, many of the compounds which dissolve into the groundwater are hazardous to health. Most concerns relate to the so-called **BTEX** compounds (i.e. benzene, toluene, ethylbenzene, and xylene). As Table 7.4 shows, drinking water limits for all BTEX compounds are low – only 10 μg/L in the case of benzene,[vi] which is the most toxic of them all. **MTBE**[vii] (methyl tertiary butyl ether) is another petroleum-associated contaminant which is problematical at very low concentrations, causing taste and odor problems at concentrations in excess of only 20 μg/L. Now suspected to be a carcinogen, MTBE is highly soluble and is about ten times more mobile in groundwaters than the BTEX compounds.

Dense nonaqueous phase liquids (**DNAPLs**) are another category of manmade organic compounds commonly originating as point sources of pollution (Fetter 1999). The most common DNAPLs are chlorinated hydrocarbons, such as perchloroethene[viii] (PCE) and trichloroethene (TCE), which are widely used as solvents (and are for this reason sometimes referred to as **chlorinated solvents**). Spills of these liquids have occurred over the years at many industrial premises. Because DNAPLs are denser than water, upon entering an aquifer they sink through the saturated zone and accumulate in "pools" on top of the underlying aquitard (Figure 9.3c). The DNAPLs will flow downslope until they occupy depressions at the base of the aquifer, and there they will sit for centuries or even millennia, slowing releasing a range of slightly more soluble **volatile organic compounds (VOCs)**[ix] into the overlying groundwater. The VOCs released into groundwater from DNAPL pools are chlorinated hydrocarbons, including dichloroethane, dichlorethene, dichloromethane, chloroform, and vinyl chloride, all of which are toxic and/or carcinogenic. These chlorinated hydrocarbons have been reported to be the most commonly detected organic contaminants in water supply wells in the USA, and the same is likely true in much of the world. Failure to realize that many of the chlorinated hydrocarbons found in groundwater actually originate from *in situ* degradation of DNAPL pools within aquifers (rather than from direct inputs from surface) long misled many investigators. Because DNAPL pools can be very large and the rates at which they release VOCs to water are very slow, many such aquifer pollution problems are likely to persist for centuries or millennia in the absence of measures to remove/encapsulate the DNAPL pools (Mackay 1998).

Many other historical industrial processes have resulted in extensive contamination of soils, which remain in place long after the demise of the responsible industry, slowly releasing contaminant to any underlying aquifer. Such bodies of **contaminated land** are an important, if highly diverse, category of long-term point sources of groundwater pollution (e.g. Lerner and Walton 1998). A bewildering range of contaminants can be released from these sites, depending on their former industrial use. Amongst the organic contaminants the following compounds are frequently found as follows:

- NAPLs are very common on a wide range of sites, given the ubiquitous use of automotives.
- Chlorinated hydrocarbons are common on sites of former mechanical workshops, laundries, clothing manufacturers, and other operations which used solvents/de-greasing agents.
- A range of coal-tar byproducts such as **polycyclic aromatic hydrocarbons (PAH)** (e.g. benzo[*a*]pyrene, naphthalene) are common at former sites of town gasworks and cokeworks.
- A particularly degradation-resistant category of compounds known as **polychlorinated biphenyls (PCBs)** is often associated with electrical components; PCBs are not very mobile in groundwaters due to their great affinity for sorption sites.
- **Tributyl tin oxide (TBTO)** was long an active ingredient of barnacle-resistant boat paints, and as such it is common in sediments of former boatyards and docks.

Inorganic contaminants associated with contaminated land include a wide range of potentially (eco)toxic metals, such as Cd, Zn, Pb, Ni, Cr, etc., plus the metalloid As. These are found in many situations, but are especially prominent in leachates from some former mine waste heaps and metalworking sites, as well as (in the case of Ni, Cd, and Pb) any operations which used large quantities of DC power. Cyanide is a common contaminant on old gasworks and cokeworks sites. Extremes of pH are associated with leachates draining from mine wastes (*p*H often <4) and old steelworks slags (*p*H 9 to ≤14).

9.3.4 Diffuse sources of aquifer pollution: land-use impacts

The distinction between point and diffuse sources can be fuzzy at times; for instance, a large number of septic tanks in a small area will give rise to so many overlapping plumes that the overall pollution effect on the aquifer will be pervasive and difficult to trace to individual sources. For this reason, the literature sometimes even refers to LUST and landfills as contributors to diffuse pollution. However, the unequivocal diffuse sources of groundwater pollution relate to widespread agricultural land-use practices, most notably pest control and soil fertilization.

A wide range of organosulfur, organophosphorous, and organochlorine compounds effective as **pesticides**[x] have been used in arable agriculture (and for dipping sheep and cattle) for many decades. Similar compounds are also used for weed control along railways and roads, and as wood preservatives (by making the wood toxic to pests). The scale of agricultural pesticide application can be truly staggering, especially when airborne applications to prairie corn fields are contemplated. Even with more modest back-of-tractor or hand-spraying applications of these compounds, the loadings arriving at the soil surface can be substantial. Although some degradation of pesticides does occur, and many are highly prone to sorption onto soil organic matter, they are now very widespread groundwater contaminants in many parts of the world. Prominent pollutant pesticides include the "drins" (aldrin, dieldrin, endrin), aldicarb, atrazine, simazine, mecoprop, and lindane, all of which are subject to strict limitations in drinking water (Table 7.4). Although changes in pesticide application practices can lessen the loadings of these compounds entering the subsurface, very long lag times can be expected before this results in substantial changes in dissolved concentrations arriving at wells.

A second major diffuse source associated with agriculture is **nitrate** contamination. Although often ascribed simply to excessive application of *artificial* fertilizers by arable farmers, the true story of how excess nitrate can become available for leaching into underlying aquifers is rather more complex (Addiscott 1988). It has been shown, for instance, that many of the heaviest loadings of nitrate pollution entering aquifers in England are due to initial ploughing up of old grasslands that had not been tilled for decades or even centuries. Even in well-established arable fields, nitrate leaching is not a simple matter of overdosing with synthetic nitrogen fertilizers. For a start, "organic" fertilizers (e.g. piggery wastes) are at least as likely as the synthetic variants to yield soluble nitrate. In addition, the nature and quantity of soil organic matter exerts a powerful influence on nitrate mobility (such that using carbon-rich "organic" fertilizers may sometimes be disadvantageous in sensitive areas). Finally, the timing of ploughing and planting of crops may be more important than the quantities of fertilizer applied: if fields are ploughed and left bare just before the onset of the rainy season, nitrate leaching will ensue; however, if they are planted with wet season crops (e.g. winter wheat in northern Europe), much of the free nitrate will be used up by the seedlings. To date, most concerns over nitrate as a groundwater pollutant have related to the notion that it is directly responsible for an illness called methaemoglobinemia, which in rare cases is lethal to babies under 6 months old. A theoretical link to stomach cancer has also been suggested in the past. Both of these health concerns have now been thoroughly discredited, and some health *benefits* of nitrate in drinking water have been identified (Addiscott and Benjamin 2004): for instance, transformations of nitrate on the human tongue inhibit dental caries, and are crucial to the functioning of defence mechanisms against the microbes which cause gastroenteritis. However, if large quantities of nitrate are introduced to the sea in discharging groundwaters extensive algal blooms can develop, with wide-ranging negative impacts on marine ecosystems.

9.3.5 Quality degradation as a side-effect of excessive abstraction

We have already seen how excessive pumping of coastal aquifers can induce marine groundwaters to penetrate inland, rendering well waters too saline for potable supply (see Section 7.2.4 and Figure 7.2). Such saline intrusion risks are not wholly restricted to coastal aquifers, however.

(a) Natural conditions

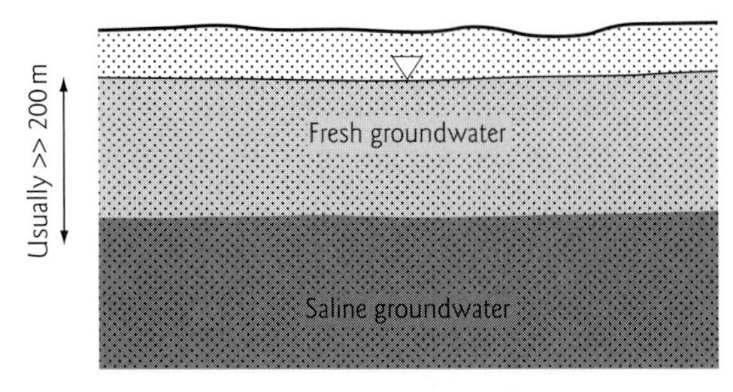

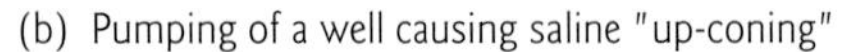

(b) Pumping of a well causing saline "up-coning"

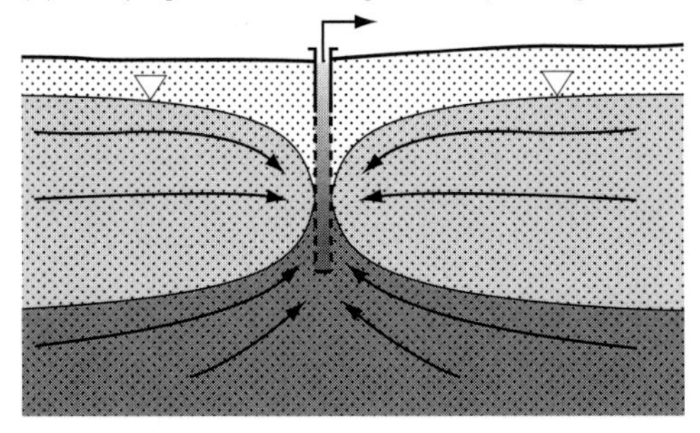

Fig. 9.4 The phenomenon of saline up-coning. (a) Under natural conditions, saline water underlies the fresh groundwater at depth. (b) Vigorous pumping of fresh groundwater induces the fresh/saline water interface to cone upwards, eventually leading to saline water ingress to the well. (cf. Figure 7.2.)

Saline groundwater is commonly present at depth in major inland aquifers, and excessive abstraction can lead to "up-coning" of this saline water to wells (Figure 9.4). While up-coning is by no means an ubiquitous problem, it has been estimated that more than two thirds of the US continental interior is underlain by saline groundwater which could potentially up-cone if groundwater abstraction were sufficiently vigorous (Rhoades et al. 1992).

Another way in which drawdown can induce a deterioration in water quality is encountered in those parts of the world in which **acid-sulfate soils** are naturally present. These are soils which are rich in the mineral pyrite (FeS_2), the presence of which indicates that the host soil was saturated with sea water in prehistoric times. As such, acid sulfate soils are generally found in coastal plains which were formerly submerged. As long as these soils remain water-logged, the pyrite will remain stable, and no acidity will be released. Drawdown of the water table (by pumping and/or excavation of deep ditches) exposes the pyrite to air, in which it rapidly oxidizes, releasing sulfuric acid. Although raising the water table once more can help to redress the problem, it takes much longer to arrest acid release than it takes to initiate it (e.g. Johnston et al. 2004).

Where pyritic strata are present at depth and pumping causes drawdown on a very large scale, it is possible for the acidic products of pyrite oxidation to accumulate in the unsaturated zone in the form of hydroxysulfate salts. When pumping finally ends and the water table rises once more (cf. Section 8.3.5), these acidic salts can dissolve in the rising groundwaters, which then become acidic and heavily enriched in iron and other metals. This has happened to some extent in central London, as rising groundwaters in the Chalk aquifer have reached a bed known as the Basal Sands, in which pyrite oxidation has been taking place during more than a century of drawdown (Mühlherr et al. 1998). On a much larger scale, the withdrawal of dewatering pumps from very large systems of mine workings has led to the generation of vast quantities of acid, metal-rich groundwaters, which often cause ecological devastation, and sometimes even jeopardize water supplies, after they begin to discharge to rivers (e.g. Younger et al. 2002a).

Endnotes

i Ongoing depletion of groundwater storage is sometimes referred to as **groundwater mining**.

ii For extensive discussion and documentation of the data illustrating global climate changes, the interested reader is advised to consult: http://www.ncdc.noaa.gov/oa/climate/globalwarming.html

iii This is particularly true for regions with complex climatic dynamics, sensitive to interannual changes in atmospheric warming and cooling in the Pacific Ocean (e.g. the southeastern USA and southern Chile).

iv Adams and MacDonald (1998) purposely chose the term "susceptibility" to avoid confusion with "vulnerability," which is well established as the term describing the risk of aquifer pollution (Section 9.3.2).

v The term "accidentally" in this context includes pollution sources attributable to culpable ignorance and stupidity.

vi The US EPA drinking water limit for benzene (i.e. 5 μg/L) is even lower than the WHO limit; on the other hand the US EPA limits for toluene and ethylbenzene equal or exceed those given in Table 7.4.

vii MTBE was originally considered a "green" additive to vehicle fuels, designed to abate air pollution by improving oxygen transfer during combustion.

viii Perchloroethene (PCE) is also commonly referred to as tetrachloroethene; see Table 7.4.

ix It should be noted that the term "VOC" also includes petroleum hydrocarbons (associated with LNAPLs) in addition to the chlorinated hydrocarbons which are released into groundwater from DNAPL pools.

x The term "pesticide" actually includes all herbicides, fungicides, insecticides, nematicides, and acaricides.

10
Modeling Groundwater Systems

If the organism carries a "small-scale model" of external reality and of its own possible actions within its head, it is able to try out various alternatives, conclude which is the best of them, react to future situations before they arise, utilize the knowledge of past events in dealing with the present and future, and in every way to react in a much fuller, safer, and more competent manner to the emergencies which face it.

(Kenneth J.W. Craik, 1943, *The Nature of Explanation*)

Key questions

- What is a model?
- Why model groundwater systems?
- What is a conceptual model?
- How can we handle aquifer boundaries in models?
- How can we represent our conceptual model mathematically?
- What is a black box model?
- What is an analytical model?
- What is a physically based model?
- What are finite difference methods?
- What are Monte Carlo methods?
- How can we model contaminant movement in groundwater?
- Can geochemical reactions be modeled realistically?
- How is groundwater modeling carried out in practice?

10.1 Why simulate groundwater systems?

10.1.1 Models – what and why?

A colleague and myself once felt obliged to attend a low-key social function where neither of us knew many people. In fact, when we arrived there was only one person there whom I recognized, so I made a beeline for him and introduced my colleague. "What do you do for a living?" my old acquaintance asked my colleague. "Modeling" he replied. I remember the poorly concealed look of bewilderment on the face of my acquaintance which betrayed his opinion that, whatever his other virtues, my colleague merits no space on the catwalk! While it's true that my colleague is no picture postcard, he's very good at identifying the essential features of aquifers and translating these into useful mathematical descriptions, which we can then encode in computer programs and use to answer all sorts of interesting questions about the movement of water and solutes through the subsurface. In other words he is excellent at *groundwater* modeling. But what is modeling in this context?

A model is a simplified representation of reality. We simplify reality by distilling the complexity of a real system down to its essentials, which we can summarize as a list of justified assumptions about the overall nature of the system. This list of assumptions is our **conceptual model**, as it enshrines our concepts of how the system works. Often there is no need to take the process of modeling any further. In other cases, there may be important puzzles over the future behavior of the system which we need to solve: this is when the next step in modeling comes into its own. We find a way of formulating our conceptual model in mathematical terms: we "put our numbers where our mouth is." This process of translation results in our **mathematical model** of the system. We now need to decide how to solve the equations which make up our mathematical model. Depending on the complexity of the equations and on the quantity and quality of data from the real system with which we might compare the results obtained by solving them, the **solver** that we choose might be a manual calculation, a few manipulations using a pocket calculator, a simple spreadsheet, an elegant analytical solution (i.e. a direct mathematical solution of a complicated equation), or a complex numerical algorithm. The last three are all implemented using computers, and hence are sometimes referred to as **computer models**. Similarly, we often talk loosely of **analytical models** or **numerical models**, though the adjectives in these phrases refer to the method of solving a mathematical model, not to the nature of the model itself.

In summary, modeling is a three-step process: developing a conceptual model, translating this into its mathematical equivalent, and solving this mathematical model. In essence, all mathematical modeling boils down to a formalized, quantitative assessment of the consistency between our concepts of system behavior and the data upon which these concepts are based (Konikow 1981). As such, the solution of mathematical models is always the servant of the ongoing process of refining our conceptual understanding of system behavior.

Modeling according to the three-step procedure just outlined is ubiquitous in science and engineering. In all spheres of endeavor, modeling enables us to:

- Assess the credibility of alternative explanations for observations.
- Harness our knowledge of the past to equip us both to understand present conditions and to predict possible future developments before they arise.

For further thought-provocation on the theory and practice of modeling, the text of Nordstrom (2003) is recommended. In terms of geological philosophy, modeling is consistent with the principle of **uniformitarianism**,[i] which states that "the present is the key to the past" (i.e. that processes currently operative on Earth can be invoked to explain features seen in the ancient rock record). When we use our knowledge of past conditions to develop a model to predict the future behavior of a groundwater system, we are implicitly applying the principle of uniformitarianism inversely.

10.1.2 Motives for aquifer modeling

There are numerous reasons why the simulation of aquifer behavior is often desirable. Mostly, the motives are entirely practical, for instance:

- Wanting to predict the possible consequences of a proposed new abstraction (typically in terms of drawdown or changes in aquifer outflows).
- Designing wellfields for purposes of dewatering construction or mining sites.
- Deducing the pros and cons of alternative aquifer management strategies, which simply cannot be tested at full scale (as only one of them can ever be implemented).

- Understanding the relationships between groundwater flow patterns and surface ecosystems.
- Assessing the long-term safety of subsurface storage/disposal of radioactive wastes.
- Predicting the possible impacts of climate change on groundwater resources.

Such applications of groundwater models are amongst the most valuable tools at the disposal of aquifer managers (Section 10.6; see Chapter 11) and are probably the single greatest application of advanced numerical analysis to the resolution of practical problems in any field of engineering.

In addition, simulation of groundwater systems is occasionally undertaken for purely scientific purposes, such as:

- Understanding how groundwater flow processes affect landform development, accumulation of ores, and other geochemical processes (e.g. Back et al. 1988).
- Simulating the coupled flow and mineral dissolution processes responsible for the formation of cave systems (e.g. Liedl et al. 2003).
- Reconstructing the behavior of aquifers during the cold periods of the Quaternary Era (e.g. Hiscock and Lloyd 1992).

Perhaps the most exotic use of groundwater simulation software is in astrophysics (e.g. Zhu et al. 1999): it turns out that certain aspects of galactic behavior can be explained by analogy to porous medium flow! Nevertheless, call me boring, but the remainder of this chapter will focus solely on mainstream applications of groundwater modeling.

10.2 Conceptual models

The term **conceptual model** has a formal meaning in hydrogeology, having been defined by Bear and Verruijt (1987) as "a set of [rigorously justified] assumptions which represent our simplified perception of a real system." In terms of groundwater *flow*, a conceptual model will be a suite of assumptions that summarize our current understanding of "how water enters an aquifer system, flows through the aquifer system and leaves the aquifer system" (Rushton 2003). Similar comments can be made with regard to water quality: the conceptual model comprises simplifying assumptions concerning the origins, transport, and fate of specific substances (dissolved and/or colloidal) present in the groundwater. The existence of a conceptual model allows others "to assess critically the current thinking and to provide further insights" (Rushton 2003). Box 10.1 presents a typical conceptual model, comprising a list of assumptions and their justifications.

Conceptual modeling should always precede any attempt to mathematically model a groundwater system (see Rushton 2003). On the other hand, conceptual modeling does not necessarily have to be followed by mathematical modeling at all. Rather, conceptual models are largely ends in themselves. They represent the current consensus on system behavior, whether this is informed by direct interpretation of field and laboratory data alone, or whether further meaning has been extracted from these data by mathematical modeling. Recalling the maxim of Konikow (1981) that mathematical modeling amounts to assessing the consistency between our data and our concepts, once we have applied a mathematical model, we should always return to our conceptual model and amend it as appropriate. Throughout the entire modeling process, the conceptual model should always remain supreme.

Box 10.1 illustrates some of the key decisions that need to be made about any aquifer system when constructing a conceptual model. These include:

- Specifying the modes of recharge to the aquifer and estimating their magnitudes (see Chapter 2).
- Specifying the magnitude of K (or T) and S (see Chapter 3) and deciding whether the degree to which these vary from place to place within the aquifer is sufficiently great that it ought to be considered.
- Defining the boundaries of the aquifer system, in terms of both their locations and their properties.

Few topics in hydrogeology cause more headaches to the novice than the definition of

Box 10.1 Example of a conceptual model: flow and solute transport in a floodplain river–aquifer system.

The table below summarizes a conceptual model that was successfully used as the basis for simulations of a river–aquifer system in western Europe. The aquifer is a body of sand and gravel of Quaternary age (formed under periglacial conditions). Near the river channel, Holocene alluvium locally overlies the aquifer and forms the river banks. The streambed is lined with up to 0.5 m of silty/muddy sediments. The full conceptual model statement contained tables of physical and hydrochemical data, maps, cross-sections, and other forms of evidence to substantiate the justifications offered for each of the assumptions.

Assumption . . .	. . . justification
All groundwater flow is laminar and Darcy's Law applies	The granular nature of the aquifer, and the lack of gradients steep enough to provoke turbulent flows
The aquifer is surrounded by zero-flux boundaries at its base and along the valley flanks, but the boundary with the river is a head-dependent flux	The aquifer is underlain by dense mudstones of Tertiary age, which also outcrop in the valley flanks where the sands and gravels feather-edge; the river partially penetrates the aquifer and can only interact with the aquifer via its bed sediment
The sand and gravel aquifer is unconfined except near the river, where it is locally overlain by silty alluvium	Pumping tests show high values of *S* consistent with unconfined conditions; piezometry beneath the alluvium indicates local confinement, in that head in wells piercing the gravels rises above the elevation of the base of the alluvium
The sand and gravel aquifer is homogeneous, and isotropic in the horizontal plane, but vertical *K* is less than *K* horizontal	Although the aquifer is composed of interleaving lenses of sand and gravel, the variations in *K* associated with these occur over distances of only a few meters, with repetition of patterns on scales of tens of meters; stratification of the sands and gravels implies K vertical $<$ K horizontal
The streambed sediment is homogeneous, isotropic, of low *K*, and very low *S*	Sampling of streambed sediment revealed it all to be of fine silt/mud grade (hence low *K*); there is no evidence that it varies dramatically from place to place; as the streambed is always fully saturated, it can only have "confined"-type storativity (i.e. very low values for *S*)
All flow in the streambed sediment is vertical	Because the alluvium forms the banks of the river in almost all places, and the aquifer is much more permeable than either the alluvium or the streambed sediment, and because the channel gradient is very gentle, the law of groundwater refraction suggests that flow in the streambed sediment will be predominantly vertical almost everywhere

Assumption . . .	. . . justification
Water temperature is consistent throughout the river–aquifer system	Although not strictly true at any one instant, it is demonstrably true **on average** (and it helpfully simplifies mathematical modeling of both flows and geochemistry!)
Salinity is never high enough to influence the density of the waters	This is true; no conductivity measurements in the study area exceed 1000 μS/cm, which is a typical freshwater value
For the solutes of interest (chloride and lindane) sorption is the most important geochemical process, and precipitation/dissolution, redox transformations, and biodegradation are unimportant	This is readily demonstrable from the literature for these two solutes. Chloride always behaves in a conservative manner, and lindane is a very refractory organic compound, which is nevertheless highly prone to sorption onto organic matter and certain mineral surfaces
Dispersion can be modeled as a Fickian[vi] process; it is predominantly mechanical in nature in the aquifer, but occurs predominantly by molecular diffusion in the streambed sediments	While there are valid criticisms to be made of Fickian[vi] models for mechanical dispersion, no practical alternative formulation yet exists for deterministic applications; molecular diffusion is well described by Fick's Law;[vi] the fine grain size of the streambed sediment ensures it dominates

model boundaries. Furthermore, inadequate knowledge (or misinterpretation) of aquifer boundary properties is one of the main sources of error in groundwater modeling. Some explanation is therefore warranted here. Put simply, groundwater flow system boundaries are surfaces which define the limits within which flow is deemed to be possible. When we define a boundary, we are effectively saying that, whatever happens in the universe beyond our boundary, we can reduce its effects to certain stated conditions without affecting our ability to accurately account for real processes occurring within our defined area of interest. Evidently, there is a certain amount of arrogance inherent in the definition of boundary conditions, but it's a kind of arrogance which has to be learned by anyone aspiring to be a successful modeler.

In hydrogeological terms, boundaries can be permeable or impermeable. Boundaries can be *physical* (i.e. coincident with real geological or landscape features) or *conceptual* (in that they are defined on the basis of assumed hydraulic characteristics). Before considering examples of these two types of boundary, the generic hydraulic properties of boundaries must be clarified.

The hydraulic characteristics associated with a particular boundary are formally known as **boundary conditions**, and correct assignment of these is a crucial skill for the groundwater modeler to acquire. Whether physical or conceptual in nature, all boundaries must either be permeable or impermeable in nature. Impermeable boundaries have a relatively simple definition: as no groundwater can cross them, any groundwater flow lines which approach them must swing and run parallel to them. Permeable boundaries offer a greater variety of possibilities. Three principal possibilities exist:

1 The boundary transmits groundwater into or out of the flow system at a rate which is independent

of the head distribution within the modeled area; this is called a **specified flux boundary**. (In many ways, impermeable boundaries are simply a special case of specified flux conditions; for this reason the term **zero flux boundary** is used as a synonym for impermeable boundary).

2 The boundary transmits groundwater into or out of the flow system at a rate which is *dependent* on the adjoining groundwater head inside the modeled area; this is called a **head-dependent flux boundary**.

3 The boundary transmits groundwater into or out of the flow system at any rate compatible with the difference between the value of head specified on the boundary and the head distribution within the modeled area; this is called a **specified head boundary**. (Popular alternative terms for this type of boundary, such as "constant head" or "fixed head," are potentially misleading due to ambiguities; this is because they can be taken to imply that the value of head is the same all along the length of the boundary (which is usually *not* the case), and/or that the head values associated with the boundary are fixed irrespective of the period of time being simulated (often not the case either).

Examples of *physical* aquifer boundaries include:

- The outcrop area of the aquifer strata. As shown in Figure 10.1a–c, an outcrop area can be assigned one of three different types of boundary conditions, depending on local circumstances and the scale of the modeled area.
- A fault bringing the aquifer rock mass into contact with essentially impermeable rock; this can only be designated zero flux boundary conditions (Figure 10.1d). (It should be noted that not all faults act as impermeable boundaries – some fault planes are sufficiently permeable that they themselves behave as head-dependent flux boundaries or even specified head boundaries; Figure 10.1e).
- A river cutting into an aquifer (Figure 10.2a–c). In general, the conditions along a river–aquifer interface should be represented as a head-dependent flux (10.2a). A specified flux might be a good alternative where the baseflow in the stream is well known, and the period of simulation is brief (10.2b). The poorest choice for a river boundary is specified head conditions (10.2c), because these make it very difficult to realistically incorporate the moderating effect of fine-grained streambed sediments, they effectively divide the aquifer into two isolated areas (either side of the river), and they can easily give rise to spurious quantities of groundwater inflow/outflow if used carelessly.
- The coastline along which an aquifer meets the ocean (see Figures 3.5, 7.2). If any boundary merits the specified head condition, this is it (Box 10.2). Head-dependent flux conditions are frequently also applicable. Particular care needs to be taken in modeling this type of boundary where salt water intrusion is being considered, in which case density contrasts between fresh and marine groundwaters need to be taken into account (see Sections 3.3.1, 7.2.4 and 9.3.5).

Examples of *conceptual* boundaries often used in practical modeling exercises include:

- A groundwater divide, i.e. a line across which the water table gradient changes direction, which is common in recharge zones (Figure 10.2d); these can be assigned zero flux conditions as long as their position would not be expected to migrate laterally over great distances during the period to be simulated.
- A flow line. As defined on a groundwater flow net (see Section 10.3.3, Figures 3.8, 10.4), a flow line can be conceived of as being "impermeable" inasmuch as the waters either side of the line should not mix significantly. The assignment of zero flux boundary conditions to a flow line might well be reasonable, provided the flow line in question is not expected to migrate laterally over great distances during the period to be simulated.

Although seldom discussed as such, the upper and lower surfaces of aquifers are also important boundaries. In almost all cases, the lower boundary will be assigned zero flux conditions, and these will be taken into account in the way the flow patterns are translated into mathematical equivalents. The upper surface can be specified as the water table (in an unconfined aquifer; see Figure 1.4) or the base of the overlying

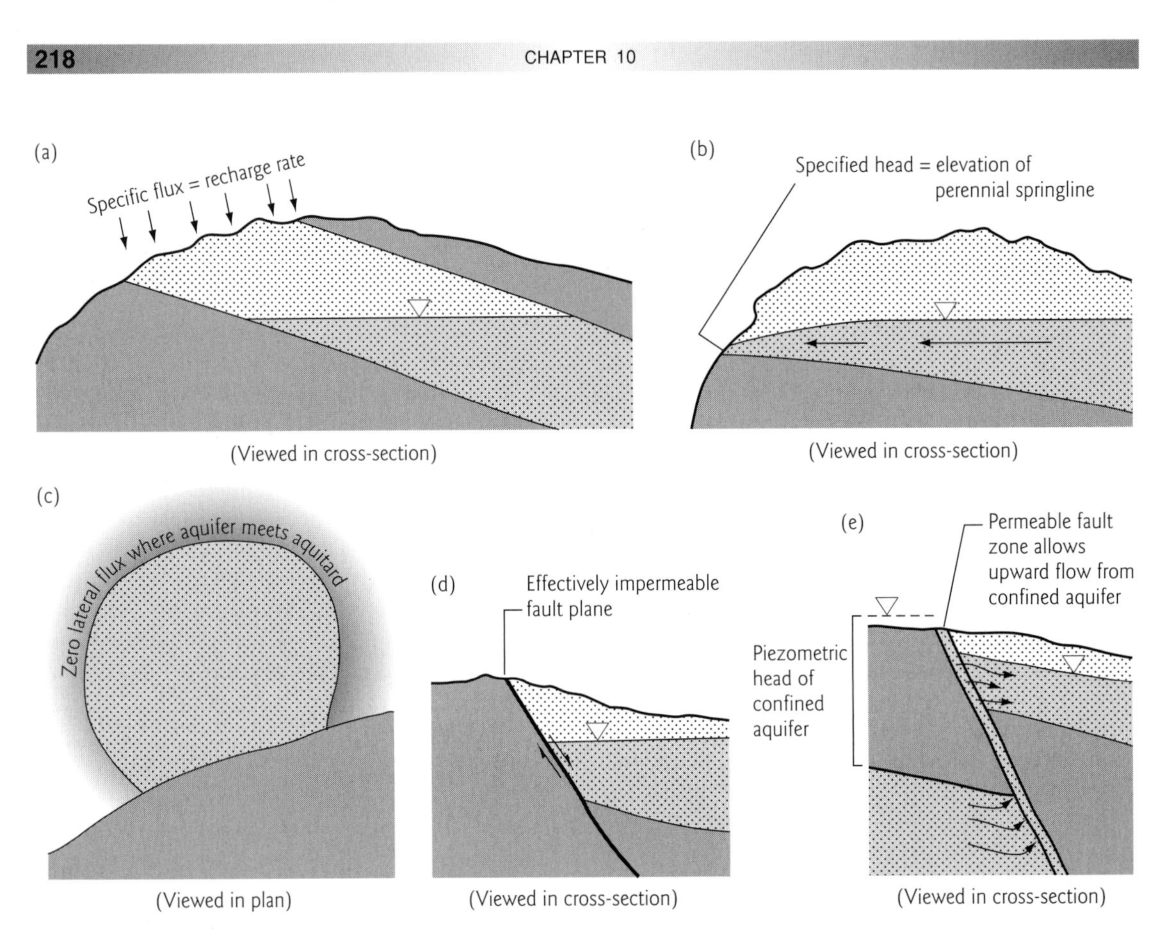

Fig. 10.1 Geologically determined aquifer boundaries. (a) The outcrop area of an aquifer, receiving recharge which can be represented as a specific flux boundary. (b) A specified head boundary identified with the elevation of a line of contact springs (which effectively mark the "outcrop" of the water table). (c) Identification of the edge of the outcrop of an aquifer as a zero flux boundary. (Compare with (a) in which a similar situation is shown in cross-section.) (d) A zero flux boundary formed by a fault plane filled with impermeable gouge, bringing an aquifer (right) into contact with an aquitard (left). (e) A permeable fault plane hydraulically connecting a deep confined aquifer and a shallow unconfined aquifer. If the head in the shallow aquifer varies considerably over time, this situation could be represented as a head-dependent flux boundary; otherwise, if water levels are fairly stable over time it might be more simply represented as a specified flux.

aquitard (in a confined aquifer; Figure 1.5). Special care must be exercised when it is possible that the piezometric surface might fall sufficiently during the period to be modeled that an aquifer which was originally confined (see Figure 1.5b, case (ii)) becomes unconfined (see Figure 1.5b case (i)).

Where discretion exists, it is always advisable to impose boundary conditions as far away from the main area of interest as possible, to avoid them dominating local hydrological behavior.

10.3 Representing the conceptual model mathematically

10.3.1 Peering through the darkness: "black box" models

The simplest approach to mathematically representing aquifer dynamics is to assume that the entire aquifer is a simple water tank, which receives water from a single source (thus lumping all forms of recharge together) and releases it via a single tap

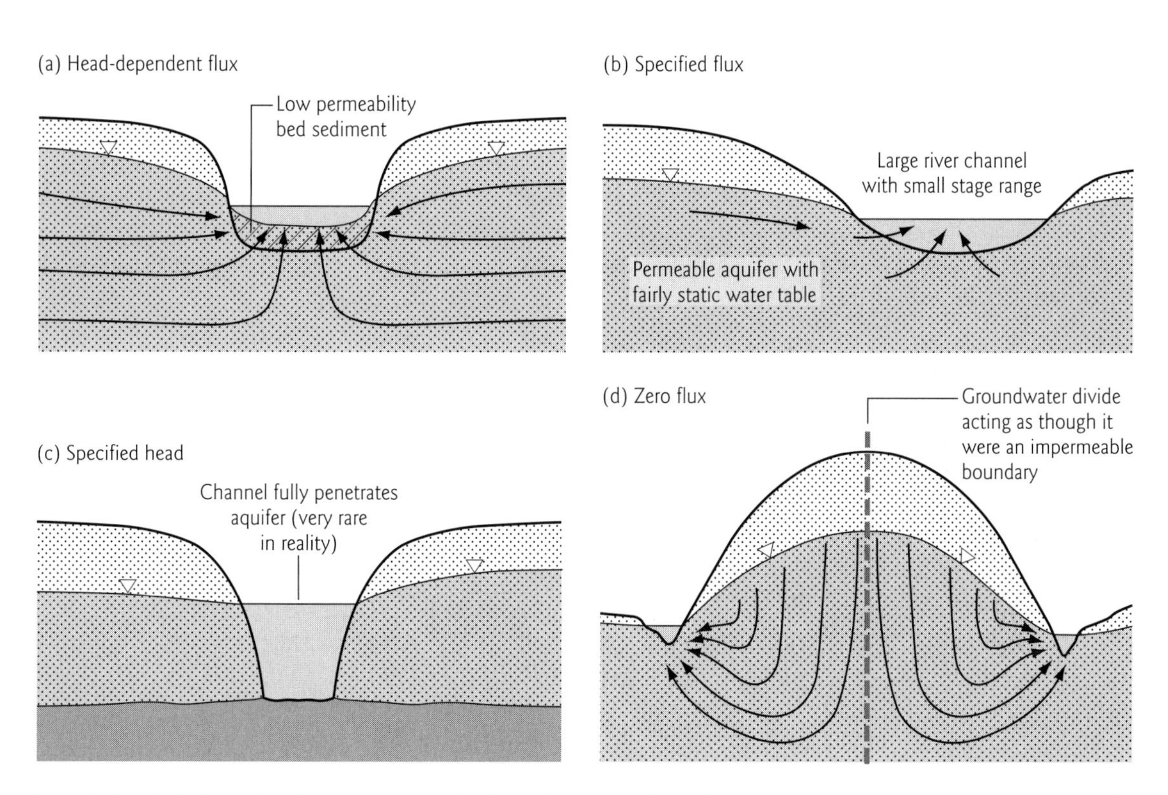

Fig. 10.2 Hydrologically determined aquifer boundaries. (a) Boundary between an aquifer and a river which partially penetrates the saturated thickness of the aquifer. In this case, the stream bed is lined with low permeability sediment, so that the rate of groundwater flow into the river is governed both by the difference in head between the aquifer and the river stage and by the permeability of the stream bed. (b) If the water table in the aquifer and river stage are both fairly static, then groundwater discharge to a partially penetrating river can be represented as a specified flux. (c) The usually fictitious (but widely used!) case of a fully penetrating river acting as a specified head boundary. (d) A typical groundwater divide in a recharge area, which can be viewed as a zero flux boundary, given that groundwater flows away from it on both sides. Caution must be exercised in designating groundwater divides as zero flux boundaries, for natural drawdown (e.g. during an extended drought) or drawdown caused by pumping of wells can shift groundwater divides laterally over large distances.

(lumping all groundwater discharge pathways together), with the difference in rate between the two at any one time being accommodated by changes in water level within the tank (representing groundwater storage). A simple model like this is one example of what are termed **black box models**. The use of the phrase "black box" acknowledges that the inner workings of the aquifer are deliberately ignored. An approximate synonym for "black box model" is **lumped-parameter model**, alluding to the ways in which inflow, storage, and outflow properties are "lumped" together into single factors. Adopting such simplistic representations inevitably ignores the actual dynamics of groundwater flow patterns and storage changes within the aquifer. In some types of investigation, the processes occurring within an aquifer might truly be of little interest, in which case the ease of use of black box models can offer great advantages over more sophisticated alternatives.

Black box models come in many shapes and sizes. The very simplest are based on simple correlations between rainfall and spring discharges (e.g. Ford and Williams 1989) and/or groundwater levels (e.g. Bloomfield et al. 2003). A little

Box 10.2 A simple analytical solution.

Imagine a body of beach sand, nestling at the foot of vertical cliffs carved into virtually impermeable metamorphic rocks. Any rain falling on top of the cliffs drains inland. The lowest level to which groundwater can drain within the sand body is set by the low-tide mark. Only direct recharge (from rainfall onto the sand) can add any water to the system. In cross-section the scenario looks something like this:

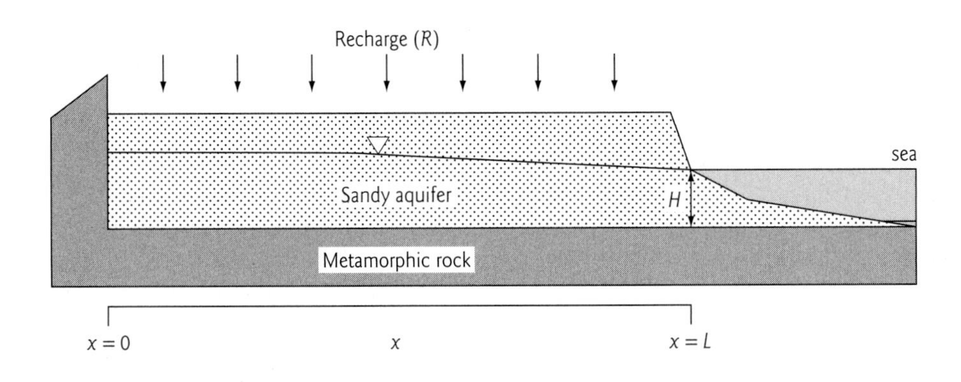

If we need only to know the long-term average head distribution in this aquifer (and not how levels change over the seasons), then the groundwater flow equation presented by Hubbert (1940) simplifies to: $T(\delta^2h/dx^2) = -R$ (where T is transmissivity (m^2/day) and h is the head (meters above base of aquifer) at any point along the x-axis shown on the sketch above). The boundary conditions relevant here are defined by the fact that the water table ends abruptly against the cliff (so that the water table gradient ($\delta h/dx$) becomes zero at that point (i.e. $x = 0$), and $h = H$ at the coastline). Then, all we need to do to solve this model analytically (Rushton and Redshaw 1979) is to integrate the flow equation twice and evaluate the coefficients of integration using our two boundary conditions. This yields the following expression:

$$h = H + (0.5R\,(L^2 - x^2)/T)$$

This is the *exact* solution of the groundwater flow equation for the system described. This means that provided we know R (m/day), L (m), T (m^2/day), and H (m above base of aquifer), then as long as we specify where we are in the aquifer (i.e. give a value for x), we can always calculate the head at that point. Assuming $R = 0.001$ m/day (i.e. around 370 mm/yr), $L = 500$ m, $T = 100$ m^2/day, and $H = 10$ m, then we can calculate the following values:

For $x = 15$ m, $h = 11.25$ m
For $x = 215$ m, $h = 11.02$ m
For $x = 387.2$ m, $h = 10.50$ m

and so on. (If you take a copy of this box along to the beach next time you go, you'll find it easy to make friends by predicting the depth to water table beneath each sunbather! Indeed, combined with the activities suggested in Box 3.2, it'll liven up your day considerably.)

more realism can be incorporated into the model by assuming that the storage–outflow relationship for an aquifer will mimic the exponential form typical of baseflow recession curves (see Section 5.3.1). These correlation-based black boxes are sometimes referred to as **transfer functions**, for they are literally mathematical functions which transfer information on aquifer inputs into estimates of aquifer outflows.

Most black box models assume an aquifer to be a single entity, giving another shade of meaning for the word "lumped." However, in some cases field evidence indicates that distinct compartments of the aquifer display a degree of independent behavior: this is often the case in karst systems, for instance, and in flooded underground mine workings. For such aquifers, it is sometimes feasible to create a number of partially interconnected boxes, each with its own hydraulic properties (see Adams and Younger 2001a). Multi-box models are more realistic than singular black boxes, and they are partly based on real field evidence of aquifer functioning, attributes which arguably qualify them as "gray boxes" (Ford and Williams 1989).

Paradoxically, simplistic black/gray box models find some of their most important application "niches" in the simulation of large and highly complex groundwater flow systems for which data are either sparse or literally uncollectable (e.g. for possible conditions in aquifers in the distant future, after major climatic change). Examples include:

- Modeling the flows of major springs draining karst aquifers in mountainous terrains, for which meaningful observation borehole data are seldom available (e.g. Ford and Williams 1989).
- Modeling flows within and between very large systems of interconnected underground mines, especially during the process of flooding after abandonment (e.g. Adams and Younger 2001a).
- Making estimates of the possible doses of radiation which might ultimately be experienced by humans and animals in the event of future leakage from radioactive waste respositories (e.g. Birkinshaw et al. 2005).

Very few studies have attempted to compare black box models with the output from more sophisticated models for the same systems. A recent rigorous intercomparison of the two kinds of model has shown that simplification of flow system dynamics needs to be undertaken very carefully if it is not to result in irrational predictions (Birkinshaw et al. 2005).

10.3.2 Pursuing realism: system dynamics models

The term **system dynamics** is used to denote an approach to the study of complex systems in which complex feedbacks occur. **Feedback** in this context refers to situations in which one attribute of a system (X) affects another (Y), whilst Y in turn affects X. Before the advent of system dynamics modeling in the early 1960s, these sorts of feedback-dominated systems were very difficult to analyze. They are particularly prevalent in many socioeconomic situations, and it is in that sphere that system dynamics modeling was pioneered. In recent years, however, applications of the system dynamics modeling approach to environmental systems have begun to be reported (Ford 1999). System dynamics modeling represents a significant advance over cruder black box models (Section 10.3.1), and has considerable potential for modeling complex systems for which insufficient data exist to support the application of fully physically based models (Section 10.3.3).

Many groundwater systems possess strong feedback mechanisms. In many unconfined sand and gravel aquifers, for instance, drawdown decreases the saturated thickness, which in turn decreases the transmissivity; when transmissivity decreases, hydraulic gradients tend to steepen, moving more water towards the zone of low saturated thickness; thus head and transmissivity are linked by feedback loops in many thin unconfined aquifers. This effect is of great practical importance in the design of construction dewatering systems (see Section 11.4.4). System dynamics models clearly have great potential for application to such cases, although at the time of writing, groundwater systems are only just beginning to be analyzed in this manner (Plemper 2004).

10.3.3 Flow nets as mathematical models

Although not often considered as mathematical models, flow nets are indeed graphically constructed groundwater flow models. The basic rules for flow net construction have already been given (see Section 3.3.2), and Figure 3.8 illustrates flow nets in both plan and cross-sectional views. We have also seen how a modified version of Darcy's Law (see equation 3.3) can be used to quantify the rates of groundwater movement through individual flow tubes. By summing the flows in individual tubes, we can easily arrive at a total flow rate for a given aquifer (or part of an aquifer). If we compare this figure with our estimate of recharge rate (see Chapter 2), we are beginning to close the "water balance" for the aquifer, which is fundamentally an expression of the law of conservation of mass. This is the same law that is invoked in the derivation of sophisticated physically based models of groundwater flow systems (Section 10.3.4). It is therefore not surprising that in addition to being amenable to manual construction (Section 3.3.2), flow nets can also be generated automatically from the output of physically based models. This is in fact a very common use of the latter category of models; it provides the estimates of the directions and velocities of groundwater flow upon which models of solute transport (Sections 10.5.2 and 10.5.4) are based.

10.3.4 What's really going on? – Physically based models

Among the major milestones in the development of groundwater analysis techniques, the derivation of the full equations of groundwater flow by Marion King Hubbert in 1940 stands alongside the discovery of Darcy's Law (1856) and the development of transient-state test pumping interpretation methods (Theis 1935). M. King Hubbert[ii] (1903–1989) was one of the most brilliant and visionary earth scientists of his generation, and is widely remembered as having been the first person to predict (in 1949) that the era of abundantly available oil and gas would be of very short duration. His place in posterity has been assured by the accuracy that his 1949 predictions can now be seen to have possessed. His bequest to hydrogeology was no less auspicious: the paper in which King Hubbert expounded his unifying theory of groundwater motion (Hubbert 1940) has been endlessly rehearsed in more recent literature, all too often without due acknowledgement of his pioneering achievement. If hydrogeology needs a monarch, King Hubbert fits the bill by name and nature: his insights paved the way for the modern groundwater modeling industry more surely than any other development. Hubbert himself once remarked: "Our ignorance is not so vast as our failure to use what we know." It is one of life's ironies that, for three decades, leading hydrogeologists were only too aware of the power possessed by Hubbert's groundwater flow equations without possessing the wherewithal to apply them in the service of aquifer management. It was only with the advent of the first generation of industrial mainframe computers that it became possible to solve the equations of Hubbert (1940) as they apply to aquifers in the real world (Prickett and Lonnquist 1971).

What do the equations of Hubbert (1940) express? They essentially apply the law of conservation of mass to aquifers, stating that at any point in any aquifer, the quantities of groundwater flowing to and from that point must either be equal in magnitude, or else the difference between them must equal the change in groundwater storage at that point. In the form of an equation, we can write:

(rate of water inflow) – (rate of water outflow) = (rate of change in volume of water in storage)

(The three terms in this equation all have units of "volume per unit time," with m^3/day being a typical practical choice). There are three possible scenarios in relation to this equation:

1 Rate of inflow = rate of outflow: in this case there will be no change in storage.

2 Rate of inflow exceeds rate of outflow: there will be an *increase* in storage.
3 Rate of outflow exceeds rate of inflow: there will be a *decrease* in storage.

We learned in Chapter 3 that rates of groundwater flow can be calculated using Darcy's Law (see equation 3.1). Thus if we know the value of K and the hydraulic gradient in a given portion of an aquifer, we can easily use Darcy's Law to evaluate the inflow and outflow rates in the above equation over any time interval of interest.[iii] Furthermore, if the S value for the aquifer is also known, we can relate the change in volume of storage to changes in water levels within the aquifer (see Section 1.4.4). Conversely, if we have an independent estimate of the volumes of water entering and leaving a given portion of aquifer, then because Darcy's Law and the calculations involving S both involve values of groundwater head, it is possible to calculate head patterns in the aquifer at any point in time as long as K and S are known. By the time we have substituted Darcy's Law and S values into the simple mass conservation equation given above, it will have become a complex formula of a type known as **partial differential equations**. While techniques for solving such equations had been developed long before the publication of the work of Hubbert (1940), the laborious nature of the calculations (for all but the most simplified versions of the equations; see Section 10.4.2) largely precluded their use outside of research institutions prior to the advent of digital computers. Thereafter, the implementation of powerful numerical methods (Section 10.4.3) first became feasible (Prickett and Lonnquist 1971) and eventually commonplace (e.g. Anderson and Woessner 1992).

One key attribute of the equations of groundwater flow expounded by Hubbert (1940) is that they encapsulate the application of the fundamental laws of physics to the analysis of groundwater flow. For this reason, we refer to models based on the solution of these equations as being **physically based**. In scientific terms, physically based models represent the gold standard: ultimately, it is possible to argue with the mathematical formulation of any black box model or system dynamics model; but unless the Laws of Physics themselves be disproved, the equations of Hubbert (1940) will stand, immune from criticism, amenable only to simplification and manipulation to suit particular applications.

Before solving the groundwater flow equations it is necessary to specify the boundaries within which a solution is to be sought. In practice this means defining the boundaries to the aquifer system under analysis (Section 10.2) and specifying the period in time to be represented in the simulation. The area contained within the model boundaries is often termed the **model domain**. Within the model domain, it is important to specify whether aquifer parameters (K (or T) and S) are the same everywhere (i.e. **homogeneous**) or whether they vary from one place to another (i.e. **heterogeneous**). Less commonly, it may also be necessary to specify whether the magnitude of K (or T) at any one point is the same in all directions (i.e. **isotropic**), or whether it varies systematically depending on the direction in which it is measured (**anisotropic**). Decisions over homogeneity versus heterogeneity and isotropy versus anisotropy are generally made at the conceptual modeling stage (Section 10.2; Box 10.1).

10.4 Ways of doing the sums: solving physically based models

10.4.1 Options, pros, and cons

Before digital computers became widely available, there were only two feasible techniques for indirectly solving physically based mathematical models for large, heterogeneous aquifer systems:

- Manual construction of flow nets (see Sections 3.3.2 and 10.3.3).
- Physical analog models, of which the most widespread were electrical analog models, consisting of networks of resistors and capacitors (see Section 3.3.2 and Rushton and Redshaw 1979). Other

physical analog models included viscous fluids flowing between parallel transparent plates ("Hele–Shaw models"), sand tanks, and heated plates (Todd 1980).

While manual flow net construction remains a useful hydrogeological skill, physical analog models are now obsolete (save for some (largely untapped) value as teaching aids). With the exception of some simple procedures which are amenable to the use of hand-held calculators (Box 10.2) and graphical methods, the solution of physically based mathematical models nowadays is almost invariably carried out by computer. There are two basic alternatives for solving physically based mathematical models: analytical solutions and numerical solutions. These two approaches should be seen as complementary rather than competing, for the following three reasons:

- Analytical solutions are exact and mathematically rigorous, whereas numerical solutions are only *approximations* to the exact solution. Analytical solutions therefore provide a very important "reality check" on the performance of numerical solutions while the latter are under development.
- Many analytical solutions are easy to apply, even using manual calculations, whereas the implementation of numerical solutions is very time-consuming and demanding of computing power, even for relatively straightforward applications.
- Analytical solutions can only be applied to relatively simple model domains, in which all aquifer parameters are homogeneous, and in which boundary conditions are simple (generally rectilinear and constant in value), whereas numerical solutions can be applied to heterogeneous domains with very complex boundary conditions. As such, numerical solutions are applicable in many more practical situations than most analytical solutions.

10.4.2 Analytical solutions

We have already encountered analytical solutions in Section 3.4, where test-pumping interpretation techniques were introduced. The methods of Theis (1935) and Cooper and Jacob (1946) (see Box 3.3) and the many later techniques derived from their work (Kruseman and de Ridder 1991) are all analytical solutions to the one-dimensional version of the groundwater flow equation, as expressed in radial coordinates. In the context of groundwater modeling, an **analytical solution** can be defined as an exact solution to a given version of the groundwater flow equation, applicable at any space–time point within a specified model domain, which is derived using the methods of advanced calculus and allied mathematical techniques. Because analytical solutions have to apply to any point within a model domain, they are not applicable to heterogeneous domains. Furthermore, because their derivation tends to be very demanding, they are generally only applicable to domains with relatively simple boundary conditions. To give a flavor of the nature of analytical solutions, Box 10.2 presents a simple analytical solution to the groundwater flow equation for a one-dimensional flow field.

10.4.3 Numerical solutions

Over the years, analytical solutions have been derived for a wide range of groundwater flow and pollutant transport problems. For instance, elegant analytical solutions have now been derived for problems as complex as the migration of reactive solutes within two-dimensional groundwater flow fields. However, for them to be amenable to analytical solutions, these problems must be posed in rather idealized forms, which will seldom be encountered in real field settings. For the most part, then, analytical models are of more academic than practical use (with the notable exception of the radial-flow solutions which form the basis of test-pumping interpretation). Because they are useful in verifying the performance of inherently approximate numerical solutions, however, analytical solutions remain important to the practical modeler.

In the real world, aquifers tend to be irregularly shaped; they are frequently heterogeneous in their properties, and may also be anisotropic; they may be contaminated with a number of solutes which interact both with the aquifer materials and with each other; nonaqueous phase liquids (NAPLs) may be present, as well as pockets of gas, so that the system is an amalgam of multiple, interacting phases. The only mathematical techniques powerful enough to accommodate such complexities are numerical solutions. In groundwater modeling, a **numerical solution** can be defined as an *approximate* solution to a given version of the groundwater flow equation, which is obtained by replacing those terms in the equation that contain derivatives (i.e. differential expressions such as $\delta h/\delta x$) by simple algebraic expressions. These algebraic expressions are calculated on and between specific (but arbitrarily selected) points in the overall simulation domain (termed **nodes**), which together form a model grid. The evaluation of these algebraic expressions is usually achieved using successive approximations, i.e. repetitive calculations for all of the nodes in a given grid. Each computational sweep of the model grid is termed an **iteration**, and a number of iterations are typically calculated until the changes in values between successive iterations become negligible, at which point the solution is said to have **converged**.

A number of different approaches can be taken to implement numerical solutions. Details of the various approaches are well beyond the scope of this book, and only a brief introduction is given below. (Readers requiring further details are referred to Huyakorn and Pinder (1983) and Anderson and Woessner (1992).) The most widely used groundwater modeling codes obtain numerical solutions using **finite difference methods**. Figure 10.3 shows a typical finite difference grid superimposed upon a base map of an area underlain by an aquifer. Most finite difference models perform their calculations for the central points in each of the grid squares, and are therefore called **block-centred** finite differences. In many ways, finite difference methods can be likened to coupling hundreds (sometimes even thousands) of tiny black box models together: each grid cell in the model calculates the Hubbert (1940) mass conservation equation for that tiny part of the aquifer:

Water in – water out = change in storage

Because calculating this equation for any one cell means updating it for its four neighbors, the calculations proceed cell by cell through the grid, iterating until the entire grid has converged. There are a number of different techniques available to perform these iterative calculations (e.g. line-successive over-relaxation (LSOR); strongly implicit procedure (SIP); alternating direction implicit (ADI) method), each of which has its own strengths and weaknesses for certain configurations of boundary conditions and internal parameter variations. If many model runs are envisaged (for instance because the code is to be run for a large number of input data sets), it may well be worth spending some time on trial runs to identify the best solver for that particular grid.

Second in popularity in widely used groundwater modeling codes are **finite element methods**. As for finite difference solutions, finite elements require the subdivision of the model domain into a grid of contiguous cells. One of the advantages of finite element methods over finite differences is that the cell shape can be triangular, or almost any polygonal shape; indeed, the cell shape can be allowed to vary from cell to cell. In certain finite element implementations, the cells can be allowed to distort automatically during model execution (in which case they are known as **deformable elements**). For certain hydrogeotechnical applications, such as cross-sectional analyses of subsidence, this can be a very handy feature. Other numerical methods sometimes used in groundwater flow analyses include **integrated finite difference methods** (which also offer the option of polygonal cells) and **finite volume methods** (which can be very useful in three-dimensional analyses and river–aquifer modeling). Several numerical methods are almost wholly restricted to solving solute transport problems, such as the **method of characteristics** and **particle tracking** (also known as the **random walk method**; see Prickett et al. 1981). Discussion of these techniques is beyond the scope of this text; the interested reader should refer to Huyakorn and Pinder (1983) and Prickett et al. (1981).

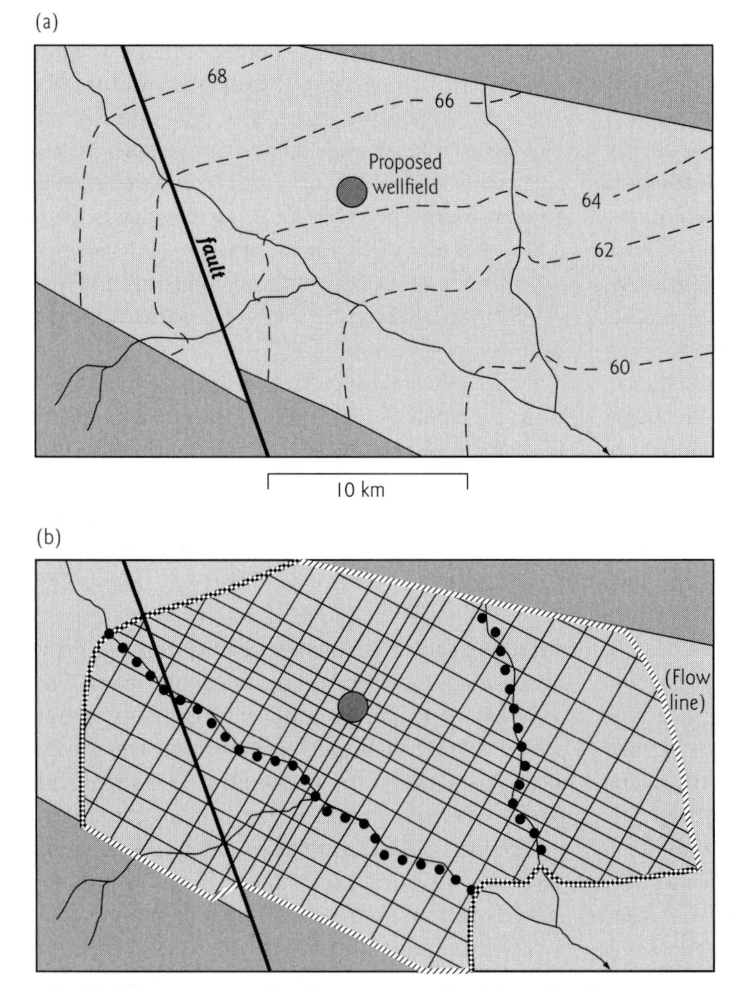

Fig. 10.3 Overlaying a finite difference model grid on a hydrogeological base map. (a) Hydrogeological map, showing outcrop of aquifer (light shaded area), aquitard (darker shading), a river system, water table contours (dashed lines with elevation relative to sea level in metres) and the location of a proposed wellfield. (b) Grid developed for area shown in (a), specifically designed for the purposes of simulating the effects of wellfield pumping on groundwater levels and flows. The area enclosed within the outermost boundaries is called the "model domain". Grid lines divide the domain into a large number of quadrilaterals ("cells"). Notice that the subdivision (i.e. "discretisation") of the domain is finest in the vicinity of the proposed wellfield, as it is expected that hydraulic gradients will need to be modeled most accurately there. Three types of boundary conditions are shown. The specific flux boundary conditions are user-defined, and are therefore set as far from the wellfield as possible; they are made to correspond to water table contours, along which calculations of flows entering/leaving the incoming flow can be most easily made. The zero flux boundary conditions are in part geologically determined (aquifer/aquitard boundary), but at the right hand edge of the domain a flow line (drawn perpendicular to the contours shown in (a)) is used as an artificial zero flux boundary. The river-aquifer boundary is internal to the domain, and is represented as a head-dependent flux (cf. Figure 10.2a), on the safe assumption that the river is partially-penetrating.

Further information on how numerical models (especially finite difference models) are used in practice is given in Section 10.6 below.

10.4.4 Playing dice: probabilistic modeling

One philosophical problem which we have not yet broached in our discussion is the issue of **nonuniqueness**. When groundwater models began to become widely used in the 1980s, many commentators began to point out that the uncertainties which hinder our understanding of real aquifers are sufficient to cast doubt on the practice of developing a single model for any one domain (e.g. Gutjahr and Gelhar 1981; Dagan 1982; Marsily 1986). It is relatively easy to demonstrate that the same set of results can be obtained using very different combinations of boundary conditions, T and S values, and recharge distributions. Where the different combinations of parameters are equally credible (which many experienced practitioners would argue is not so often as the critics liked to suggest), then on what basis can we offer a single set of model results as our unique answer to a given problem? This classic "single input/single output" approach is often referred to as **deterministic modeling**, to distinguish it from the alternative approach of **probabilistic modeling** (also known as **stochastic modeling**).

The proponents of probabilistic groundwater modeling credibly argue that it is more reasonable to conceive of our input data as probability distributions rather than single numbers. If we then process these probabilistic inputs through numerical modeling programs, the output will also be delivered in the form of probability distributions. There are a number of procedures for doing this (Marsily 1986). One of the most popular approaches is called **Monte Carlo modeling**, the name deliberately alluding to the roulette tables of that European capital of *soi-disant* respectable gambling. ("Las Vegas modeling" might have been the name if US researchers had got there first). In this approach, the probability distributions describing T, K, S, recharge and other input variables are assembled, and then randomly sampled at the start of each model run. Each run of the model yields one **realization** (i.e. possible answer) to the problem. After 1000 or more realizations have been accumulated, there should be sufficient output data to allow assembly of credible probability distributions for variables of interest. The trouble with Monte Carlo modeling, of course, lies in the heavy demands it places on computer run-times. Given that each model grid has to be run more than 1000 times, the computing power needs to be hefty, especially where the model grid has thousands of nodes and where aquifer properties are highly variable. Even with modern computers, the burden often remains daunting for real-world problems. While many more sophisticated and computationally efficient approaches to probabilistic modeling have been devised (e.g. Gutjahr and Gelhar 1981; Dagan 1982, 2004; Marsily 1986), the set-up and execution of these programs is still far more demanding than the simple deterministic approach.

With the fervor of converts, by about 1990 many opinion-formers in the world of groundwater modeling could be heard predicting the imminent supremacy of probabilistic modeling as the only defensible option for mainstream practice, with poor old deterministic modeling becoming at least marginalized, if not disappearing altogether. The prophets of the ascendancy of probabilistic modeling argued that the then-anticipated exponential improvements in digital processing power during the 1990s would remove the last practical barrier to the widespread use of their favored approach. As we shall see in Section 10.6, things have not really turned out as these prophets anticipated, with deterministic modeling still dominating practice (e.g. Rushton 2003) in all but a handful of specialized, research-oriented niches, such as in radioactive waste management (e.g. Selroos et al. 2002). There are a number of reasons why, at the start of the twenty-first century, probabilistic modeling still does not predominate. First, as computing power has increased, deterministic modelers have tackled ever-more complex problems, which even now would be prohibitively expensive to analyze probabilistically. Second, decision-makers rarely like to be presented with probability density functions rather than "headline" numbers. Third, the application of the more efficient probabilistic techniques demands a level of mathematical skill to which few practitioners can honestly lay claim. Finally, many hydrogeologists find the exercise of judgment (which is paramount in deterministic modeling) to be more intuitively comfortable than relinquishing control to a random number generator. In other words, heuristic knowledge can go a long way towards narrowing probabilities in a manner that is not necessarily amenable to formal statistical description.

10.5 One step beyond: simulating groundwater quality

10.5.1 Groundwater quality modeling: a challenging endeavor

Prediction of changes in groundwater quality is often at least as desirable as being able to predict flow system changes. In any case, the groundwater quality manager doesn't really have the luxury of ignoring flow modeling at the expense of groundwater quality simulations. This is because it is impossible to model spatial and temporal patterns of groundwater quality without first taking into account groundwater velocities and mechanical mixing processes. In other words, a good flow model is a prerequisite for an adequate groundwater quality simulation. As if that wasn't bad enough, groundwater quality modeling isn't simply twice the work of flow modeling: it's at least three times the work and can easily be ten times! As a rough rule-of-thumb, for each solute specified as being of interest, one can expect the workload to increase by an amount equal to that required to develop the initial flow model. Groundwater quality modeling is difficult, expensive, and often difficult to defend in full. Despite being fraught with uncertainties, the more comprehensive groundwater quality modeling exercises are so complex that the probabilistic modelers have yet to raise anything other than a white flag in this territory: it remains a bastion of determinism. On the plus side, groundwater quality modeling is invariably fascinating, sometimes very useful, and often deeply satisfying, at least after the toil is complete.

Groundwater quality modeling originated in two quite distinct strands of activity, which have lately begun to intertwine and produce a precocious new hybrid. **Solute transport modeling** grew out of physically based modeling of groundwater flows. It is concerned with explaining and predicting the distribution of specific solutes within groundwater flow systems (Section 10.5.2). The second strand, **hydrogeochemical modeling**, grew out of the wider field of low-temperature geochemistry, and as such it is concerned with how dissolved substances are actually present in solution, how they interact with each other, and how they react with solid mineral phases (Section 10.5.3). Through most of their evolutionary history, up until the early 1990s, these two traditions were largely distinct, being implemented by different communities of researchers. Largely thanks to the leadership provided by the US Geological Survey (e.g. Nordstrom 2003), the two traditions began to merge to the benefit of both: solute transport models began to incorporate more rigorous and realistic representations of geochemical reactions, and hydrogeochemical models began to incorporate more sophisticated models of groundwater flow and mixing processes (Appelo and Postma 1993). The result is a rapidly expanding hybrid research field known as **reactive transport modeling** (Walter et al. 1994a,b; Section 10.5.4).

10.5.2 Solute transport models

The fundamental principles of solute movement in groundwater can be summarized in terms of the following four processes:

- Most solute movement occurs due to the bulk movement of the groundwater, carrying its solute load with it; this process is termed **advection**, and

it is quantified simply by estimating the velocity[iv] of groundwater flow.
- Mixing of groundwaters occurs during the advective transport of solutes. In effect, solutes spread out from areas of high concentration, invading surrounding areas of groundwater at lower concentrations. This process is called **dispersion**.
- Rock–water interactions (various types of geochemical reactions; see Section 4.4.2), which generally slow down (**retard**) rates of solute transport. Sorption processes are generally the main cause of **retardation**.
- For certain types of solutes (radionuclides and organic compounds) natural processes of **degradation** (radioactive decay and biodegradation respectively) can deplete the absolute mass of a given solute during transport (though the concentrations of "daughter products" will increase concomitantly).

Advection is by far the dominant process in all reasonably permeable aquifer materials. This suggests that scoping calculations of solute transport using groundwater flow data alone will provide most of the information we need when assessing a groundwater pollution problem. As a corollary the quantification of dispersion usually has a lower priority (Lehr 1988). This is comforting, for dispersive processes in aquifers are not nearly so well understood as is advection.

Dispersion of solutes during groundwater flow has two main components:

- **Molecular diffusion**: this is a slow process, occurring by Brownian motion,[v] which only becomes an important component of total dispersion in low-permeability rocks.
- **Mechanical mixing**: this arises from small-scale variations in permeability which result in a considerable spread of groundwater velocities (in both magnitude and direction) around the mean value which is used to quantify advection.

For the purposes of modeling, it is common to resolve the overall effects of dispersion into two components: dispersion in the direction of advection is called **longitudinal dispersion**, whereas sideways displacements are referred to as **transverse dispersion**. Figure 10.4 illustrates how advection and dispersion in these two directions are manifest in the migration of pollutants from a point source. In most cases, the magnitude of these effects is calculated using a well-known formula called Fick's Law, which was originally developed to describe molecular diffusion.[vi] In applying Fick's Law, we need to use a factor known as the **dispersion coefficient**, which is obtained by multiplying the local velocity of groundwater flow by a constant known as the **dispersivity**. In theory, dispersivity is a "characteristic length" which somehow typifies the propensity of a given portion of an aquifer to generate mechanical mixing effects in proportion to the velocity of the groundwater moving through it. In reality, it has proven impossible to identify unique values of dispersivity for given portions of aquifers; everything seems to depend on the nature of the field data from which the dispersivity is calculated (e.g. short-term tracer test vs. evaluation of a long-established plume of pollutants). Consequently, the use of Fick's Law and dispersivity to quantify dispersion in aquifers has long been controversial (Marsily 1986). In practice, most solute transport modelers bear the shortcomings of the approach in mind, but cautiously apply it nonetheless, for want of any better approach of comparable simplicity.

The four processes that govern the movement of solutes in aquifers can be summarized in the form of a single equation, which is often called the **advection-dispersion equation**[vii] (**ADE**). The ADE is another example of a partial differential equation. While a number of analytical solutions of the ADE have been developed for systems with simple boundary conditions (see Fetter 1999), for most practical applications it has to be solved numerically. Although finite difference and finite element solutions can be applied to the ADE, without very cautious handling these two techniques give rise to errors which manifest themselves in a spurious increase in the effects of dispersion on solute concentration distributions. These errors are termed **numerical dispersion**, and their avoidance has been the focus of much mathematical research. It is precisely because they are free from numerical dispersion that the

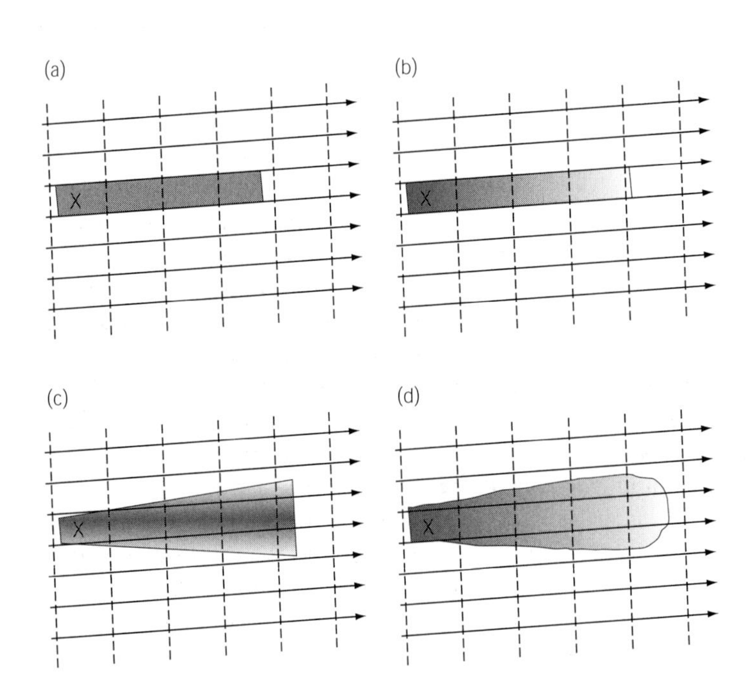

Fig. 10.4 The effects of advection and dispersion on the migration of a pollutant plume in an aquifer. A portion of a flow net is shown, with water table contours depicted by dashed lines and flow lines by solid arrows. A point source of pollution (e.g. a leaking landfill) is marked at point X, and a plume of pollutants is shown migrating down-gradient within the aquifer, with the intensity of shading being proportional to pollutant concentrations within the plume. (a) If transport were solely by advection, with zero dispersion, then pollutant concentrations would remain the same everywhere within the plume, the plume would have a very sharp front, and the pollutants would not be present outside the flow tube within which X lies. (b) If advection were accompanied by dispersion in the longitudinal direction only (i.e. dispersion in the same direction as advection), then concentrations would be lower at the leading edge of the plume than they would be closer to the source; all pollutants remain within the same flow tube as X however. (c) If there were no longitudinal dispersion, but only *transverse* dispersion, the plume is still sharp fronted, but the pollutants are now spreading into the neighboring flow tubes as well. (d) The most realistic pattern in reality: advection with both longitudinal and transverse dispersion. While most pollutants remain within the same flow tube as X, some spreading into neighboring tubes has occurred, the leading edge of the plume is no longer a sharp linear feature, and pollutant concentrations close behind the leading edge and around the outer boundaries of the plume are lower than they are close to the source.

method of characteristics and particle tracking techniques are preferentially applied to solute transport problems (Prickett et al. 1981; Huyakorn and Pinder 1983; Marsily 1986). Brief notes on the practical use of solute transport models are given in Section 10.6.

10.5.3 Hydrogeochemical modeling: speciation, mineral equilibria, and reactions

"It isn't rocket science!" How many times have you heard this expression? Well, there's at least one area of groundwater modeling to which this phrase is definitely inapplicable. This is because one of the first practical applications of the thermodynamic modeling techniques which now underpin all hydrogeochemical modeling was in the design of NASA's rocket fuel systems in the 1970s. The successful modeling of solution chemistry and reactions is one of the triumphs of groundwater science; it is also one of the few strongholds of rigor in an otherwise hand-wavy field.

The first step in all hydrogeochemical simulations is the modeling of **speciation**, which is defined as "the distribution of dissolved components among free ions, ion pairs, and complexes" (Nordstrom 2003). In other words, while a simple water analysis tells us the total amount of a given substance present in a liter of water (see Section 4.1.3), the manner in which that substance actually occurs in solution is only rarely revealed by laboratory analyses. For instance, an analysis might report the occurrence of 50 mg/L of Fe in solution. But how much of this is present as the free ion Fe^{2+}, how much as free Fe^{3+}, and how much as various ionic complexes (in which iron is paired with various anions, such as $FeSO^0_{4(aq)}$, $Fe(OH)^{2+}_{(aq)}$, and $FeSO^+_{4(aq)}$)? It is possible to use thermodynamic theory to develop models of ionic interactions. However, because the answer to our question requires that we also know about other occurrences of the OH^- and SO_4^{2-} anions in the solution, the necessary calculations are complex; they amount to a large number of simultaneous equations, the solution of which requires the use of numerical methods. Once speciation has been modeled, it is possible to use the calculated **activities** (i.e. effective concentrations) of the various ions to calculate how the water is likely to react with respect to a wide range of minerals. It may be that the ionic activities exceed the thermodynamic thresholds above which the water is likely to begin precipitating a given mineral. In this case, the water is said to be **supersaturated** with respect that mineral. In other cases, the activities of one or more of the ions which are present in a given mineral may be so low that dissolution of that mineral would occur (were the water to encounter it). In this case the water is said to be **undersaturated** with respect to that mineral. In rare cases, waters and minerals are at equilibrium (see Section 4.4.2), i.e. the water is perfectly **saturated** with respect to a given mineral, so that it would neither precipitate nor dissolve. (Thorough explanations of these phenomena are given by Lloyd and Heathcote (1985), Appelo and Postma (1993), Nordstrom (2003), and Hiscock (2005), amongst many others).

Mineral saturation calculations tell us *whether* a given reaction might be expected to occur; they do not tell us anything (directly) about the rates of such reactions, or their consequences for other aspects of solution chemistry. There are two ways to begin to delve into these questions. One is to reconstruct mass balances of ions added to/lost from solution along well-defined groundwater flowpaths (cf. Section 4.4.4). This is an example of what is known as **inverse modeling**, where a mathematical simulation reconstructs known end-results (Parkhurst and Appelo 1999). An alternative approach is needed where predictions of hydrogeochemical evolution are required: **forward modeling** of changes in groundwater quality involves specifying ambient conditions (temperature, pressure), relative proportions of any mixing waters, reactants of interest (solutes, minerals, gases), and reaction rates (kinetic coefficients). This type of geochemical reaction modeling is a complex and challenging task (Bethke 1996).

10.5.4 Reactive transport modeling

Increasing computational power is now making it possible to combine geochemical reaction modeling with realistic simulations of solute transport: **reactive transport modeling (RTM)** (Appelo and Postma 1993; Walter et al. 1994a,b; Parkhurst and Appelo 1999). The earliest RTM initiatives involved coupling equilibrium geochemical models to two-dimensional groundwater flow models. This approach necessitates assuming that geochemical reactions are everywhere in equilibrium. While this may be justifiable in cases where groundwater flow is slow and mineral dissolution/precipitation reactions are rapid (cf. Section 4.4.2), in other situations this so-called **local equilibrium assumption** may well be violated. However, the development of nonequilibrium reactive transport models, though viable, remains an area of active research (e.g. Molins et al. 2004), which has yet to become routine practice in the investigation of real-world problems.

10.6 Groundwater modeling in practice

10.6.1 The craft of groundwater modeling

Although it is possible to set up and run fairly complicated groundwater models using spreadsheet programs (Olsthoorn 1985; Anderson and Woessner 1992), the majority of groundwater modeling practitioners now use publicly or commercially available software.[viii] The most widely used flow modeling program is MODFLOW, a quasi three-dimensional finite difference code developed by the US Geological Survey (McDonald and Harbaugh 1988; Harbaugh and McDonald 1996). This code has proved exceptionally robust in use and has withstood scrutiny in many court cases. It remains the code of choice for many public organizations in North America (Anderson and Woessner 1992) and Europe (Hulme et al. 2002). A number of solute transport codes have been successfully developed to complement MODFLOW, including a particle tracking code (MODPATH; Pollock 1994) and a three-dimensional method of characteristics program (MOC3D; Konikow et al. 1996).

Customary practices for the use of such codes are now well established (e.g. Anderson and Woessner 1992; Hulme et al. 2002; Rushton 2003), if not devoid of their critics (Konikow and Bredehoeft 1992). The typical sequence of activities is summarized in Figure 10.5. In applying the routine of Figure 10.5 to groundwater flow modeling, many practitioners first "calibrate" a so-called **steady-state model**, i.e. a version of the model which purports to represent some "average" set of head conditions reflecting the balance between long-term recharge and discharge. Because real aquifers are *never* in such a state, this practice requires acceptance of an inherent element of fiction. However, because it allows testing of alternative transmissivity values before adding the complication of variations in storativity,[ix] steady-state calibration remains popular in practice. The next step requires running the code in full **transient** simulation mode, i.e. allowing for heads to vary over time. It is usual to undertake **history-matching**, in which recharge estimates for the relevant period are input to the model, and the model output values of head are compared with field records of variations in borehole water levels. (The same sort of exercise can also be undertaken using other time-varying parameters, such as spring flow rates.) Transient simulations require storativity values to be specified. These are usually adjusted until good agreement is obtained between observed and modeled data. The language used to describe this stage in model development has proven controversial over the years: the common terms "validation" and "verification" imply some degree of certainty beyond that which is really justifiable in scientific terms (Konikow and Bredehoeft 1992). Rushton (2003) advocates the more realistic term **model refinement**, which in turn implies that history-matching is a prelude to adjustment of the conceptual and mathematical models to achieve greater convergence between concepts and data (cf. Konikow 1981). As Nordstrom (2003) puts it: "Having obtained results from a . . . mathematical

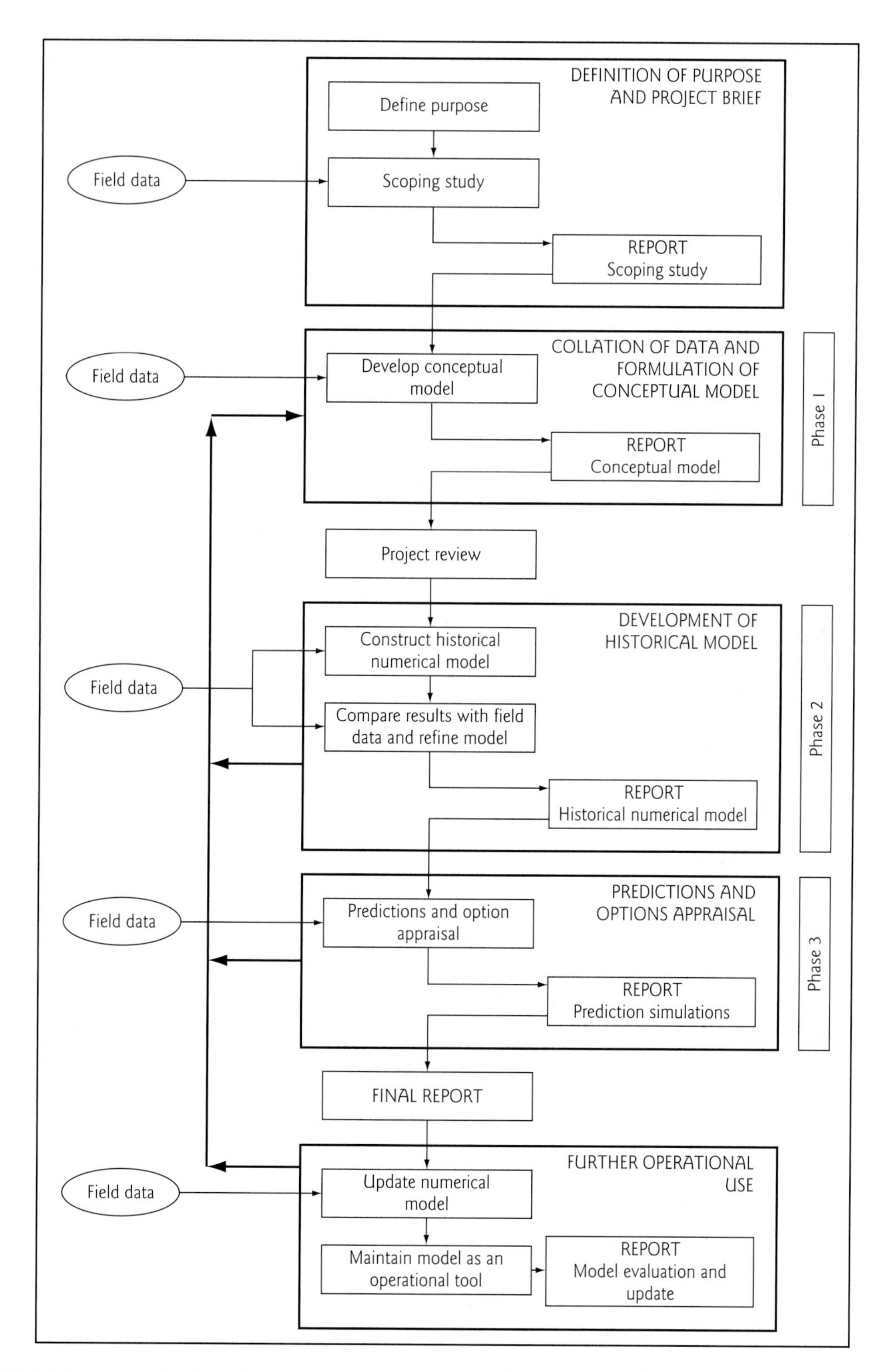

Fig. 10.5 Flow chart showing the typical sequence of steps taken in a groundwater modeling exercise. Although originally developed for modeling flow, the same sequence of steps should also be followed when modeling pollutant migration and/or geochemical processes in aquifers (mass balance or reactive transport). (After Hulme et al. 2002.)

model, some unexpected results often occur which force us to change our original conceptual model. This . . . demonstrates how science works; it is an ongoing process of approximation that iterates between idea, theory, observational testing of theory, and back to modifications of theory or development of new theories" (Nordstrom 2003). All too often, this iterative process is neglected or undertaken half-heartedly; when this happens, opportunities to improve conceptual understanding are lost, which is always regrettable.

Given the difficulties inherent in flow modeling, it is perhaps unsurprising that many solute transport modelers (including those in the vanguard of reactive transport modeling) prefer to assume that the groundwater flow system is in the (fictitious) steady state, avoiding the need to simultaneously solve for both transient groundwater flow and advection/dispersion. It is rarely acknowledged how limiting this assumption of steady-state hydraulics actually is; in many groundwater systems, changes in the direction of groundwater flow occur seasonally, no doubt giving rise to far more "dispersion" than we might ever predict by multiplying a static groundwater velocity by some spurious estimate of dispersivity.

The process of **parameterization**, whereby we assign aquifer properties (*T*, *S*, base and top elevations, etc.), initial water levels, and other data to a model grid such as that shown in Figure 10.3, is notoriously tedious and time-consuming. Advances in information technology are gradually removing much of the drudgery from this essential task. In particular:

- Data entry and scrutiny of output has been greatly eased since the advent of a number of **graphical user interfaces** (**GUIs**; such as Groundwater Vistas™, Visual MODFLOW,™ and GMS™), most of which facilitate straightforward transfers of data to/from **geographical information systems** (**GIS**).
- A number of very efficient methods have been developed for estimating parameter values by interpolation between scarce measurement points; foremost among these are the various techniques of **geostatistics** (most notably the technique known as kriging; Kitanidis 1997), which offer robust and unbiased estimation of "regionalized variables" which are common in hydrogeology (in essence, the closer a given point is to the location of a measured value, the more likely it is to display a similar value).

The modeling of groundwater flow is now a mature and fairly stable industrial practice. Since about 1990, when MODFLOW began to acquire pre-eminence, little has changed in the manner in which groundwater flow modeling is implemented. GUIs have gradually improved over time. This has been a mixed blessing. While it has undoubtedly made life easier in routine modeling jobs, it has introduced a certain distance between the user and the algorithm, which has increased the risk of misapplications of codes, due to misunderstandings of the powers and limitations of numerical methods.

10.6.2 Future horizons

New developments in computer programing techniques offer the potential of restoring an appropriate degree of interaction between algorithm and user, without losing the advantages of the present generation of GUIs (i.e. ease of use and vivid graphics). **Object-oriented programing** (OOP) is now transforming the way in which virtually all new computer programs are written, and though the groundwater software houses have been rather resistant to change, there are clear signs that the advantages of OOP are beginning to be applied to simulation of both flow and transport (e.g. Gandy 2003).

Given the enthusiasm with which three-dimensional imagery has been harnessed in all popular GUIs, it seems certain that the groundwater modeling industry will eventually also begin to make more use of **virtual reality** and related, life-like graphical interaction programming techniques. The ability to climb around within, say, a simulated pollutant plume, looking at the distribution of redox zones and amending the conceptual model as you go, is very appealing. As more and more virtual reality suites are constructed and the costs of using them decrease, it is hard to imagine groundwater professionals not taking advantage of the new horizons opened up by this technology.

Endnotes

i Uniformitarianism is one of the founding principles of geology. The brainchild of James Hutton (1726–1797), it provided the theoretical basis for the stratigraphic interpretation of the entire ancient rock record (cf. Oldroyd 1996).

ii Faced with two potentially embarrassing Christian names, Hubbert wisely chose the second for his given name.

iii If the terminology and abbreviations used here seem unfamiliar, Chapter 3 should be read before continuing.

iv If we know Q by applying Darcy's Law (Equation 3.1), then velocity = Q/(effective porosity $\times$ A)

v Brownian motion refers to the physical phenomenon that solutes move about randomly within a fluid. As all particles are doing this, eventually they will spread themselves out approximately evenly throughout the fluid.

vi Molecular diffusion is generally represented using **Fick's Law**, which summarizes the common observation that molecules tend to spread evenly throughout a solution until they are present throughout the solution. A model is said to be **Fickian** if it follows the mathematical form of Fick's Law (whether or not it is used to model molecular diffusion).

vii In older literature the ADE is sometimes called the "convection–diffusion equation," which is the name invariably given to a mathematically identical formula used in modeling solute transport in surface waters (James 1992).

viii Most of these can be accessed via the International Ground Water Modeling Center (http://typhoon.mines.edu/).

ix Storativity is effectively set to zero in steady-state simulations, as there are no time-dependent head changes.

11

Managing Groundwater Systems

Cha bhi fios air math an tobair gus an tràigh e. *(Gàidhlig–Scotland)*
(How good the well is won't be known until it dries up.)
Chan fhuil meas air an uisce go dtriomaithear an tobar. *(Gaeilge–Ireland)*
(No respect is given to the water until the well dries up.)

(Ancient Gaelic Proverbs)

Key questions

- Do groundwater resources really need to be actively managed?
- How can a permitting system protect the rights of groundwater users?
- Does centralized planning and management have a role?
- Do current groundwater management strategies protect freshwater ecosystems?
- What is "sustainable development," and how does it relate to groundwater use?
- Can aquifers be fairly shared between neighboring countries?
- What steps can be taken to minimize groundwater-related hazards in the construction and mining industries?
- How can we protect groundwater quality from degradation?
- Is remediation of polluted groundwaters a feasible proposition?

11.1 Approaches to groundwater resource management

11.1.1 "Might is right": *laissez faire* resource exploitation

Mankind had been exploiting groundwater for millennia before the need to actively manage aquifers ever became a matter of public debate. As long as water was withdrawn from wells using only buckets or hand-pumps, issues of resource availability seldom arose. All that began to change with the advent of steam-driven pumps in the eighteenth century. Although originally developed for mine dewatering purposes (see Younger 2004b), by the mid nineteenth century large beam engines were pumping groundwater for public supply purposes at many locations in Europe and North America. For the first time in history, single-point groundwater withdrawals of several megaliters per day had become possible; these brought with them regional-scale drawdowns. As cones of depression expanded, established groundwater users found their old, shallow wells beginning to dry up, and the influence of the newcomers was easy to infer. The existing law was ill-prepared to cope with the adjudication of resource disputes of this type. Under English Common Law[i] water beneath the ground was held to be the absolute property of the landowner, which meant that well owners could pump as much water as they liked without regard to its effects on their neighbors, irrespective of who started to pump first. As long as the occurrence of underground water was regarded as "occult and mysterious" (an infamous phrase actually used in some early legal judgments over groundwater disputes), this dispensation held. However, as enlightenment gradually spread even to the world of groundwater engineering, pressure mounted for more logical legal codes, in which the lateral continuity of aquifers and the principle of prior rights (i.e. "first come, first served") are taken into account.

Nevertheless, in most underdeveloped countries *laissez-faire* approaches to groundwater abstraction rights are still the norm, often to the ultimate disadvantage of poor individuals and communities: those who can afford the most powerful pumps can cause the most drawdown (see Box 11.1).

11.1.2 Abstraction permits: protecting prior rights and reactively managing aquifers

It was in the 1960s that most developed countries began to enact laws which afforded some measure of protection to the rights of abstractors, such that they could begin to regard their established pumping operations as enjoying legal safeguards against new abstractions nearby. Typically, these legal codes are enforced by some kind of permitting system, in which approval for a proposed new abstraction will not be granted unless the proposer can satisfy some public authority that their new pumping operations will not adversely affect the continued availability of water to established abstractors in the vicinity. All such laws have their loopholes, of course, such as exemptions which effectively allow certain types of operations (e.g. military, mining, or construction dewatering) to proceed without any need for a permit, or even a need to make some prior assessment of the likely consequences of their actions. Occasional problems therefore still arise, even in countries with long-established abstraction permitting regimes, in which the "small guy" finds their well drying up due to the actions of someone else, and they are powerless to mount any defence under the law.

Abstraction permitting regimes are essentially *reactive* groundwater management tools, in that the water management authorities simply respond to proposals made by others. Where extensive aquifers with high transmissivities are located in humid, temperate regions, this purely reactive approach may be sufficient to ensure that all demands for groundwater are met without giving rise to any of the symptoms of aquifer overexploitation (cf. Sections 9.2.2 and 9.2.4). However, in drier regions, especially where arable production is an important economic activity, simple "first come, first served" permitting systems may not be sufficient to ensure that abstraction rights are distributed such that they ensure satisfaction of

Box 11.1 Deadly serious: *laissez-faire* groundwater development in India.

The use of groundwater for irrigation has been steadily increasing in many parts of India for more than a century. In some Indian states, the last decade of the twentieth century saw a veritable explosion in the numbers of deep boreholes fitted with powerful electric submersible pumps. In some cases, this explosion has been prompted by the provision of cheap (and sometimes even free) electricity supplies. In the absence of any effective regulatory regime for groundwater abstractions, a number of serious consequences are now coming to light. For instance, studies in Karnataka and Andra Pradesh have revealed dramatic declines in river baseflows and plummeting water table levels. The old, shallow wells, worked by hand-pumps, which have supplied village water needs for decades, are most vulnerable to a sudden drop in the water table. Water tankers now supply many poor people, selling them water that until recently they would have obtained for free (apart from maintenance costs) from their own wells. Because the rising demand for water means that shallow wells have to be replaced by deep boreholes that require machinery and funds to drill, the gap between rich and poor is widening: more affluent people have the resources to continue exploiting the diminishing water supplies, whereas the poor are stripped of what little self-reliance they once possessed. As the most vulnerable farmers run out of irrigation water, and even of the water they require to maintain livelihoods and their basic water and sanitation needs, they are thrust into a vicious circle of debt, in which they have to borrow increasing amounts to extract reducing quantities of water. The result is now evident in the high rate of suicides amongst farmers, which has recently become a political issues in India. While suicides are obviously prompted by many factors, a former member of the National Planning Commission reported in 2004 that many suicides have been committed by farmers spending tens of thousand of rupees on digging wells, finding no water, and then borrowing to dig further in desperation. At the Conference on the Groundwater Crisis in Anantapur District, Andra Pradesh, India, August 19, 2004 (attended by over 1500 farmers, NGOs, and government officials), Sri YV Malla Reddy, Director of the Ecology Center of the Rural Development Trust reported that 75% of 400 farmer suicides in the area were attributable mainly to failure of irrigation boreholes. The promotion of irrigation that involves groundwater overexploitation, within an economic setting which greatly exacerbates the gap between rich and poor, is unsustainable in the long term (Calder 1999). In economic terms, it leads to "boom" and "bust" cycles in agricultural production, resulting in steep rises in price inflation. Despite the gravity of these issues, far more international attention has focused on allegations that global soft drinks companies operating in parts of India are causing the depletion of aquifers. The evidence on the latter cases is equivocal; the suicide of hundreds of farmers whose livelihoods have been destroyed through unrestrained irrigation by their more affluent neighbors has an unmistakable and grim eloquence of its own.

specific water demands which have political priority due to their socioeconomic importance. In principle, there is no reason why trading of permits could not be used to redress such imbalances, with market forces ensuring that those who can extract most value from a given volume of groundwater can purchase the right to use more of it from other users. In practice permit trading systems are still rare, for at least two reasons:

- Inherent difficulties in coming up with compatible suites of tariffs, checks, and balances which make such systems appealing both to long-term investors and regulatory authorities.

- Concerns over the potentially oppressive dynamics which they might introduce, in which desperately poor individuals and communities may be forced into parting with their water supply life-line in return for short-lived financial rewards.

11.1.3 Active groundwater management: securing supply in scarcity

Moving beyond abstraction permitting altogether, some countries implement *active* groundwater management regimes, in which a public authority is given sole rights to abstract groundwater throughout a given country, and it simply allocates the pumped water between different end-users in accordance with nationally agreed priorities. Such "command-and-control" approaches to groundwater management were typical of communist governments in the Cold War era. To this day they are also typical of most densely populated countries in arid and semi-arid areas, where the strategic importance of water to maintaining social stability far outweighs any predilection for free trade solutions.

Active management of aquifers might include measures such as aquifer storage and recovery (see Section 7.4.4), river augmentation schemes (see Section 7.4.3), regional water transfers (such as the Great Man-Made River Project in Libya; see Box 9.1), and ordinances forbidding abstraction altogether in the event that symptoms of overexploitation emerge (see Section 9.2.4). Examples of these types of approaches may be cited from countries all over the world (e.g. Box 9.1), with all complexions of political regimes. Nevertheless, it goes without saying that the scope for implementing the most stringent command-and-control measures increases in proportion to the authoritarianism of the government in question. Paradoxically (though not surprisingly) it is precisely the world's most authoritarian regimes which have the poorest records in sustainably managing aquifers for the benefit of all citizens and with respect for the needs of ecosystems.

11.1.4 Holistic groundwater management: respecting ecology and human rights

So far we have focused on groundwater resource management systems that are designed to ensure the rights of human water users; but what about ecosystems. There is no doubting the major advances which groundwater resource management systems have yielded over the last four decades. As we saw in Chapter 6, however, many freshwater ecosystems are critically dependent on the availability of abundant, clean groundwater. While it is entirely feasible to incorporate ecological conservation goals into the decision-making processes of abstraction permitting or command-and-control regimes, it is easy to identify multiple examples of fully developed aquifers in which no attention was paid to the ecological consequences of maintaining large drawdowns (cf. Sections 9.2.2 and 9.2.4). The lesson to be learned from history is that when we focus solely on supplying human demands for water, we have a tendency to incrementally diminish or destroy formerly diverse ecosystems, upon which we ultimately depend for valuable genetic resources, the fertilization "services" provided by insects and birds, and the uplifting recreational benefits of access to unspoiled natural landscapes. Self-interest, narrowly defined, will ultimately be our undoing. *Enlightened* self-interest, in which we ensure that we meet our water needs without irretrievably damaging freshwater ecosystems, is the only sensible course if we wish to secure the future of our own species.

To complete the evolution from *laissez-faire* to holistic groundwater management, we need to go beyond the simple assertion of the prior rights of existing abstractors, and begin to ask:

- How can we ensure equitable allocation of water resources, according to the true needs of individuals and communities?
- How can we meet present-day demands without removing options for future human generations to do the same?
- How can we meet present and future human demands for water without destroying freshwater ecosystems?

The first of these questions arises from the notion that access to water is a basic human right. The United Nations Committee on Economic, Cultural and Social Rights declared as much in 2002: "Water is fundamental for life and health. The human right to water is indispensable for leading a healthy life in human dignity. It is a prerequisite to the realization of all other human rights." The corollary to this is that no system of groundwater management should be tolerated which is not capable of ensuring the satisfaction of the basic human need for water. In trying to evade the simplicity of the UN message, the perpetrators of inequitable water allocation policies have become fluent at obscuring the issue in arguments over detail ("It's enough that they have access to water for 2 hours per day"; "They only have a right to enough water to keep them alive – anything else is negotiable"). In some parts of the world inequitable allocation is a result of deliberate political decisions (see Trottier 1999); alternatively, it may be an "unintended" (if utterly predictable) outcome of political neglect or *laissez-faire* policies. Whatever the cause, in many underdeveloped countries it commonly falls to women and children to carry as much as 15 liters of water every day over distances of many hundreds of meters; back-breaking work, indeed, but necessary to ensure even the most minimal access to water for their families. Altogether, more than one billion humans (or 15% of the world population) still lack access to a reliable source of water in the year 2005; in Africa, fully 40% of the population lack such access (Clarke and King 2004).

The second of the above questions encapsulates the concept of "intergenerational solidarity." This concept has long had currency in various indigenous cultures, in which respect for ancestors and concern for descendants are prized virtues. However, it has only recently entered the mainstream of western political thought, in the wake of the first Earth Summit, held in 1992 in Rio de Janeiro. Efforts to move towards political implementation of intergenerational solidarity have since been initiated world-wide, albeit with only limited success to date in the water sector.

The third question, how we can meet human demands without destroying freshwater ecosystems, amounts to *interspecies solidarity*. Its pursuit is at least as challenging as the other two. The challenges relate to:

- The difficulties in achieving the necessary advances in social consciousness and political will, especially given the relatively slow response/recovery times for groundwater flows and ecosystems in comparison to human life-times, and
- Our relatively poor understanding of the dynamics of coupled hydro-ecological processes in many groundwater-fed ecosystems (see Chapter 6 and Petts et al. 1999).

In the following section, we will consider how these, and related, questions are now being addressed in the pursuit of "sustainable development" of groundwater resources.

11.2 Towards sustainable groundwater development

11.2.1 Sustainable development: the Brundtland definition and Agenda 21

The term "sustainable development" was first formally defined in 1987 by a committee of the World Commission on Environment and Development chaired by the then Prime Minister of Norway, Ms Gro Harlem Brundtland, as follows:

> A *sustainable development* is a development which meets the needs of the present without compromising the ability of future generations to meet their own needs. (World Commission on Environment and Development 1987)

The document containing this definition has come to be known as the Brundtland Report. This definition implicitly advocates equitable access to resources ("meets the needs of the present") while promulgating the concept of intergenerational solidarity. The missing element in this definition is interspecies solidarity: as it stands, the Brundtland definition can be construed as being anthropocentric in nature. This is a point to which we shall return.

Having been adopted formally by world leaders at the 1992 Earth Summit in Rio, an action plan for sustainable development named "Agenda 21" was proposed. (The number 21 refers to the twenty-first century.) In framing Agenda 21, world leaders took cognizance of a number of stark facts:

- Almost 20% of the world population is destitute.
- 90% of world population growth is taking place in underdeveloped countries.
- 20% of the world population use 80% of the planet's natural resources.
- For the first time in history, more than half of the world's population is now living in urban areas, many of which are now mega-cities with populations in excess of 5 million.

A number of themes for action were identified in Agenda 21, relating to prosperity, justice, housing, soil fertility, international cooperation, and environmental protection. While the agenda-setting was global, it was recognized that actions necessarily take place at local level. Thus was born the movement known as "Local Agenda 21," in which partnerships between governmental, private sector, and/or community organizations aim to implement sustainable development at the local or regional scale. Ten years on from the Rio Summit, a second global summit was held in Johannesburg, at which the commitment to Agenda 21 was re-affirmed and the UN Commission on Sustainable Development[ii] was charged with developing and implementing a 15-year development program. The very first thematic priorities in this program include water.

11.2.2 Sustainable water resources development: principles and governance issues

A number of definitions for "sustainable water resources development" have been proposed in the years since the Rio Earth Summit. Table 11.1 summarizes a few of these, and offers some critique of their scope. Probably the most rounded of these definitions is that enshrined in the South African Water Act of 1998. Futhermore, unlike some of the others, the South African definition is not simply rhetoric, but public policy. In the explanatory memorandum of the Water Act 1998, the following statement is made: "Sustainability is not an end in itself, but a critical approach to ensure the renewable supply of water for present and future generations. Key to this is the 'reserve' – the water needed to supply human needs and protect ecosystems." In accordance with this principle, in implementing their Water Act the government of South Africa abolished all pre-existing water rights in the country, and established the "reserve" as the only water right. Definition of the reserve is a multifaceted task, which is being addressed on a catchment-by-catchment basis, by a process which interweaves hydrological analysis, ecological investigations, and social considerations. The latter are evaluated not only by dispassionate observers, but also through community participation.

The South African Water Act 1998 is arguably the most progressive water management legislation currently being enforced anywhere in the world. The principles which it embodies are in harmony with subsequently published global guidelines on the application of the principles of sustainable development to the water sector, developed by the World Humanity Action Trust (WHAT 2000). Table 11.2 summarizes some of the Trust's key recommendations and comments on their particular applicability to groundwater systems. Implementation of the kinds of practical measures listed in Table 11.2 pre-supposes that the system of governance in a given country (or group of neighboring countries) is capable of enforcement. This in turn requires that "those managing water catchments have the accountability, professional competence and legal authority to carry out their duties, and [that legislation exists which] should make possible the meaningful participation of all interested parties" (WHAT 2000). The key goal is to ensure that public values are pre-eminent, irrespective of the absolute ownership of wells, pumps, and other infrastructure.

Table 11.1 Various definitions of "sustainable water resources development."

Definition	Source	Critique
Sustainable water resources development ensure that the benefits of use of a hydrological system will meet present objectives of society without compromising the ability of the system to meet future objectives	UK Groundwater Forum/ Foundation for Water Research	This definition mirrors Brundtland, but is anthropocentric rather than ecocentric; it presumes we can accurately pre-judge "future objectives"
A balance between the benefits of environmental protection and the costs of achieving them	Scottish Environment Protection Agency	This definition amounts to anthropocentrism in a cost–benefit framework, which is itself arbitrary due to the subjectivity inherent in valuing ecological "goods" and specifying discounting period durations
The sustainable use of water requires two conditions: (i) no loss of the potential functions of the hydrological system; (ii) preservation of ecosystems and biodiversity	Netherlands Public Health Institute	This definition seems to overlook issues of equitable allocation (which is admittedly not a huge problem in the Netherlands)
The maintenance and protection of the (ground) water resource to balance economic, environmental, and human (social) requirements	Hiscock et al. (2002)	A robust definition specifically developed with groundwater resources in mind
"Some for all, for ever" – equitable allocation and utilization of water for social and economic benefit, and through environmentally sustainable practices	South African Water Act of 1998	Perhaps the best single definition in use anywhere, especially given the expansion on the various components of the definition given in the Act

In particular, the World Humanity Action Trust advocates the use of a permitting approach (cf. Section 11.1.2) for both abstraction *and* discharges to water bodies (including aquifers), in which strict limits will be imposed on the duration, volume, and quality of all abstractions/discharges, and in which the tariff structure is designed to encourage sustainable management practices (WHAT 2000).

Design of appropriate tariff structures is a challenging activity. As for many other natural resources, the economic valuation of groundwater is not straightforward, due to the need to simultaneously evaluate the value of water in use, and the value of "unused" groundwater left *in situ*, where it performs numerous roles of ecological and social importance. While much progress has been made in the development of appropriate economic approaches applicable to aquifer management (e.g. National Research Council 1997), a wider debate still rages over the validity of attempting to compare sets of values which cannot reasonably be expressed in the same units of measurement (Martinez-Alier 2002).

11.2.3 How sustainable is a given groundwater abstraction?

The constraints on groundwater utility were described in detail in Section 7.2. For any one

Table 11.2 Selected recommendations of the World Humanity Action Trust (WHAT 2000, www.stakeholderforum.org/policy/governance/future.pdf) for sustainable development of water resources, and comments on their applicability to groundwater systems.

Recommendation . . .	Groundwater issues . . .
Water management at national, regional, and international levels must be based on the catchment	From an ecological perspective this makes sense, though many major aquifers underlie more than one surface catchment, and therefore intercatchment groundwater flows must be adequately accounted for and managed
In managing water resources, institutions and individuals must take into account the impacts of their activities on ecosystems and the precautionary principle	This certainly applies to groundwater resources, though its practical application needs special care in view of the time-lags inherent in groundwater flow systems
Governments must actively encourage a greater awareness of sustainable water use and water issues at all levels of society	Educational initiatives are particularly challenging in relation to groundwater systems, given the widespread lack of a basic appreciation of groundwater occurrence
Governments should prepare legislation immediately to ensure that full cost recovery is achieved with a tariff structure designed to increase efficiency of water use	Save in emergency situations, charging people the true cost of water abstraction and supply is an important first step towards the development of an adequate perception of the true worth of groundwater in many societies
Subsidies, existing or proposed, should be carefully evaluated to ensure that they accomplish the socioequity goals advanced as their justification and do not impose unacceptable environmental impacts	This is a particular issue in relation to the uncritical approval of the use of groundwater for large-scale irrigation in semi-arid and arid countries (see Box 9.1)
Investment in water projects in international catchments should encourage co-operation between catchment countries	The same applies to transboundary aquifer systems (Section 11.2.4); in fact the problems can be even more insidious where transboundary groundwater systems are concerned, as pumping from a "downstream position" can cause drawdown "upstream" (unlike in rivers)
All financial investment should require proof that sustainable and efficient water use will be guaranteed	Besides applying this principle to direct investments in the water sector, it is also important that large "accidental" users of groundwater (such as forestry, which depletes recharge, and mining, which directly intersects vast quantities of groundwater) be subject to the same level of scrutiny
Assumptions that water will be provided for all new developments must cease	It is good practice to specify and enforce absolute limits on abstraction from many aquifers; if that prompts the siting of water-intensive industries elsewhere, so be it. Only when the political will exists to honestly make such hard (but necessary) decisions will there ever be a prospect of achieving sustainable groundwater management

abstraction (or closely spaced cluster of abstractions, as in a wellfield) it is important to be able to assess the likely sustainability of a current or proposed rate of pumping with considerable precision and clarity. To do this, we make use of the two factors "deployable output" (DO) and "potential yield" (PY), which are defined in Section 7.2.2. Having quantified these two factors, paying particular attention to the ecological elements of the "environmental issues" cited in the definition of DO, then the decision logic for assessing the sustainability of a given groundwater abstraction can be summarized very simply, as follows (Younger 1998):

DO > PY: The abstraction is not sustainable, and the aquifer is probably overexploited (cf. Sections 9.2.2 and 9.2.4).
DO = PY: Although the groundwater resource is fully developed, the abstraction is probably sustainable.
DO < PY: The abstraction is sustainable and the groundwater resource may well have spare capacity for further abstraction.

It should be noted that identification of an excess of DO over PY is not necessarily a bad thing as long as "it is not permanent. It may be a step towards sustainable development . . . the term aquifer overexploitation is mostly a qualifier intended to point to a concern about the evolution of the aquifer-flow system from some specific, restricted points of view . . . implementing groundwater management and protection measures needs quantitative appraisal of aquifer evolution and effects based on detailed multidisciplinary studies, which have to be supported by reliable data" (Custodio 2002).

11.2.4 Transboundary groundwaters

Particularly difficult problems arise in those parts of the world where major aquifers extend across one or more international frontiers. In such cases, it may be a waste of time for an individual country to enact laws to enforce sustainable management of groundwaters unless their neighbors do so too. Some of the potential problems associated with transboundary groundwaters include (Figure 11.1):

- Excessive abstraction either up-gradient (Figure 11.1a) or down-gradient (Figure 11.1b) across the international frontier can diminish the available groundwater resources in the neighboring country.
- Inadequate groundwater protection measures in an up-gradient country can lead to contamination of valuable groundwater resources in the down-gradient country (Figure 11.1c).

Shared aquifers are a major bone of contention in arid regions of North Africa and the Middle East. For instance, the gigantic Nubian Sandstone Aquifer underlies parts of Chad, Sudan, Libya, and Egypt, all of which are countries that suffer frequent surface water droughts. Of particular concern are the various aquifers that underlie parts of both Israel and the occupied Palestinian territories of the West Bank (limestones) and Gaza Strip (unconsolidated sands) (e.g. Trottier 1999). In that particular case, international peace negotiations are unlikely to reach a satisfactory conclusion unless the equitable use of these transboundary aquifers can be agreed. Elsewhere, the management of transboundary aquifers is thankfully not usually a matter of war and peace, though careful international negotiations are often necessary to avoid political hostilities developing. Since the year 2000, an initiative called ISARM (International Shared Aquifer Resource Management) has been under development by various UN agencies and the International Association of Hydrogeologists, with the aim of promoting conflict reduction through facilitating joint studies of shared aquifers. Since May 2003, the world's first major transboundary aquifer management concordat has been under negotiation between the governments of Argentina, Brazil, Paraguay, and Uruguay to facilitate collaborative management of one of the world's biggest aquifers: the Guaraní Aquifer. With an extension of 1.2 million square

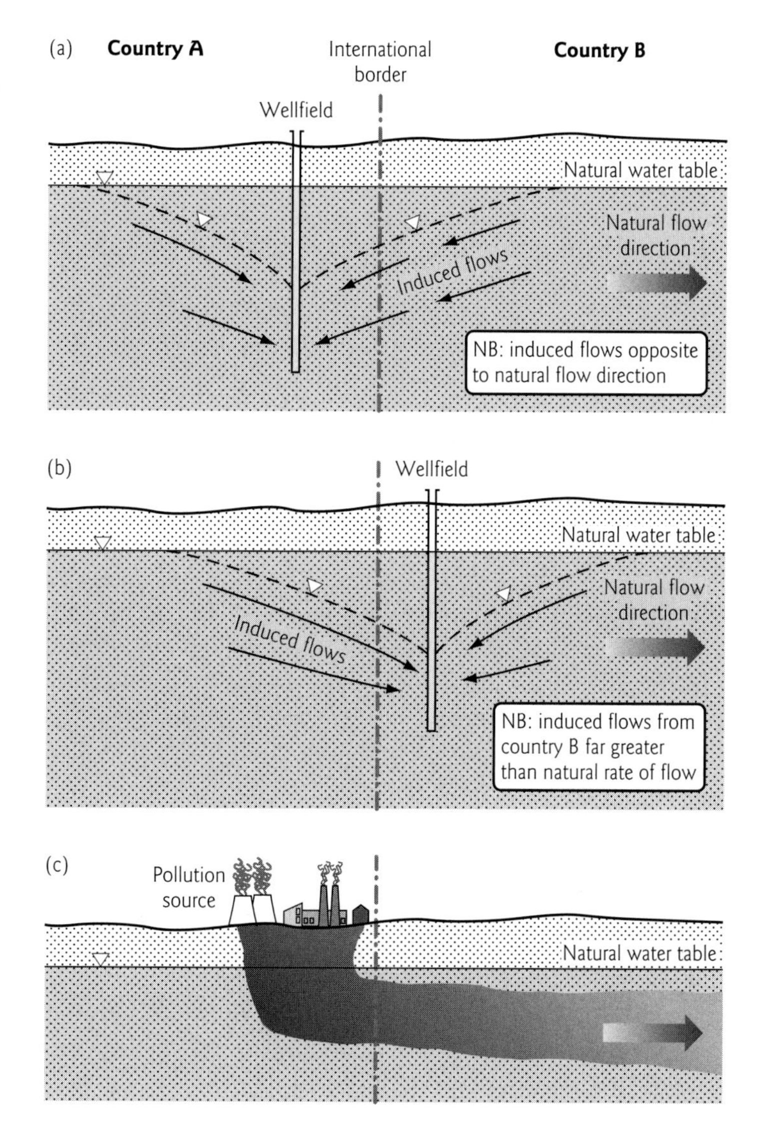

Fig. 11.1 Transboundary aquifers: issues of groundwater interception and pollution across international borders. Dashed line indicates water table induced by pumping on one or other side of the border between two countries. (a) Abstraction upgradient of an international border. In this case, the abstractors in Country A are intercepting groundwater which would have flowed naturally from their country into Country B, and are also inducing inflow of a further complement of water which originated as recharge in Country B. (b) Abstraction downgradient of an international border. Although Country B always received inflow from Country A even under natural conditions, a significantly greater inflow is now being induced by pumping. (c) Pollution of groundwater from a point source upgradient of an international boundary. Because of the natural flow direction of groundwater in this aquifer, a pollution source located as shown in Country A contaminates far more groundwater in Country B than in its country of origin.

kilometers, this major sedimentary aquifer was only relatively recently recognized as a transboundary resource. In the process of developing an agreement for the management of this aquifer, the basic principles espoused by the ISARM initiative have been vindicated, though aquifer-specific characteristics are obliging all stakeholders to develop innovative measures (legal, institutional, technical, scientific, social, and economic) to ensure harmonious sustainable management of the Guaraní system in decades to come (Kemper et al. 2003).

11.3 Groundwater control measures to mitigate geohazards

11.3.1 Introduction

The various geohazards associated with natural and artificially altered groundwater flow systems were described in Chapter 8. Here we consider some of the common measures taken to minimize the risks posed by these geohazards. As in the case of groundwater pollution, the chief maxim is "prevention is better than cure."

11.3.2 Stabilizing slopes by groundwater control

Depending on the scale of the slippage, the failure of slopes can have a range of consequences ranging from minor inconvenience for livestock, through the disruption of important rail and road networks, to massive loss of human life. In almost all cases, the likelihood of slope failure can be greatly reduced by means of pore water control (cf. Section 8.3.2). Because slope stabilization is meant to produce stable landscape features which may stand largely unattended for decades or more, preferred methods of pore water control all rely on the use of **gravity drainage** rather than pumping. At its simplest, pore water control can be achieved by preventing infiltration on or near the slope by means of interceptor drains (Figure 11.2). A cut-off drain set just back from the crest of the slope is indispensable. This will usually be concrete-lined and/or filled with cobbles to prevent it becoming a focus for erosion. It is also common to install a dendritic network of cobble-filled drains on the slope face itself (Figure 11.2a), which will result in rapid drainage of any incipient overland flow to a toe trench, which must be engineered to prevent undercutting of the base of the slope. Where it is necessary to lower the water table well below the surface of the slope, then galleries or subhorizontal boreholes can be driven in from the toe of the slope to intersect all potential zones of water-logging (Figure 11.2b). In extreme cases, it may be worthwhile drilling vertical wells (which can be back-filled with sand or gravel) to connect perched lenses of saturated material to the gallery or subhorizontal borehole at depth (Figure 11.2c).

11.3.3 Dealing with the risk of subsidence due to groundwater abstraction

Subsidence is typically irreversible. *Post hoc* responses to subsidence are really a matter of damage limitation and safeguarding public safety. The case of Florida, where all householders are required by state law to take out insurance policies against the risks of subsidence, has already been mentioned (see Section 8.3.3). Such policies can provide the wherewithal for activities such as grout injection to stabilize subsurface voids, capping of dolines with reinforced concrete plinths, and underpinning of buildings. By the time such measures can be implemented, however, the physical and psychological damage will already have been done. It is therefore preferable to address potential subsidence risks before they arise. This is best approached by means of governmental ordinances to control locations and/or technologies of construction in subsidence-prone areas. Hazard mapping can be used to identify high-risk areas (e.g. Gutiérrez-Santolalla et al. 2005). Where construction *must* proceed in such areas, for instance where linear transport routes cross them, it is possible to incorporate structural safeguards into the designs of roads, bridges, and rail embankments, so that they will remain serviceable even if dolines develop immediately beneath them (e.g. Jones and Cooper 2005).

11.3.4 Groundwater control in the construction industry

The hazards posed to shallow excavations by groundwater have been outlined in Section 8.3.3. There are two principal approaches to the mitigation of these hazards: physical exclusion of groundwater and artificial groundwater lowering. It is important to note that these two approaches are generally complementary rather than competitive.

Perhaps the most common groundwater exclusion method is the installation of sheet-piling, which is basically a system of interlocking steel sheets driven vertically into the ground. Sheet-piling is principally used to support soil so that deep excavations can be made without having to excavate huge areas to obtain slopes which are sufficiently gentle that they will stand unsupported. However, because the piles interlock, they often impede (though do not entirely stop) the flow of groundwater towards the excavated area. A greater degree of water-tightness can be achieved by using variants of piling technology that incorporate the injection of cement- and/or

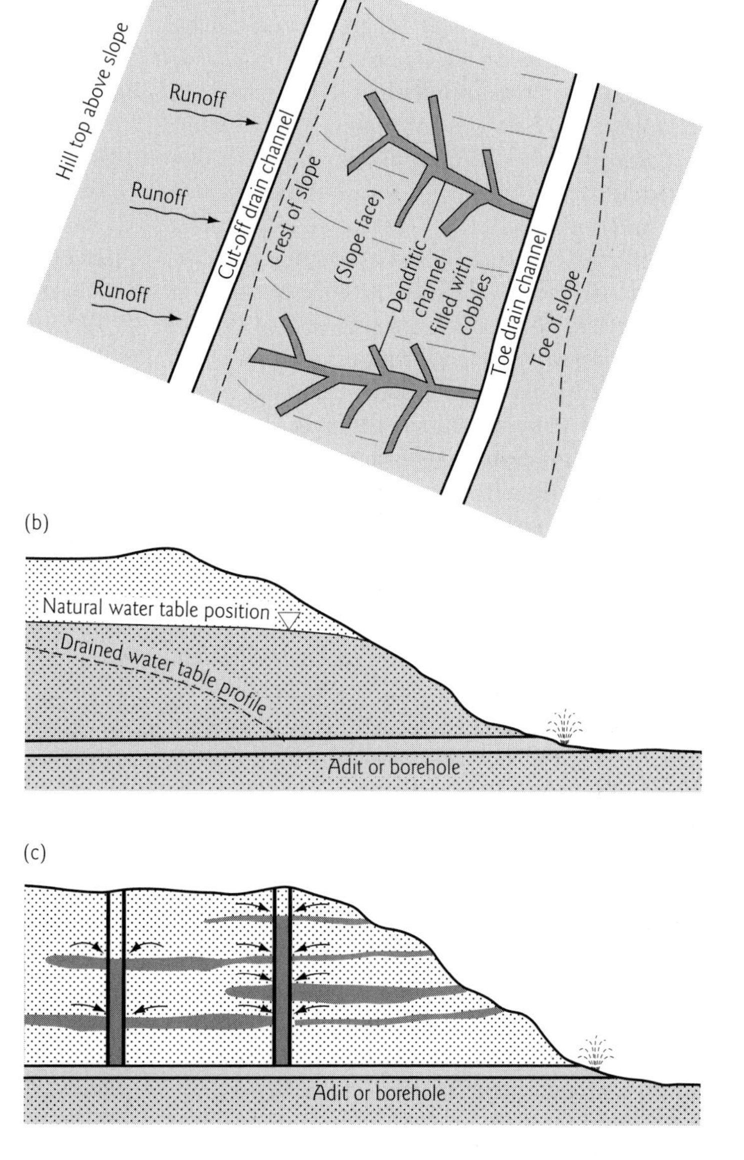

Fig. 11.2 Common strategies for slope stabilization by groundwater control. (a) Plan view of typical slope stabilization measures in materials prone to failure when subject to high pore water pressures. Runoff originating on the hill top above the slope is intercepted by a cut-off drain immediately above the crest of the slope. Runoff which overflows this channel in extreme events, plus any localized recharge originating due to rainfall directly onto the slope face, will be intercepted by the dendritic channels filled with cobbles, cut into the slope face itself. These cobble-filled channels, plus any runoff on the slope face which does not enter them, discharges into the toe drain, which rapidly moves water away from the vulnerable slope toe area. (b) Cross-section showing the lowering of a naturally high water table in the vicinity of a vulnerable slope by installation of an adit (i.e. a drainage tunnel) or a horizontal borehole. This approach may suffice in relatively massive, homogeneous aquifer materials. (c) In cases where the aquifer material is heterogeneous, such that perched aquifers develop at distinct horizons (cf Figure 1.6), the approach shown in (b) above may be insufficient to ensure slope stability; vertical wells back-filled with freely-draining material (sand or gravel as appropriate to ground conditions) can ensure that the perched lenses of groundwater are drained down to the base level drain (i.e. the adit or horizontal borehole).

bentonite-based grouts. In extreme cases, temporary groundwater exclusion can be achieved by means of ground freezing, allowing conventional excavation against vertical walls of frozen soil and the installation of impermeable walls using ordinary building techniques (e.g. Bell 2004). These more elaborate techniques are significantly more expensive than sheet-piling. For this reason it will often be more cost-effective to combine sheet-piling with groundwater lowering than to go for extensive grouting or ground-freezing.

Groundwater lowering is achieved using temporary wellfields (Powers 1992; Preene et al. 2000; Cashman and Preene 2002). To understand how these wellfields are designed it is necessary to recall the cone of depression which develops around any pumping well (see Section 3.3.3; see Figures 3.9, 3.10). While a single well *may* produce enough drawdown to lower the water table below the planned maximum excavation depth, the exponential shape of distance–drawdown curves (Figure 3.9b) is such that two or more wells can usually obtain greater drawdown at any one point for a given pumping rate. In other words, the total drawdown at any one point due to a number of wells pumping in consort equals the sum of the individual drawdowns which any one of the wells would cause were it pumping on its own (Figure 11.3a). As drawdowns due to additional wells are simply superimposed on the existing total, this "law" of wellfield design is known as the **principle of superposition**.

Fields of conventional water wells similar to those used for water supply purposes (e.g. Figure 3.7), from which the groundwater is removed using electric submersible pumps, may be used where the problem aquifer is thick and extensive. Often, such deep-seated aquifers will underlie the soils into which shallow excavations are to be sunk, and the aim of groundwater lowering will be to de-pressurize the underlying aquifer so that floor heave is prevented (Figure 11.3b). In the construction industry, applications of this type are known as **deep well** installations.[iii]

To control groundwater within the shallow soils themselves, **wellpoints** are used. A single wellpoint array will comprise between four and twelve shallow (<10 m total depth), small-diameter (50 mm) wells all connected to a common surface collector pipe, which is placed under suction using a single large pump (Figure 11.3c). Wellpoints can often be installed very cheaply, using a simple technique known as **jetting**, in which the erosive power of water is used to create a hole into which the well components can easily slide. In firmer soils, it may be necessary to use more conventional drilling techniques (augers, percussion, or rotary) to install the wellpoints. Due to the inherent physical constraints on suction lift devices, the theoretical maximum depth to water in a pumping wellpoint system operating close to sea level is 8 m below ground surface,[iv] though taking into account mechanical inefficiencies and friction losses, a realistic limit closer to 6 m obtains in most practical applications. However, greater drawdown levels can be achieved by a multistage design, in which concentric arrays of wellpoints are installed on benches of decreasing altitude towards the center of the excavation (see Powers 1992; Preene et al. 2000; Cashman and Preene 2002). One of the principal drawbacks of wellpoints is the clutter associated with the indispensable surface pipework immediately surrounding an excavation.

Where groundwater must be lowered further than 6 m below ground, and especially where the shallow aquifers include low-permeability materials such as silts or muds, **ejectors** are often the best option. Ejectors occupy a niche in the dewatering industry where pumping water levels are too deep for wellpoints, but well yields are too low to allow the use of electric submersible pumps (as in classic "deep wells"). Ejector wells are drilled, just like deep wells: they differ from the latter simply in the manner of water extraction. Unlike electric submersible pumps, which quickly burn out if run in a dry borehole, ejectors can pump air/water mixtures without any problems, and if the top of the borehole is sealed, the air pressure within the well will drop below atmospheric pressure, thus allowing direct drainage of water from the unsaturated zone. As such, ejectors can be used to improve the stability of low-permeability silts and fine sands. Because they use decompression of introduced water to lower air pressure at the base of a borehole, ejectors can operate to far greater depths than wellpoint suction pumps; pumping water levels of 30 m depth are common, and with careful design and installation ejectors can even pump from depths as great as 50 m.

Both wellpoints and ejectors can be used in conjunction with sheet-piling (Figure 11.3d). Besides preventing lateral inflows of groundwater through gaps in the sheet piles (which are generally more of a nuisance rather than a danger), arrays of wellpoints or ejector wells can

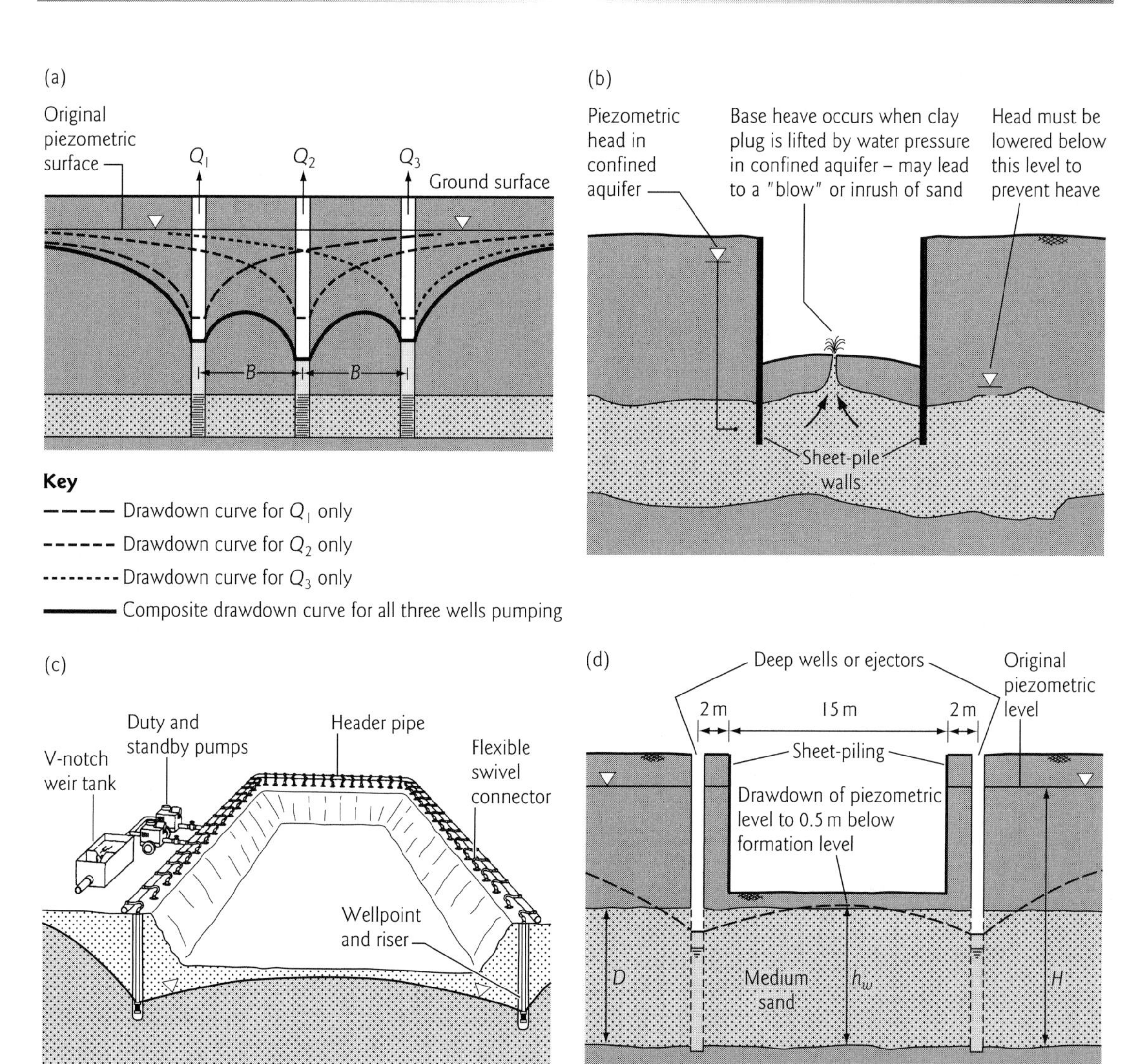

Fig. 11.3 Groundwater control for construction projects. (a) The Principle of Super-position: the drawdown achieved by a number of wells pumping together is the sum of the drawdown they would each cause if operating solo (adapted after Kruseman and de Ridder 1991). (b) Typical wellpoint dewatering system for a shallow excavation. The system comprises a large number of individual shallow wells (wellpoints), each containing a "riser" (narrow plastic pipe, perforated only near its base), which are all connected via flexible swivel connectors to a common header pipe. The header pipe is placed under vacuum by suction provided by the duty pump. (Note that although the perspective of the diagram allows the subsurface components of only two wellpoints to be shown, every flexible swivel connector shown on the diagram has its own wellpoint – a total of 45 in this example; adapted after Preene et al. 2000.) (c) The logic of de-pressurization of a deep confined aquifer to prevent base heave in the sole of an excavation. If the hole is excavated within the sheet-piling without lowering the head in the underlying sand aquifer, base heave is likely to occur (adapted after Cashman and Preene 2002). (d) Combined use of sheet piling and deep wells or ejector to prevent base heave in a deep excavation (adapted after Cashman and Preene 2002).

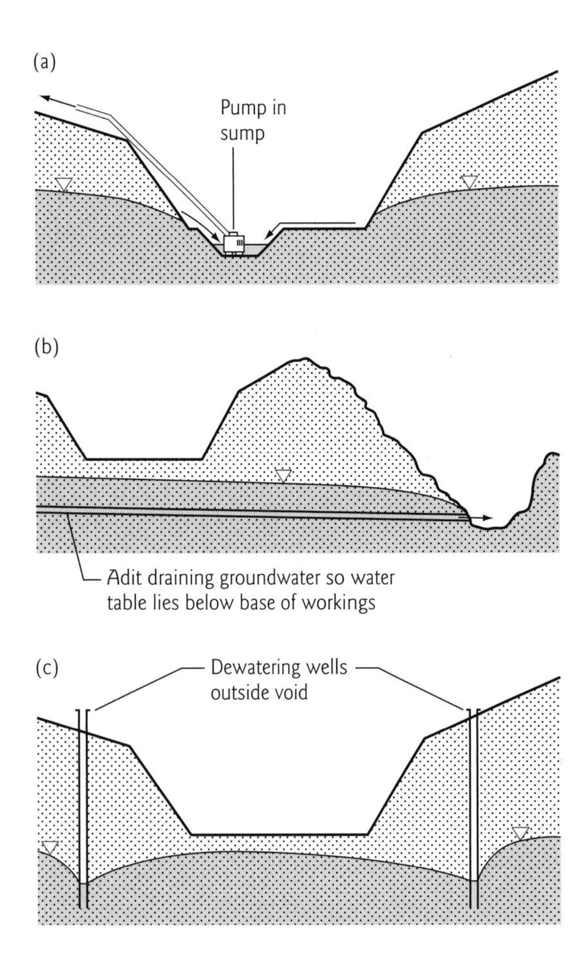

Fig. 11.4 Mine dewatering: the three principal strategies (after Younger et al. 2002). (Note that although the examples shown relate to surface mines, exactly the same principles apply to the dewatering of underground mine workings.) (a) Sump dewatering, in which no dewatering infrastructure is installed outside of the working voids, and all incoming groundwater is directed to one or more storage ponds ("sumps"), from which it is pumped out of the mine. (b) Adit dewatering, in which a drainage tunnel (adit) is driven below the projected sole of the deepest workings, to underdrain the ground scheduled for mining. (c) External dewatering, in which dewatering wells located outside of the working area (or sometimes both inside and outside of the working area) are pumped sufficiently that the net drawdown lowers the water table below the sole of the working area. It should be noted that strategies (a) through (c) are not necessarily alternatives; it is often necessary to use some combination of all three to achieve adequate groundwater control throughout a mined area.

also be used to prevent the development of quick conditions or floor heave (Section 8.3.3) in the bottom of the excavation.

11.3.5 Groundwater control in the mining industry

The hazards posed to mining operations by groundwater have been outlined in Section 8.3.4. Whether the aim is to minimize the risk of potentially fatal flooding of underground workings, or to prevent production losses due to slope failures in open pits, three basic approaches to dewatering are used by the mining sector worldwide (Younger et al. 2002a):

- **Sump dewatering** (Figure 11.4a), in which any water which enters the mine workings is diverted to one or more low-points in the workings (sumps), whence it is pumped out of the mine.
- **Adit dewatering** (Figure 11.4b), in which one or more long, near-horizontal tunnels (adits) are driven in to the mined ground to an artificial high-permeability drainage pathway below the natural level of groundwater discharge.
- **External dewatering** (Figure 11.4c), in which one or more pumping boreholes (or, in some cases, old mine shafts) located outside of the current zone of mineral extraction are used to intercept groundwater which would otherwise have flowed into the active mine-workings.

At many mines, some combination of all three approaches will be used. For instance:

- Even where effective external dewatering is taking place, some sump dewatering is nearly always necessary to prevent flooding of isolated low points in the workings in low-permeability strata.
- Where an adit system is in use, it is often still necessary to pump water from some parts of the workings before they can drain away freely through the adit.

The craft of mine dewatering is complex and generally sparsely documented. The interested reader is referred to Younger et al. (2002a) for an overview.

11.4 Preventing groundwater contamination

11.4.1 Principles of groundwater protection

The concept of aquifer vulnerability was outlined in Section 9.3.2 and Figure 9.2, and various approaches to mapping vulnerability were described. By applying such approaches, it is possible to classify the Earth's surface into zones of greater or lesser vulnerability to groundwater pollution. Armed with such a classification, it is possible to identify areas in which specific industrial, agricultural, or municipal activities may be undertaken without any risk to groundwater quality, and other areas in which careless land use could lead to intractable problems of aquifer pollution. Depending on the system of governance in a given country, it may be possible to ensure that land-use practices harmonize with the aim of protecting groundwater quality. In North America and the European Union, for instance, public bodies are responsible for permitting (or forbidding) proposed developments by individuals or companies. If the officials who make such decisions take aquifer protection considerations into account, then long-term safeguarding of groundwater quality can be achieved. In most jurisdictions, these safeguards are implemented at two scales of resolution: generalized *aquifer protection*, in all areas underlain by aquifers (whether or not they are currently pumped in that vicinity) and *source protection*, in the vicinity of existing authorized abstractions.

11.4.2 Vulnerability maps

Vulnerability maps provide the principal tool for ensuring aquifer protection. Such maps are published by governmental agencies in many countries. Figure 11.5a shows an extract from one such map, in which areas with different levels of vulnerability are clearly marked. Before using such maps in the decision-making process, it is essential that reference is made to the specific vulnerability mapping method used to derive the map (e.g. GOD, DRASTIC, or some other approach; Section 9.3.2). For instance, in the case of the Republic of Ireland, vulnerability is defined on the assumption that any pollution source will be located 2 m below the ground surface (a depth chosen to cover the typical depths of most building foundations and buried pipelines, etc.). Where a proposed development would entail excavating to a greater depth, it is essential that the published vulnerability map be re-interpreted accordingly (Kelly C., 2005, Geological Survey of Ireland, personal communication). It is also important to bear in mind that vulnerability maps can only provide high-level guidance on the likely compatibility between a given activity and a given site; it will never be a substitute for site-specific investigation and adequate precautionary design.

11.4.3 Source protection zones

The natural filtration and biogeochemical transformation processes which occur in most aquifers are a key reason why groundwaters are often preferred to surface waters as sources of potable supply (cf. Sections 7.1.1, 7.2.3 and 9.3). Obviously, the further a given drop of water moves through an aquifer before reaching a well, the less likely it is to contain contaminants by the time it arrives. On this basis, it is logical to be especially vigilant about potentially polluting activities occurring close to a wellhead (or a spring), whilst allowing oneself to become progressively less paranoid the further away the activities occur. Applying this logic, source protection zones tend to be defined as concentric rings (normally skewed in the up-gradient direction; Figure 11.5b), with strictest land-use controls applying to the innermost rings. The inner source protection zone is often defined on the basis of protection against pathogen contamination. We have already seen that most pathogen contamination of springs and wells is introduced by users themselves (see Section 7.3), so that it is important that the activities of the well owners are as tightly regulated as those of other people. Natural rates of bacterial die-off during flow through porous media are reasonably well known, from studies of slow sand filters which are widely used in water treat-

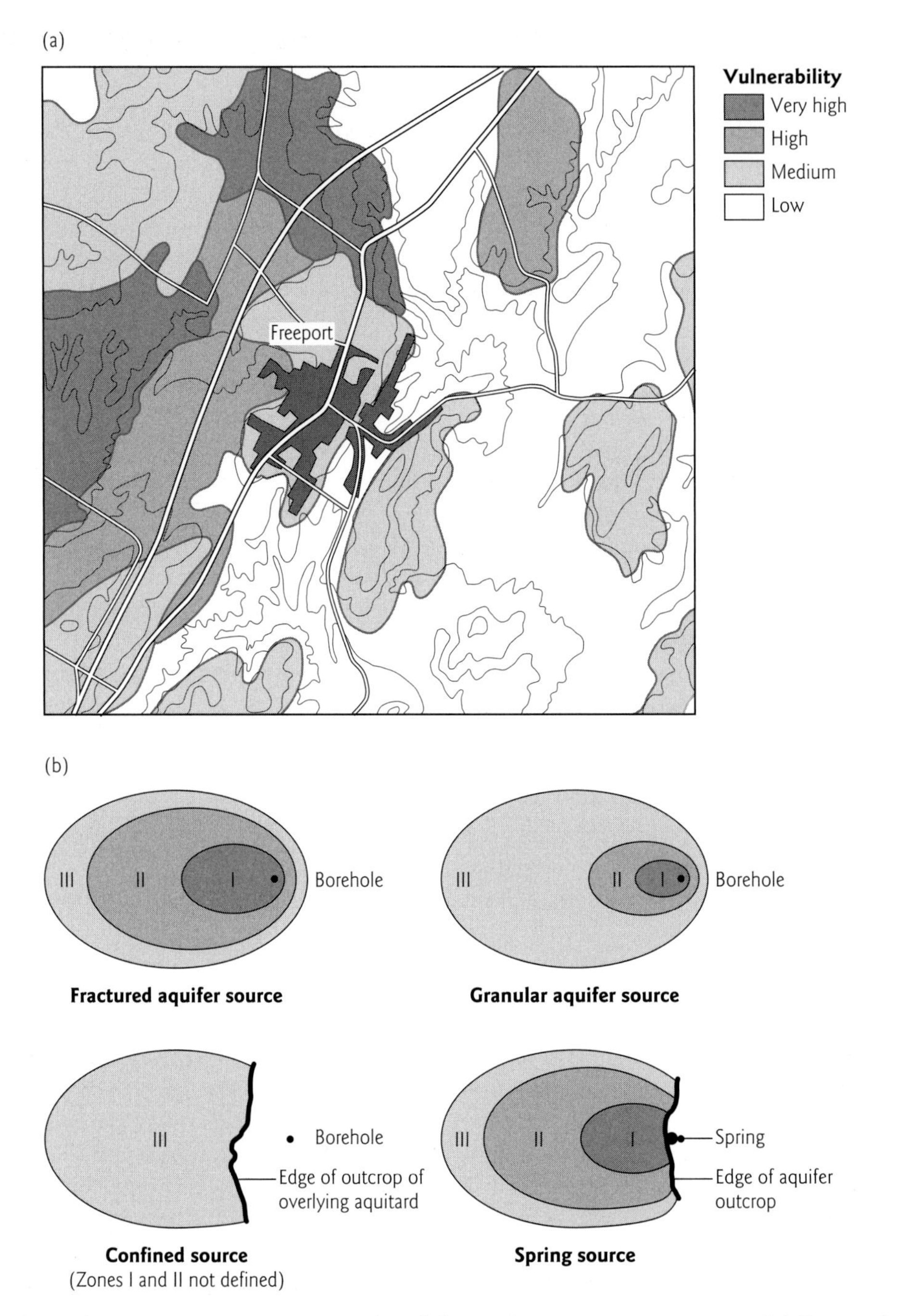

Fig. 11.5 Groundwater protection: mapping vulnerability and source protection zones. (a) Extract from a vulnerability map for a small area in the eastern United States, in which the land surface is categorized according to the degree of vulnerability of groundwater underlying it (adapted after Aller et al. 1987). (b) Source protection zones: schematic plans of the anticipated layout of the inner (I), intermediate (II), and entire (III) source protection zones anticipated in a range of hydrogeological settings (adapted after NRA 1992). Groundwater flow direction is left to right in all cases, which explains why none of the zones are concentric, but skewed upgradient into ovoid shapes. The first two cases are for wells in unconfined aquifers: the difference in shape between the "fractured" and "granular" reflects the generally higher effective porosity of the latter. The third case, for a well piercing an overlying aquitard to reach a deeper confined aquifer, assumes that aquifer to be granular. The final case is for a spring draining an unconfined aquifer. It should be noted that real source protection zones rarely turn out to be so ideally ovoid in plan, due to the complexities of aquifer heterogeneity and spatial variations in recharge.

ment (e.g. Binnie et al. 2002). On this basis, it is usually safe to assume that groundwater which take more than 50 days to reach a pumping well is likely to be free from pathogens. The inner source protection zone is usually defined accordingly, as an area with a notional 50-day travel time to the well, or a 50 m radius, whichever is the greater.

Beyond the inner protection zone, the only other zone which can be defined with any clarity is the outer limit zone, corresponding to the entire area from which the well draws water (i.e. the entire borehole catchment). This can be defined in various ways, including comparing the pumping rate with the recharge rate and calculating the necessary contributing area. Some source protection zone policies settle for an inner and outer source protection zone alone; this is so in the Irish policy, for instance (Misstear and Daly 2000).

Many other policies seek a more nuanced approach, distinguishing up to four intermediate "rings" between the inner and outer protection zones. The notion is that processes such as dilution, biodegradation, sorption, and precipitation are all more likely to reach completion the further the water has to travel to the well. However, the rates of these processes vary greatly between hydrogeological settings and (save for dilution) between different solute types. In the absence of obvious generalized attenuation rates applicable to all contaminants, many jurisdictions arbitrarily define intermediate source protection zone rings on the basis of a range of notional days of travel time to source (200 days, 400 days, etc.). Besides the arbitrary nature of these subdivisions, they also suffer from the problem that groundwater travel times are notoriously difficult to estimate with any great certainty. Nevertheless, many public authorities prefer to deal with these arbitrary subdivisions rather than be accused of imposing overly strict land-use limitations over very wide areas.

The establishment of source protection zones has been underway in many countries for more than 20 years now. Have they worked? This is a very difficult question to answer, for the only proof lies in an absence of evidence. However, it is the view of most "old-timers" in the hydrogeological profession that the establishment of source protection zones has resulted in far fewer instances of well contamination than they used to experience.

11.5 Remediating contaminated groundwaters

11.5.1 Making decisions about groundwater remediation

Sometimes all preventative actions fail, and we have a body of contaminated groundwater to deal with. Before considering what technical options might exist for remediating the contamination, it is important to ask a number of more fundamental questions, such as:

- What is at risk if we do *not* remediate the groundwater?
- What level of clean-up would be acceptable, given the possible risks?

The word "risk" arises in both of these questions. Many technical specialists draw a distinction between "risk" and "hazard." A hazard is generally accepted to be something which is *potentially* dangerous, but this only becomes a *risk* if there is a credible mechanism for this danger to be realized. A good example from the natural world would be cadmium. This metal is inherently hazardous to human health, but as long as it remains buried at great depth it poses us no risk. Once exposed by miners, a serious risk of exposure to cadmium arises. Using this type of reasoning, groundwater specialists routinely use assessments of risk to inform decision-making over remediation. The logical framework is summarized in the following sequence:

SOURCE ⇒ PATHWAY ⇒ RECEPTOR

In making decisions about groundwater remediation, this logical framework can be applied at a number of scales:

- Groundwater itself can be regarded as a "receptor," the pollution of which is unacceptable. Using this approach, the "source" is typically a surface (or near-surface) point or diffuse source of pollutants (see Section 9.3), the "pathway" is the unsaturated zone and the "receptor" is the saturated zone. This approach is consistent with the argument that sustainable management of groundwater precludes us allowing contamination of groundwaters at present, even if they are currently unused, as this effectively eliminates choice for future generations (Section 11.2.1).
- A well, spring, or groundwater-fed ecosystem can be regarded as the "receptor," in which case the "source" may be a surface or subsurface source of pollution, and the "pathway" includes both the unsaturated and saturated zones.
- If the original source of pollution at the ground surface has been removed, a body of polluted groundwater can itself be regarded as the "source," with the pathway being a well or spring, and the receptor being one or more humans, animals, or plants.

In whatever manner a particular risk assessment is framed, the key to risk *minimization* lies in eliminating the source and/or breaking the pathway. **Source elimination** tends to be more easy to achieve for surficial point (as opposed to diffuse) sources of groundwater pollution. Physical removal of a pollutant source may be feasible in the case of small scale spills or LUST (see Section 9.3.3); in the case of old landfills or extensive bodies of contaminated land, it may be more appropriate to fit the source with an impermeable cover, and possibly also to inject grout to impermeabilize the soils beneath it, to prevent any further release of pollutants.

Breaking the pathway between a source and a receptor generally requires either *in situ* or *ex situ* remediation.

11.5.2 *Ex situ* remediation technologies

The principal *ex situ* approach is **pump-and-treat**, in which one or more pumping wells abstract the contaminated groundwater, and conventional water treatment technologies are used to strip the contaminants from the water before it is either re-injected, dispatched for use elsewhere, or discharged to a surface water course (US EPA 1996). The selection of treatment technologies is highly dependent on the form of pollution present (Nyer 1992; US EPA 1995). For instance, a wide range of organic compounds, trace metals, and metalloids are susceptible to irreversible sorption onto **activated carbon**. In the case of volatile organic compounds (VOCs; see Section 9.3.3), it is usually cost-effective to first displace these into the vapor phase (by cascading the contaminated groundwater through a tower filled with baffles, a process known as **air stripping**) before passing the VOC-carrying air through an activated carbon filter. LNAPLs are generally amenable to physical separation from pumped water using simple segmented tanks. Many metals and metalloids are amenable to precipitation and sedimentation processes, with sorptive and electrochemical processes sometimes proving useful for selective removal of small dissolved quantities of toxic substances.

For certain pollution problems it is possible to house the treatment element of a pump-and-treat facility below the ground surface, directly above a single well which serves both to pump the polluted water and reinject the treated water. Known as **recirculating wells**, these compact pump-and-treat installations are sometimes referred to as *in situ* technologies, though this is not strictly the case, given that water is still pumped out of the saturated zone for treatment.

A particular problem with pump-and-treat is the so-called **rebound** of contaminant levels in the groundwater after the end of pumping operations (Mackay 1998). Typical circumstances are that the concentrations of contaminants in the raw groundwater (both pumped and in nearby observation wells) decline throughout the period of pump-and-treat operations. Eventually the apparent contaminant levels are well below target concentrations, and the pump-and-treat operation is suspended. After natural hydraulic gradients have become re-established, sampling of the monitoring wells shows contaminant concentrations once more above acceptable limits; they are said to have "rebounded." The explanation for this phenomenon varies from one type of contaminant to another. For instance, in relation to LNAPLs, it may be due to "stranding" of free product in the unsaturated zone as the water table dropped during pumping, only for the contaminants to be mobilized again after water table recovery. Similarly, contaminants sorbed to aquifer materials before the water table was drawn down may desorb after water levels rise once more. In the case of DNAPLs and other contaminants, rates of release which were too slow to alter the dissolved load of chlorinated hydrocarbons as long as groundwater velocities were artificially high during pumping may be enough to raise concentrations significantly under the far slower flow regime re-established after recovery of water levels.

An emerging category of *ex situ* remediation technologies which are typically applied to discrete surface discharges of polluted groundwaters are so called **passive treatment processes**. (The name distinguishes them from the more "active" treatment processes just described.) To date, passive treatment systems have found most application in the remediation of polluted groundwaters emanating from flooded mine workings or mine waste depositories, but they also have substantial potential for application to landfill and contaminated land leachates and associated groundwaters. Passive treatment has been formally defined as follows: "Passive treatment is the deliberate improvement of water quality using only naturally available energy sources (e.g. gravity, microbial metabolic energy, photosynthesis), in systems which require only infrequent (albeit regular) maintenance in order to operate effectively over the entire system design life" (PIRAMID Consortium 2003). As they do not require ongoing inputs of electricity or chemical reagents, and are generally constructed using nonhazardous materials (limestone, compost, etc.), passive systems can be installed in remote areas and left to operate unattended for long periods of time. Even better, it is normally possible to design passive systems so that they closely resemble natural wetlands. This not only makes them pleasant to look at; it also means that they can integrate into the surrounding landscape and merge into surrounding natural ecosystems. Indeed, in some cases, it has proved possible to achieve effective passive treatment by appropriating all or part of a natural wetland system and slightly modifying it to receive a polluted groundwater discharge. Passive systems of this nature may be regarded as instances of "enhanced natural attenuation," a remedial approach that is explored further in the following section.

11.5.3 *In situ* remediation technologies

A combination of frustrations over the poor performance of many active remediation technologies (both *in* and *ex situ*) and alarm at the enormous costs of many major ground water clean-up projects has led to a critical re-examination of the ability of natural aquifer processes to retard contaminants (Bekins et al. 2001). The outcome has been a dramatic upsurge in interest in **natural attenuation**, which is formally defined as an assemblage of "physical, chemical, or biological processes that, under favourable conditions, act without human intervention to reduce the mass, toxicity, mobility, volume, or concentration of contaminants in soil or groundwater" (Bekins et al. 2001). The relevant processes are those which have already been described in Sections 4.4.2 and 10.5.2. For natural attenuation to qualify as a remedial "technology," it must be carefully monitored, and its ability to meet pre-defined targets (to avoid damaging sensitive receptors) periodically reviewed. When it is adopted in this mode, we may speak of **monitored natural attenuation** (MNA).

The branding of natural attenuation as a remediation technology has its detractors. Environmental activists suspect that MNA is simply a handy alibi, allowing polluters to shirk their true responsibilities. A systematic investigation of the various pros and cons of MNA was undertaken by the US National Research Council between 1997 and 2000. It was found that communities are more likely to accept MNA as legitimate where the responsible parties and regulatory authorities are able to provide sound evidence that natural attenuation processes are transforming the contaminants to harmless products (rather than simply diluting them). To this end, it is essential that proponents of MNA are able to demonstrate robust cause-and-effect relationships to explain the disappearance of contaminant species and the appearance of harmless "daughter products" (National Research Council 2000). If convincing pollutant degradation pathways can be identified and credibly conveyed to stakeholders from nonscientific backgrounds, then MNA is likely to remain the predominant approach to groundwater remediation for decades to come (Rittmann 2004).

Where monitoring shows that utterly natural attenuation is *not* achieving desired goals, then it is possible to conceive of various interventions which can increase the rates and/or extend the scope of the key hydrological and biogeochemical processes; we would then be in the realm of **enhanced natural attenuation** (ENA). For organic contaminants, the aim of ENA is usually to enhance the metabolic rates of native bacteria, which are capable of catalyzing the breakdown of complex molecules into less harmful, smaller molecules. The relevant technologies are sometimes referred to collectively as **bioremediation**. Typical strategies include **air sparging**, in which air is pumped down wells into the contaminated groundwater, stimulating the activity of aerobic bacteria. This approach is provingly highly successful in the remediation of methyl tertiary butyl ether (MTBE), for instance. Similarly, injecting simple carbon compounds such as lactate and acetate into aquifers contaminated with heavy metals stimulates bacterial sulfate reduction, trapping the metals as insoluble sulfide minerals. Delivery systems for such microbial metabolites can range from simple injection boreholes to pairs of dual-screened wells fitted with rotating mixers, which force the water downwards in one well and upwards in the other. Between the two wells, a closed loop of groundwater circulation develops (Figure 11.6a). Introduction of acetate, lactate, and other compounds into the downflow well leads to intense biodegradation between the two wells. This approach, dubbed **horizontal flow treatment wells** (McCarty et al. 1998), has recently been applied to the clean-up of rocket fuel components on various military bases in the USA.

In many shallow aquifers it is possible to apply the principles of passive treatment for the purposes of *in situ* remediation. This is done using one or other variant of **permeable reactive barrier (PRB)** technology. The PRB concept is deceptively simple (Figure 11.6b): a permeable medium of geochemically appropriate material is placed in the path of the polluted groundwater in the form of a "barrier" across the flow path. As the groundwater flows through the barrier, beneficial (bio)geochemical reactions take place which result in an overall improvement in water quality, so that the groundwater flowing out of the down-gradient face of the barrier is significantly less polluted than that which entered.

Unfortunately, the simplicity of PRBs ends at this point, for both the construction and long-term deployment of PRBs are beset with considerable uncertainties, which remain topics of active research. The key issues that need to be addressed in any PRB design are nonetheless clear, and include:

1 The likelihood that water will flow *through* the PRB rather than bypassing it.
2 The degree to which geochemical reactions within the PRB will improve water quality.
3 The frequency with which the reactive medium must be replenished (or completely removed and replaced) to prevent either:
   (a) a loss of performance due to exhaustion of reactive components and/or
   (b) destruction of permeability by the clogging of pores by minerals which precipitate within the PRB substrate.

Extensive guidance on the design, installation, maintenance, and monitoring of PRBs is now available, both for organic contaminants (e.g. Gavaskar et al. 1998) and for acidic, metalliferous groundwaters associated with abandoned mine sites (e.g. Benner et al. 1997; Amos and Younger 2003; PIRAMID Consortium 2003).

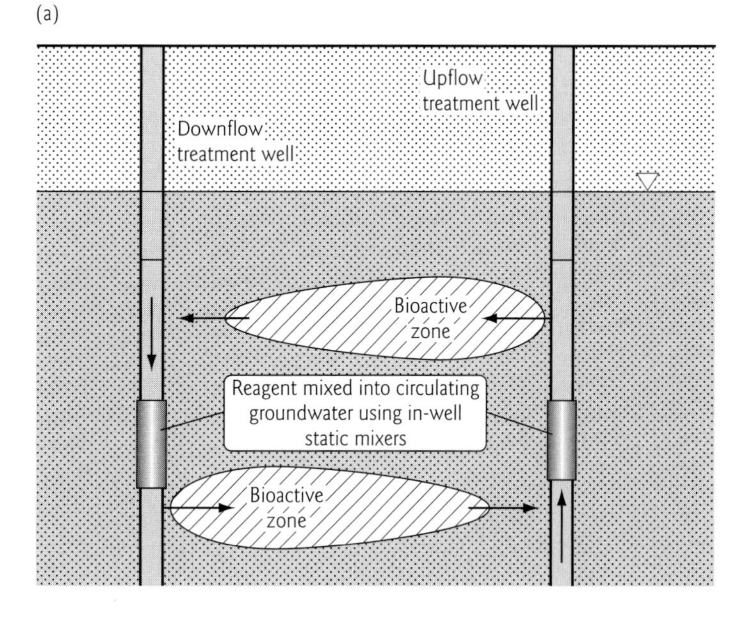

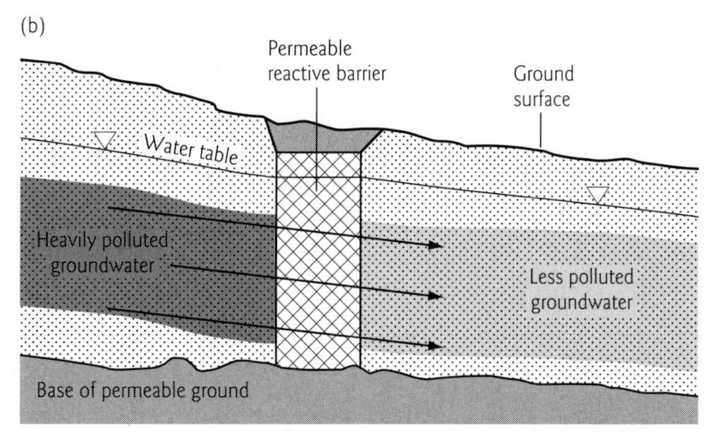

Fig. 11.6 *In situ* groundwater remediation technologies: schematic cross-sections. (a) Horizontal flow treatment wells: a pair of adjoining wells fitted with static mixers rotating in opposite directions, so that groundwater is induced to flow downwards in the left hand well, and upwards in the other (adapted after Knarr et al. 2003). A closed loop of circulation is established between the two wells, in which artificially introduced microbial metabolites foster the biodegradation of pollutants in the groundwater. (b) Permeable reactive barrier (PRB): heavily polluted groundwater passes through a trench filled with reactive media, losing pollutants to geochemical reactions, so that less polluted groundwater exits into the aquifer down gradient (adapted after Younger et al. 2002).

Endnotes

i "English Common Law" was the original basis for the legal system not only in England, but also throughout North America, Australasia, India, and many other countries which were formerly part of the British Empire.

ii The UN Commission on Sustainable Development: http://www.un.org/esa/sustdev/csd/

iii Field examples and clear illustrations are available online at: www.project-dewatering.co.uk

iv This limit is altitude-dependent, dropping to only 3 m below ground at 4000 m above sea level.

References

Abd El Samie, S.G. and Sadek, M.A., 2001, Groundwater recharge and flow in the Lower Cretaceous Nubian Sandstone aquifer in the Sinai Peninsula, using isotopic techniques and hydrochemistry. *Hydrogeology Journal*, **9**: 378–389.

Abdulrazzak, M.J., Sorman, A.U., and Al-Hames, A.S., 1989, Water balance approach under extreme arid conditions – a case study of Tabalah Basin, Saudi Arabia. *Hydrological Processes*, **3**: 107–122.

Acornley, R.M., 1999, Water temperature within spawning beds in two chalk streams and implications for salmonid egg development. *Hydrological Processes*, **13**: 439–446.

Adams, B. and MacDonald, A.M., 1998, Aquifer susceptibility to side-effects of groundwater exploitation. In: Robins, N.S. (ed.) Groundwater pollution, aquifer recharge and vulnerability. *Geological Society, London, Special Publications*, **130**: 71–76.

Adams, R. and Younger, P.L., 2001a, A strategy for modeling ground water rebound in abandoned deep mine systems. *Ground Water*, **39**: 249–261.

Adams, R. and Younger, P.L., 2001b, Simulating the hydrological effects of limestone extraction. In: Ribeiro, L. (ed.) *Proceedings of the Third International Conference on Future Groundwater Resources at Risk*. Lisbon, Portugal, June 2001, pp. 301–310.

Addiscott, T., 1988, Farmers, fertilizers and the nitrate flood. *New Scientist*, **120**(1633): 50–54.

Addiscott, T.M. and Benjamin, N., 2004, Nitrate and human health. *Soil Use and Management*, **20**: 98–104.

Aiuppa, A., Bellomo, S., Brusca, L., D'Allesandro, W., and Federico, C., 2003, Natural and anthropogenic factors affecting groundwater quality of an active volcano (Mt Etna, Italy). *Applied Geochemistry*, **18**: 863–882.

Al-Aswad, A.A. and Al-Bassam, A.M., 1997, Proposed hydrostratigraphical classification and nomenclature: application to the Palaeozoic in Saudi Arabia. *Journal of African Earth Sciences*, **24**: 494–510.

Al-Bassam, A.M., Al-Dabbagh, M.E., and Hussein, M.T., 2000, Application of a revised hydrostratigraphical classification and nomenclature to the Mesozoic and Cenozoic succession of Saudi Arabia. *Journal of African Earth Sciences*, **30**: 917–927.

Alden, A.S. and Munster, C.L., 1997, Assessment of river–floodplain aquifer interactions. *Environmental and Engineering Geoscience*, **3**: 537–548.

Al-Otaibi, M., 1997, *Artificial groundwater recharge in Kuwait: planning and management*. PhD Thesis, Department of Civil Engineering, University of Newcastle, UK.

Allen, D.M., Mackie, D.C., and Wei, M., 2004, Groundwater and climate change: a sensitivity analysis for the Grand Forks aquifer, southern British Columbia, Canada. *Hydrogeology Journal*, **12**: 270–290.

Allen, M.R. and Ingram, W.J., 2002, Constraints on future changes in climate and the hydrologic cycle. *Nature*, **419**: 224–232.

Allen, R.G., Pereira, L.S., Raes, D., and Smith, M., 1998, *Crop Evapotranspiration: guidelines for computing crop water requirements*. FAO Irrigation and Drainage Paper no. 56. United Nations Food and Agricultural Organisation (FAO), Rome, 290pp.

Aller, L., Bennet, T., Lehr, J.H., and Petty, R.J., 1987, *DRASTIC: a standardized system for evaluating groundwater pollution potential using hydrogeologic settings*. US EPA Report 600/2-85/018. United States Environmental Protection Agency, Cincinatti.

Amos, P.W. and Younger, P.L., 2003, Substrate characterisation for a subsurface reactive barrier to treat colliery spoil leachate. *Water Research*, **37**: 108–120.

Anderson, M.P. and Woessner, W.W., 1992, *Applied Groundwater Modeling: simulation of flow and advective transport*. Academic Press, San Diego.

Andreu, J.M., Pulido-Bosch, A., and Estévez, A., 2001, Aquifer overexploitation in Alicante (Spain). In: Ribeiro, L. (ed.) *Proceedings of the Third International Conference on Future Groundwater Resources at Risk–FGR '01*. 25–27 June 2001, Lisbon, Portugal, pp. 713–720.

Appelo, C.A.J. and Postma, D., 1993, *Geochemistry, Groundwater and Pollution*. A.A. Balkema Publishers, Rotterdam.

Applegate, G., 2002, *The Complete Guide to Dowsing: the definitive guide to finding underground water*. Vega Books, London.

Arlosoroff, S., Tshanneri, G., Grey, D., Journey, W., Karp, A., Langenegger, O., and Roche, R., 1987, *Community Water Supply: the handpump option*. World Bank, Washington DC.

Ayers, R.S. and Westcot., D.W., 1985. *Water Quality for Agriculture*. FAO Irrigation and Drainage Paper 29 (Revision 1). Food and Agriculture Organization of the United Nations, Rome..

Back, W., Rosenshein, J.S., and Seaber, P.R, 1988, *Hydrogeology. The Geology of North America*, Vol. O-2. Geological Society of America, Boulder, CO.

Baker, V.R., Kochel, R.C., Laity, J.E., and Howard, A.D., 1990, Spring sapping and valley network development. In: Higgins, C.G., and Coates, D.R. (eds) Groundwater geomorphology; the role of subsurface water in earth-surface processes and landforms. *Geological Society of America Special Paper*, **252**: 235–265.

Banas, K., and Gos, K., 2004, Effect of peat-bog reclamation on the physico-chemical characteristics of the ground water in peat. *Polish Journal of Ecology*, **52**: 69–74.

Banks, D., Younger, P.L., Arnesen, R.-T., Iversen, E.R., and Banks, S.D., 1997, Mine-water chemistry: the good, the bad and the ugly. *Environmental Geology*, **32**: 157–174.

Banks, D., Skarphagen, H., Wiltshire, R., and Jessop, C., 2004, Heat pumps as a tool for energy recovery from mining wastes. In: Gieré, R. and Stille, P. (eds) Energy, waste and environment: a geochemical perspective. *Geological Society, London, Special Publications*, **236**: 499–513.

Banwart, S.A. and Malmström, 2001, Hydrochemical modelling for preliminary assessment of mine water pollution. *Journal of Geochemical Exploration*, **74**: 73–97.

Banwart, S.A., Evans, K.A., and Croxford, S., 2002, Predicting mineral weathering rates at field scale for mine water risk assessment. In: Younger, P.L., and Robins, N.S. (eds) Mine Water Hydrogeology and Geochemistry. *Geological Society, London, Special Publications*, **198**: 137–157.

Barker, J.A., Downing, R.A., Gray, D.A., et al. 2000, Hydrogeothermal studies in the United Kingdom. *Quarterly Journal of Engineering Geology and Hydrogeology*, **33**: 41–58.

Barnes, I., O'Neil, J.R., and Trescasses, J.J., 1978, Present day serpentinization in New Caledonia, Oman and Yugoslavia. *Geochimica et Cosmochimica Acta*, **42**: 144–145.

Barrett, M.H., Hiscock, K.M., Pedley, S., Lerner, D.N., Tellam, J.H., and French, M.J., 1999, Marker species for identifying urban groundwater recharge sources: a review and case study in Nottingham, UK. *Water Research*, **33**: 3083–3097.

Baskin, Y. (ed.), 2003, Sustaining healthy freshwater ecosystems. *Issues in Ecology* 10. Ecological Society of America, Washington DC.

Batty, L.C. and Younger, P.L., 2002, Critical role of macrophytes in achieving low iron concentrations in mine water treatment wetlands. *Environmental Science and Technology*, **36**: 3997–4002.

Bear, J. and Verruijt, A., 1987, *Modeling Groundwater Flow and Pollution. Theory and applications of transport in porous media*. D Reidel Publishing, Dordrecht.

Beeson, S., Misstear, B.D., and van Wonderen, J., 1997, Assessing the reliable outputs of groundwater sources. *Journal of the Chartered Institution of Water and Environmental Management*, **11**: 295–304.

Bekins, B.A., Rittmann, B.E., and MacDonald, J.A., 2001, Natural attenuation strategy for groundwater cleanup focuses on demonstrating cause and effect. *Eos (Transactions of the American Geophysical Union)*, **82**: 57–58.

Bell, F.G., 2004, *Engineering Geology and Construction*. Spon Press, New York.

Benner, S.G., Blowes, D.W., and Ptaceck, C.J., 1997, A full-scale porous reactive wall for prevention of acid mine drainage. *Ground Water Monitoring and Restoration*, **17**: 99–107.

Best, M.G. and Christiansen, E.H., 2000, *Igneous Petrology*. Blackwell Publishing, Malden, MA.

Bethke, C.M., 1996, *Geochemical Reaction Modeling, Concepts and Applications*. Oxford University Press, New York.

Binnie, C., Kinber, M., and Smethurst, G., 2002, *Basic Water Treatment*, 3rd edn. Thomas Telford, London.

Birk, S., Liedl, R., and Sauter, M., 2004, Identification of localised recharge and conduit flow by combined analysis of hydraulic and physico-chemical spring responses (Urenbrunnen, SW-Germany). *Journal of Hydrology*, **286**: 179–193.

Birkinshaw, S.J., Thorne, M.C., and Younger, P.L., 2005, Reference biospheres for post-closure performance assessment: inter-comparison of SHETRAN simulations and BIOMASS results. *Journal of Radiological Protection*, **25**: 33–49.

Biswas, A.K., 1970, *History of Hydrology*. North-Holland Publishing Company, Amsterdam.

Bloomfield, J.P., Goody, D.C., Bright, M.I., and Williams, P.J., 2001, Pore-throat size distributions in Permo-Triassic sandstones from the United Kingdom and some implications for contaminant hydrogeology. *Hydrogeology Journal*, **9**: 219–230.

Bloomfield, J.P., Gaus, I., and Wade, S.D., 2003, A method for investigating the potential impacts of climate-change scenarios on annual minimum groundwater levels. *Journal of the Chartered Institution of Water and Environmental Management*, **17**: 86–91.

Bobba, A.G., Singh, V.P., Berndtsson, R., and Bengtsson, L., 2000, Numerical simulation of saltwater intrustion into Laccadive Island aquifers due to climate change. *Journal of the Geological Society of India*, **55**: 589–612.

Bord, J., and Bord, C., 1986, *Sacred Waters: holy wells and water lore in Britain and Ireland*. Paladin Publishers, London.

Boulton, A., 2000, The subsurface macrofauna. In: Jones. J.B. and Mulholland, P.J. (eds) *Streams and Ground Waters*. Academic Press, San Diego, pp. 337–361.

Bouraoui, F., Vachaud, G., Li, Z.X., Le Treut, H., and Chen, T., 1999, Evaluation of the impact of climate changes on water storage and groundwater recharge at the watershed scale. *Climate Dynamics*, **15**: 153–161.

Bouwer, H., 2002, Artificial recharge of groundwater: hydrogeology and engineering. *Hydrogeology Journal*, **10**: 121–142.

Bowell, R.J., 2002, The hydrogeochemical dynamics of mine pit lakes. In: Younger, P.L., and Robins, N.S. (eds) Mine Water Hydrogeology and Geochemistry. *Geological Society, London, Special Publications*, **198**: 159–185.

Brahana, J.V., Thrailkill, J., Freeman, T., and Ward, W.C., 1988, Carbonate rocks. In: Back, W., Rosenshein, J.S., and Seaber, P.R., (eds), *Hydrogeology. The Geology of North America*, Vol. O-2. Geological Society of America, Boulder, CO, pp. 333–352.

Bras, R.L., 1990, *Hydrology. An introduction to hydrologic science*. Addison-Wesley, Reading, MA.

Brassington, R., 1998, *Field hydrogeology*, 2nd edn. Wiley, Chichester.

Bredehoeft, J.D., Papadopolous, S.S., and Cooper, H.H., 1982, Groundwater: the water-budget myth. In: *Scientific Basis of Water Resource Management*. US National Academy of Sciences, Studies in Geophysics. National Academy Press, Washington DC, pp. 51–57.

Bredehoeft, J.D., Neuzil, C.D., and Milly, P.C.D., 1983, Regional flow in the Dakota aquifer – the role of confining layers. *US Geological Survey Water Supply Paper*, **2237**, 45pp.

Brookins, D.G., 1988, *Eh-pH Diagrams for Geochemistry*. Springer-Verlag, Berlin.

Brouyere, S., Carabin, G., and Dassargues, A., 2004, Climate change impacts on groundwater resources: modelled deficits in a chalky aquifer, Geer basin, Belgium. *Hydrogeology Journal*, **12**: 123–134.

Brown, A.G. (ed.), 1995, *Geomorphology and Groundwater*. Wiley, Chichester.

Brunke, M. and Gonser, T., 1997, The ecological significance of exchange processes between rivers and groundwater. *Freshwater Biology*, **37**: 1–33.

Bullock, A. and Acreman, M., 2003, The role of wetlands in the hydrological cycle. *Hydrology and Earth System Sciences*, **7**: 358–389.

Burgess, D.B., 2002, Groundwater resource management in eastern England: a quest for environmentally sustainable development. In: Hiscock, K.M, Rivett, M.O., and Davison, R.M. (eds), Sustainable groundwater development. *Geological Society Special Publications*, **193**: 53–62.

Burgess, W.G., Burren, M., Perrin, J., and Ahmed, K.M., 2002, Constraints on sustainable development of arsenic-bearing aquifers in southern Bangladesh. Part 1: a conceptual model of arsenic in the aquifer. In: Hiscock,

K.M., Rivett, M.O., and Davison, R.M. (eds), Sustainable groundwater development. *Geological Society of London, Special Publications*, **193**: 145–163.
Burnett, W.C., Bokuniewicz, H., Huettel, M., Moore, W.S., and Taniguchi, M., 2003, Groundwater and pore water inputs to the coastal zone. *Biogeochemistry*, **66**: 3–33.
Burt, T.P., 1995, The role of wetlands in runoff generation from headwater catchments. In: Hughes, J. and Heathwaite, L. (eds) *Hydrology and Hydrochemistry of British Wetlands*. Wiley, Chichester, pp. 21–38.
Buttle, J.M., 1994, Isotope hydrograph separations and rapid delivery of pre-event water from drainage basins. *Progress in Physical Geography*, **18**: 16–41.
Calder, I.R., 1999, *The Blue Revolution, Land Use and Integrated Water Resources Management*. Earthscan Publishers, London.
Calver, A., 1997, Recharge response functions. *Hydrology and Earth Systems Sciences*, **1**: 47–53.
Cannan, C.E. and Armitage, P.D., 1999, The influence of catchment geology on the longitudinal distribution of macroinvertebrate assemblages in a groundwater dominated river. *Hydrological Processes*, **13**: 355–369.
Capper, P.L., Cassie, W.F., and Geddes, J.D., 1995, *Mechanics of Engineering Soils*. E. and F.N. Spon, London.
Carslaw, H.S. and Jaeger, J.C., 1959, *Conduction of Heat in Solids*, 2nd edn. Clarendon Press, Oxford.
Cashman, P.M. and Preene, M., 2002, *Groundwater Lowering in Construction: a practical guide*. Spon Press, London.
Cervantes, P. and Wallace, P.J., 2003, Role of H_2O in subduction-zone magmatism: new insights from melt inclusions in high-Mg basalts from central Mexico. *Geology*, **31**: 235–238.
Chapelle, F.H., 2000, *Ground-water Microbiology And Geochemistry*, 2nd edn. John Wiley and Sons, New York.
Chapman, P., 1993, *Caves and Cave Life*. Harper Collins, London.
Chavez, A., Davis, S.N., and Sorooshian, S., 1994a, Estimation of mountain front recharge to regional aquifers. 1. Development of an analytical hydroclimatic model. *Water Resources Research*, **30**: 2157–2167.
Chavez, A., Sorooshian, S., and Davis, S.N., 1994b, Estimation of mountain front recharge to regional aquifers. 2. A maximum likelihood approach incorporating prior information. *Water Resources Research*, **30**: 2169–2181.
Chen, M., Soulsby, C., and Younger, P.L., 1999, Modelling the evolution of minewater pollution at Polkemmet Colliery, Almond catchment, Scotland. *Quarterly Journal of Engineering Geology*, **32**: 351–362.
Christensen, T.H., Bjerg, P.L., Banwart, S.A., Jakobsen, R., Heron, G., and Albrechtsen, H.-J., 2000, Characterization of redox conditions in groundwater contaminant plumes. *Journal of Contaminant Hydrology*, **45**: 165–241.
Clark, L., 1977, The analysis and planning of step-drawdown tests. *Quarterly Journal of Engineering Geology*, **10**: 125–143.
Clark, L., 1988, *The Field Guide to Water Wells and Boreholes*. Wiley, New York.
Clarke, R. and King, J., 2004, *The atlas of water*. Earthscan Publications, London.
Clesceri, L.S., Greenberg, A.E., and Eaton, A.D. (eds), 1998, *Standard Methods for the Examination of Water and Wastewater*. 20th edn. American Public Health Association/American Water Works Association/Water Environment Federation, Washington, DC.
Coates, D.R., 1990, Geomorphic controls of groundwater hydrology. In: Higgins, C.G. and Coates, D.R. (eds) Groundwater geomorphology; the role of subsurface water in earth-surface processes and landforms. *Geological Society of America Special Paper*, **252**: 341–356.
Cole, J.A., 1998, Water resources (introduction). In: Herschy, R.W. and Fairbridge, R.W., (eds) *Encyclopaedia of Hydrology and Water Resources*. Kluwer Academic Publishers, Boston, pp. 712–716.
Cole, J.A., Oakes, D.B., Slade, S., and Clark, K.J., 1994, Potential impacts of climatic change and of sea-level rise on the yields of aquifer, river and reservoir sources. *Journal of the Institution of Water and Environmental Management*, **8**: 591–606.
Coleman, N.M., 2003, Aqueous flows carved the outflow channels on Mars. *Journal of Geophysical Research–Planets*, **108**, article no. 5039.
Cook, P.G., Solomon, D.K., Plummer, L.N., Busenberg, E., and Schiff, S.L., 1995, Chlorofluorocarbons as tracers of groundwater transport processes in a shallow, silty sand aquifer. *Water Resources Research*, **31**: 425–434.
Cooper, D.M., Wilkinson, W.B., and Arnell, N.W., 1995, The effects of climate changes on aquifer storage and river baseflow. *Hydrological Sciences Journal*, **40**: 615–631.

Cooper, H.H., and Jacob, C.E., 1946, A generalized graphical method for evaluating formation constants and summarizing well field history. *Transactions of the American Geophysical Union*, **27**: 526–534.

Cripps, J.C., Bell, F.G., and Culshaw, M.G., (eds), 1986, *Groundwater in Engineering Geology*. Geological Society, London, Engineering Geology Special Publication, no. 3. Geological Society, London. .

CROGEE, 2002, *Regional Issues in Aquifer Storage and Recovery for Everglades Restoration: a review of the ASR Regional Study Project Management Plan of the Comprehensive Everglades Restoration Plan*. Committee on Restoration of the Greater Everglades Ecosystem (CROGEE), Water Science and Technology Board (WSTB) and Board on Environmental Studies and Toxicology (BEST).

Cronican, A.E. and Gribb, M.M., 2004, Hydraulic conductivity prediction for sandy soils. *Ground Water*, **42**: 459–464.

Cruden, D.M., 1991, A simple definition of a landslide. *Bulletin International Association for Engineering Geology*, **43**: 27–29.

Custodio, E., 2002, Aquifer overexploitation: what does it mean? *Hydrogeology Journal*, **10**: 254–277.

Custodio, E. and Llamas, M.R., (eds), 1996, *Hidrología subterránea*, 2nd edn. ("Underground hydrology"; bound in two volumes (*Tomo I* and *Tomo II*); in Spanish.) Ediciones Omega S.A., Barcelona.

Dagan, G., 1982, Stochastic modeling of groundwater-flow by unconditional and conditional probabilities. 1. Conditional simulation and the direct-problem. *Water Resources Research*, **18**: 813–833.

Dagan, G., 2004, On application of stochastic modeling of groundwater flow and transport. *Stochastic Environmental Research and Risk Assessment*, **18**: 266–267.

Danielopol, D.L., Griebler, C., Gunatilaka, A., and Notenboom, J., 2003, Present state and future prospects for groundwater ecosystems. *Environmental Conservation*, **30**: 104–130.

Darcy, H., 1856, *Les fontaines publiques de la Ville de Dijon*. Victor Dalmont, Paris.

Das Gupta, A. and Onta, P.R., 1997, Sustainable groundwater resource development. *Hydrological Sciences Journal*, **42**: 565–582.

Davis, S.N., 1988, Sandstones and shales. In: Back, W., Rosenshein, J.S., and Seaber, P.R., (eds), *Hydrogeology*. The Geology of North America, Vol. O-2. Geological Society of America, Boulder, CO, pp. 323–332.

Davis, W.M., 1930, Origin of limestone caverns. *Geological Society of America Bulletin*, **41**: 475–628.

de Vries, J.J. and Simmers, I., 2002, Groundwater recharge: an overview of processes and challenges. *Hydrogeology Journal*, **10**: 5–17.

Destouni, G. and Prieto, C., 2003, On the possibility for generic modeling of submarine groundwater discharge. *Biogeochemistry* **66**: 171–186.

Dickson, M. and Fanelli, M. (eds), 2005, *Geothermal Energy: utilization and technology*. Earthscan Publications/ UNESCO, London.

Domenico, P.A. and Schwartz, F.W., 1997, *Physical and Chemical Hydrogeology*, 2nd edn. Wiley, New York.

Downing, R.A. and Gray, D.A., (eds), 1986, *Geothermal Energy – the potential in the United Kingdom*. British Geological Survey. HMSO, London.

Downing, R.A., Ashford, P.L., Headworth, H.G., Owen, M., and Skinner, A.C., 1981, The use of groundwater for river regulation. In: *A Survey of British Hydrogeology*. Royal Society, London, pp. 153–171.

Downing, R.A., Price, M., and Jones, G.P. (eds), 1993, *The Hydrogeology of the Chalk of North-West Europe*. Oxford University Press, Oxford.

Downing, R.A., Oakes, D.B., Wilkinson, W.B., and Wright, C.E., 1974, Regional development of groundwater resources in combination with surface waters. *Journal of Hydrology*, **22**: 155–177.

Driscoll, F.G., 1986, *Groundwater and Wells*, 2nd edn. Johnson Filtration Systems, St Paul, MN.

Dudgeon, B.A., 2005, Hydrogeochemistry of seepage entering deep tunnels in crystalline rock: Cruachan Power Station, Scotland. PhD Thesis, School of Civil Engineering and Geosciences, University of Newcastle upon Tyne, 183pp plus appendices.

Dudgeon, C.R., 1985, Effects of non-Darcy flow and partial penetration on water levels near open-pit excavations. In: *Proceedings of the 18th Congress of the International Association of Hydrogeologists*, Cambridge, England, September 1985, pp. 122–132.

Eberle, M. and Persons, J.L., 1978, *Appropriate Well-drilling Technologies: a manual for developing countries*. National Water Well Association (USA), Worthington, OH.

Ebraheem, A.M., Riad, S., Wycisk, P., and El-Nasr, A.M.S., 2002, Simulation of impact of present and future groundwater extraction from the non-replenished Nubian Sandstone Aquifer in southwest Egypt. *Environmental Geology*, **43**: 188–196.

Eckhardt, K. and Ulbrich, U., 2003, Potential impacts of climate change on groundwater recharge and streamflow in a central European low mountain range. *Journal of Hydrology*, **284**: 244–252.

Eden, W.J., 1971, Landslides in clays. *Canadian Building Digest*, **143**: 4pp.

Edmunds, W.M. and Tyler, S.W., 2002, Unsaturated zones as archives of past climates: towards a new proxy for continental regions. *Hydrogeology Journal*, **10**: 216–228.

Edmunds, W.M., Fellman, E., Goni, I.B., and Prudhomme, C., 2001, Spatial and temporal distribution of groundwater recharge in northern Nigeria. *Hydrogeology Journal*, **10**: 205–215.

Elliot, T., Andrews, J.N., and Edmunds, W.M., 1999, Hydrochemical trends, palaeorecharge and groundwater ages in the fissured Chalk aquifer of the London and Berkshire Basins, UK. *Applied Geochemistry*, **14**: 333–363.

Elliot, T., Chadha, D.S., and Younger, P.L., 2001, Water quality impacts and palaeohydrogeology in the Yorkshire Chalk aquifer, UK. *Quarterly Journal of Engineering Geology and Hydrogeology*, **34**: 385–398.

Elliott, C.R.N., Dunbar, M.J., Gowing, I., and Acreman, M.C., 1999, A habitat assessment approach to the management of groundwater dominated rivers. *Hydrological Processes*, **13**: 459–475.

Environment Agency, 2003, *Guidance on Monitoring of Landfill Leachate, Groundwater and Surface Water*. Environment Agency (England and Wales), Bristol.

Environmental Protection Authority, 2003, *Consideration of Subterranean Fauna in Groundwater and Caves During Environmental Impact Assessment in Western Australia*. (Guidance for the assessment of environmental factors, Western Australia in accordance with the Environmental Protection Act 1986.) No. 54. Government of Western Australia, Perth.

European Commission, 2001, *EUR19400–Artificial Recharge Of Groundwater–Final Report*. Office for Official Publications of the European Communities, Luxembourg.

Ewen, J., 1996a, "SAMP" model for water and solute movement in unsaturated porous media involving thermodynamic subsystems and moving packets: 1. Theory. *Journal of Hydrology*, **182**: 175–194.

Ewen, J., 1996b, "SAMP" model for water and solute movement in unsaturated porous media involving thermodynamic subsystems and moving packets: 2. Design and application. *Journal of Hydrology*, **182**: 195–207.

Fairbridge, R.W., 1998, Water: categories. In: Herschy, R.W. and Fairbridge, R.W., (eds) *Encyclopaedia of Hydrology and Water Resources*. Kluwer Academic, Boston, pp. 687–688.

Fetter, C.W., 1999, *Contaminant hydrogeology*, 2nd edn. Prentice Hall, New Jersey.

Fetter, C.W., 2001, *Applied Hydrogeology*, 4th edn. Prentice Hall, New Jersey.

Figueroa Vega, G.E., 1984, Case History No. 9.8: Mexico, D.F., Mexico. In: Poland, J.F., (ed.) *Guidebook to Studies of Land Subsidence Due to Ground-water Withdrawal*. UNESCO, Paris, pp. 217–232.

Findlay, S. and Sobczak, W.V., 2000, Microbial communities in hyporheic sediments. In: Jones. J.B. and Mulholland, P.J. (eds) *Streams and Ground Waters*. Academic Press, San Diego, pp. 287–306.

Forchheimer, P., 1930, *Hydraulik*, 3rd edn. Teubner, Leipzig/Berlin.

Ford, A., 1999, *Modeling the Environment: an introduction to system dynamics models of environmental systems*. Island Press, Washington DC.

Ford, D.C. and Williams, P.W., 1989, *Karst Geomorphology and Hydrology*. Kluwer Academic Publishers, Dordrecht.

Fornés, J. and Llamas, M.R., 2001, Conflicts between groundwater abstraction for irrigation and wetland conservation: achieving sustainable development in the La Mancha Húmeda Biosphere Reserve (Spain). In: Griebler, C., Danielopol, D.L., Gibert, J., Nachtnebel, H.P., and Notenboom, J. (eds) *Groundwater Ecology: a tool for management of water resources*. (Document EUR 19877.) Office for Official Publications of the European Communities, Luxembourg, pp. 263–275.

Forth, R.A., 1994, Ground settlement and sinkhole development due to the lowering of the water table in the Bank Compartment, South Africa. In: Oliveira, R., Rodrigues, L.F., Coelho, A.G., and Cunha, A.P. (eds), *Proceedings of the 7th Congress of the International Association of Engineering Geologists*. Lisbon, Portugal, 5–9 September 1994, Balkema, Rotterdam, pp. 2933–2940.

Foster, S.S.D., 1998, Groundwater recharge and pollution vulnerability of British aquifers: a critical overview. In: Robins, N.S. (ed.) Groundwater pollution, aquifer recharge and vulnerability. *Geological Society, London, Special Publications*, **130**: 7–22.

Foster, S.S.D., Morris, B.L., and Lawrence, A.R., 1994, Effects of urbanization on groundwater recharge. In: *Proceedings of the ICE International Conference on Groundwater Problems in Urban Areas*. Institution of Civil Engineers, London, pp. 43–63.

Freeze, R.A., 1969, The mechanism of natural groundwater recharge and discharge, 1. One dimensional, vertical, unsteady, unsaturated flow above a recharging or discharging groundwater flow system. *Water Resources Research*, **5**: 153–171.

Freeze, R.A., 1972a, Role of subsurface flow in generating surface runoff; 1 – baseflow contributions to channel flow. *Water Resources Research*, **8**: 609–623.

Freeze, R.A., 1972b, Role of subsurface flow in generating surface runoff; 2 – upstream source areas. *Water Resources Research*, **8**: 1272–1283.

Freeze, R.A. and Cherry, J.A., 1979, *Groundwater*. Prentice Hall, Englewood Cliffs, NJ.

Gandy, C.J., 2003, *Modelling Pyrite Oxidation and Pollutant Transport in Mine Wastes Using an Object-oriented Particle Tracking Code*. PhD Thesis, School of Civil Engineering and Geosciences, University of Newcastle upon Tyne.

Gavaskar, A.R., Gupta, N., Sass, B.M., Janosy, R.J., and O'Sullivan, D., 1998, *Permeable Barriers for Groundwater Remediation: design construction and monitoring*. Battelle Press, Columbus, OH.

Gee, G.W. and Hillel, D., 1988, Groundwater recharge in arid regions: review and critique of estimation methods. *Hydrological Processes*, **2**: 255–266.

Gibert, J., Danielopol, D.L., and Stanford, J.A., 1994, *Groundwater Ecology*. Academic Press, San Diego.

Gillham, R.W., 1984, The capillary fringe and its effect on water table response. *Journal of Hydrology*, **67**: 307–324.

Gillieson, D., 1996, *Caves: processes, development and management*. Blackwell Scientific, Oxford.

Griebler, C., 2001, Microbial ecology of the subsurface. In: Griebler, C., Danielopol, D.L., Gibert, J., Nachtnebel, H.P., and Notenboom, J. (eds), *Groundwater Ecology: a tool for management of water resources*. (Document EUR 19877.) Office for Official Publications of the European Communities, Luxembourg, pp. 81–108.

Griebler, C., Danielopol, D.L., Gibert, J., Nachtnebel, H.P., and Notenboom, J. (eds), 2001, *Groundwater Ecology: a tool for management of water resources*. (Document EUR 19877.) Office for Official Publications of the European Communities, Luxembourg.

Grindley, J., 1967, The estimation of soil moisture deficits. *Meteorological Magazine*, **96**: 97–108.

Grindley, J., 1969, *The Calculation of Actual Evaporation and Soil Moisture Deficits Over Specified Catchment Areas*. Hydrological Memoir of the Meteorological Office, no. 38. Meteorological Office, Bracknell (UK).

Grischek, T., Hiscock, K.M., Metschies, T., Dennis, P.F., and Nestler, W., 1998, Factors affecting denitrification during infiltration of river water into a sand aquifer in Saxony, eastern Germany. *Water Research*, **32**: 450–460.

Groenewold, G.H. and Rehm, B.W., 1982, Instability of contoured surface-mined landscapes in the northern Great Plains: causes and implications. *Reclamation and Revegetation Research*, **1**: 161–176.

Groombridge, B. and Jenkins, M., 1998, *Freshwater Biodiversity: a preliminary global assessment*. World Conservation Monitoring Centre, World Conservation Press, Cambridge, UK.

Guest, J.E., Cole, P.D., Duncan, A.M., and Chester, D.K., 2003, *Volcanoes of Southern Italy*. Geological Society, London.

Gustard, A., Bullock, A., and Dixon, J.M., 1992, *Low Flow Estimation in the United Kingdom*. IH Report no. 108. Institute of Hydrology, Wallingford, UK.

Gutiérrez-Santolalla, F., Gutiérrez-Elorza, M., Marín, C., Maldonado, C., and Younger, P.L, 2005, Subsidence hazard avoidance based on geomorphological mapping in the Ebro River valley mantled evaportie karst terrain (NE Spain). *Environmental Geology*, **48**: 370–383.

Gutjahr, A.L. and Gelhar, L.W., 1981, Stochastic-models of subsurface flow – infinite versus finite domains and stationarity. *Water Resources Research*, **17**: 337–350.

Haines, T.S. and Lloyd J.W., 1985, Controls on silica in groundwater environments in the United Kingdom. *Journal of Hydrology*, **81**: 277–295.

Hakenkamp, C.C. and Palmer, M.A., 2000, The ecology of hyporheic meiofauna. In: Jones. J.B. and Mulholland, P.J. (eds), *Streams and Ground Waters*. Academic Press, San Diego, pp. 307–336.

Hammer, D.A., 1992, *Creating Freshwater Wetlands*. Lewis Publishers, Boca Raton.

Harbaugh, A.W. and McDonald, M.G., 1996, User's documentation for MODFLOW-96, an update to the US Geological Survey modular finite-difference ground-water flow model. *US Geological Survey Open-File Report*, **96–485**.

Hardwick, P. and Gunn, J., 1995, Landform–groundwater interactions in the Gwenlais Karst, South Wales. In: Brown, A.G. (ed.), *Geomorphology and Groundwater*. Wiley, Chichester, pp. 75–91.

Hattingh, R.P., Pulles, W., Krantz, R., Pretorius, C., and Swart, S., 2002, Assessment, prediction and management of long-term post-closure water quality: a case study – Hlobane Colliery, South Africa. In: Younger, P.L. and Robins, N.S. (eds) Mine Water Hydrogeology and Geochemistry. *Geological Society, London, Special Publication*, **198**: 297–314.

Hazen, A., 1892, Some physical properties of sands and gravels. *Massachusetts State Board of Health Annual Report 1892*, 539–556.

Headworth, H.G., 2004, Recollections of a golden age: the groundwater schemes of Southern Water 1970–1990. In: Mather, J.D. (ed.), 200 years of British hydrogeology. *Geological Society, London, Special Publications*, **225**: 339–362.

Healy, R.W. and Cook, P.G., 2002, Using groundwater levels to estimate recharge. *Hydrogeology Journal*, **10**: 91–109.

Hem, J.D., 1985, Study and interpretation of the chemical characteristics of natural water, 3rd edn. *United States Geological Survey Water-Supply Paper 2254*. United States Geological Survey, Washington DC.

Herschy, R.W., 1998, World water balance. In: Herschy, R.W. and Fairbridge, R.W. (eds) *Encyclopaedia of Hydrology and Water Resources*. Kluwer Academic, Boston. pp. 787–788.

Herschy, R.W. and Fairbridge, R.W. (eds), 1998, *Encyclopaedia of Hydrology and Water Resources*. Kluwer Academic, Boston.

Higgins, C.G., 1984, Piping and sapping: development of landforms by groundwater outflow. In: LaFleur, R.G. (ed.), 1984, *Groundwater as a Geomorphic Agent*. Allen and Unwin, Boston. pp. 18–58.

Higgins, C.G. and Coates, D.R., 1990, Groundwater geomorphology; the role of subsurface water in earth-surface processes and landforms. *Geological Society of America Special Paper*, **252**.

Higgins, C.G., Coates, D.R., Baker, V.R., et al., 1988, Landform development. In: Back, W., Rosenshein, J.S., and Seaber, P.R. (eds) *The Geology of North America*, Vol. O-2, *Hydrogeology*. Geological Society of America, Boulder, CO, pp. 383–400.

Hinderer, M. and Einsele, G., 1997, Groundwater acidification in Triassic sandstones: prediction with MAGIC modelling. *Geologische Rundschau*, **86**: 372–388.

Hiscock, K.M., 2005, *Hydrogeology. Principles and practice*. Blackwell Publishing, Oxford.

Hiscock, K.M. and Grischek, T., 2002, Attenuation of groundwater pollution by bank filtration. *Journal of Hydrology*, **266**: 139–144.

Hiscock, K.M. and Lloyd, J.W., 1992, Palaeohydrogeological reconstructions of the North Lincolnshire Chalk, UK, for the last 140,000 years. *Journal of Hydrology*, **133**: 313–342.

Hiscock, K.M, Rivett, M.O., and Davison, R.M., 2002, Sustainable groundwater development. In: Hiscock, K.M., Rivett, M.O., and Davison, R.M. (eds), Sustainable groundwater development. *Geological Society Special Publications*, **193**: 1–14.

Hobbs, S.L. and Gunn, J., 1998, The hydrogeological impacts of quarrying karstified limestone, options for prediction and mitigation. *Quarterly Journal of Engineering Geology*, **31**: 147–157.

Holmes, N.T.H., 1999, Recovery of headwater stream flora following the 1989–1992 groundwater drought. *Hydrological Processes*, **13**: 341–354.

Holzer, T.L. (ed.), 1984, Man-induced land subsidence. In: *Reviews in Engineering Geology*, Vol. VI. Geological Society of America, Boulder, CO.

Horton, R.E., 1933, The role of infiltration in the hydrologic cycle. *Transactions of the American Geophysical Union*, **14**: 446–460.

Hounslow, A.W., 1995, *Water Quality Data. Analysis and interpretation*. CRC Lewis, Boca Raton, FL.

Houston, J. and Hart, D., 2004, Theoretical head decay in closed basin aquifers: an insight into fossil groundwater and recharge processes in the Andes of northern Chile. *Quarterly Journal of Engineering Geology and Hydrogeology*, **37**: 131–139.

Hubbert, M.K., 1940, The theory of ground-water motion. *Journal of Geology*, **48**: 785–944.

Hulme, P., Grout, M., Seymour, K., Rushton, K., Brown, L., and Low, R., 2002, *Groundwater Resource Modelling: guidance notes and template project brief*. Environment Agency R&D Guidance Notes W213. Environment Agency (England and Wales), Bristol.

Hunt, C.D., Jr, Ewart, C.J., and Voss, C.I., 1988, Region 27, Hawaiian Islands. In: Back, W., Rosenshein, J.S., and Seaber, P.R. (eds), *Hydrogeology*. The Geology of North America, Vol. O-2. Geological Society of America, Boulder, CO, pp. 255–262.

Huyakorn, P.S. and Pinder, G.F., 1983, *Computational Methods in Subsurface Flow*. Academic, New York.

Institute of Hydrology, 1980, *Low Flow Studies report*. Natural Environment Research Council, Wallingford (Oxon), UK.

Jackson, I. (ed.), 2004, *Britain Beneath our Feet: an atlas of digital information on Britain's land quality, underground hazards, resources and geology*. (BGS Occasional Publication No. 4.) British Geological Survey, Nottingham.

James, A., 1992, *An Introduction to Water Quality Modelling*, 2nd edn. Wiley, Chichester.

Johnston, R.H., 1989, The hydrologic responses to development in regional sedimentary aquifers. *Ground Water*, **27**: 316–322.

Johnston, R.H., 1997, Sources of water supplying pumpage from regional aquifer systems of the United States. *Hydrogeology Journal*, **5**: 54–63.

Johnston, S.G., Slavich, P.G., and Hirst, P., 2004, The effects of a weir on reducing acid flux from a drained coastal acid sulphate soil backswamp. *Agricultural Water Management*, **69**: 43–67.

Joint Science Academies, 2005, *Joint Science Academies' Statement: global response to climate change*. National Academies of Science of Brazil, Canada, China, France, Germany, India, Italy, Japan, Russia, UK and USA. (7 June 2005). National Academy of Science, Washington DC.

Jones, C.J.F.P. and Cooper, A.H., 2005, Road construction over voids caused by active gypsum dissolution, with an example from Ripon, North Yorkshire, England. *Environmental Geology*, **48**: 384–394.

Jones. J.B. and Mulholland, P.J., 2000, *Streams and Ground Waters*. Academic Press, San Diego.

Kalin, R.M. and Roberts, C., 1997, Groundwater resources in the Lagan Valley sandstone aquifer, Northern Ireland. *Journal of the Chartered Institution of Water and Environmental Management*, **11**: 133–139.

Kaplan, L.A. and Newbold, J.D., 2000, Surface and subsurface dissolved organic carbon. In: Jones. J.B., and Mulholland, P.J. (eds) *Streams and Ground Waters*. Academic Press, San Diego, pp. 237–258.

Karami, G.H. and Younger, P.L., 2002, Analysing step-drawdown tests in heterogeneous aquifers. *Quarterly Journal of Engineering Geology and Hydrogeology*, **35**: 295–303.

Kemper, K.E., Mestre, E., and Amore, L., 2003, Management of the Guarani aquifer system – moving towards the future. *Water International*, **28**: 185–200.

Khan, L.R. and Mawdsley, J.A., 1988, Reliable yield of unconfined aquifers. *Hydrological Sciences Journal*, **33**: 151–170.

Khoury, H.N., Salameh, E., and Abdul-Jaber, Q., 1985, Characteristics of an unusual highly alkaline water from the Maqarin area, northern Jordan. *Journal of Hydrology*, **81**: 79–91.

Kiernan, K., Wood, C., and Middleton, G., 2003, Aquifer structure and contamination risk in lava flows: insights from Iceland and Australia. *Environmental Geology*, **43**: 852–865.

King, J.K., Kostka, J., Frischer, M.E., Saunders, F.M., and Jahnke, R.A., 2001, A quantitative relationship that demonstrates mercury methylation rates in marine sediments are based on the community composition and activity of sulfate-reducing Bacteria. *Environmental Science and Technology*, **35**: 2491–2496.

Kitanidis, P.C., 1997, *Introduction to Geostatistics: applications in hydrogeology*. Cambridge University Press, Cambridge.

Klimchouk, A.B., Lowed, D., Cooper, A.H., and Sauro, U. (eds), 1996, Gypsum karst of the world. *International Journal of Speleology*, **25**(3–4): 307pp.

Klimchouk, A.B., Ford, D.C., Palmer, A.N., and Dreybrodt, W. (eds), 2000, *Speleogenesis: evolution of karst aquifers*. National Speleological Society, Huntsville, AL.

Knarr, M.R., Goltz, M.N., Lamont, G.B., and Huang, J., 2003, In situ bioremediation of perchlorate-contaminated groundwater using a multi-objective parallel evolutionary algorithm. *Proceedings of the 2003 Congress on Evolutionary Computation (CEC 2003). Canberra, Australia, 8–12th December 2003*. IEEE Press, Piscataway (NJ). 3: 1604–1611.

Koch, N.K., 1994, *Geohazards: natural and human*. Prentice-Hall, Englewood Cliffs, NJ.

Konikow, L.F., 1981, Role of numerical simulation in analysis of ground-water quality problems. *Science of the Total Environment*, **21**: 299–312.

Konikow, L.F. and Bredehoeft, J.D., 1992, Ground-water models cannot be validated. *Advances in Water Resources*, **15**: 75–83.

Kooi, H. and Groen, J., 2001, Offshore continuation of coastal groundwater systems; predictions using sharp-interface approximations and variable-density flow modelling. *Journal of Hydrology*, **246**: 19–35.

Korim, K., 1994, The hydrogeothermal systems in Hungary. In: Risler, J.-J. and Simers, I. (eds) *Hydrogeothermics*. International Association of Hydrogeologists, International Contributions to Hydrogeology, Vol. 15. Verlag Heinz Heise, Hannover, pp. 43–55.

Kruseman, G.P. and de Ridder, N.A., 1991, *Analysis and Evaluation of Pumping Test Data*, 2nd edn. ILRI Publication No. 47. International Institute for Land Reclamation and Improvement, Wageningen, Netherlands.

Kulkarni, H., Deolankar, S.B., Lalwani, A., Joseph, B., and Pawar, S., 2000, Hydrogeological framework of the Deccan basalt groundwater systems, west-central India. *Hydrogeology Journal*, **8**: 368–378.

Kuma, J.S. and Younger, P.L., 2001, Pedological characteristics related to ground water occurrence in the Tarkwa area, Ghana. *Journal of African Earth Sciences*, **33**: 363–376.

Kuma, J.S. and Younger, P.L., 2004, Water quality trends in the Tarkwa gold-mining district, Ghana. *Bulletin of Engineering Geology and the Environment*, **63**: 119–132.

LaFleur, R.G. (ed.), 1984, *Groundwater as a Geomorphic Agent*. Allen and Unwin, Boston.

Lamont-Black, J., Younger, P.L., Forth, R.A., Cooper, A.H., and Bonniface, J.P., 2002, A decision-logic framework for investigating subsidence problems potentially attributable to gypsum karstification. *Engineering Geology*, **65**: 205–215.

Langmuir, D., 1997, *Aqueous Environmental Geochemistry*. Prentice-Hall, New Jersey.

Larson, C. and Larson, J., 1990, *Lava Beds Caves*. ABC Publishing, Vancouver, WA.

LeBlanc, M., Casiot, C., Elbaz-Poulichet, F., and Personné, C., 2002, Arsenic removal by oxidizing bacteria in a heavily arsenic-contaminated acid mine drainage system (Carnoulès, France). In: Younger, P.L. and Robins, N.S. (eds) Mine Water Hydrogeology and Geochemistry. *Geological Society, London, Special Publications*, **198**: 267–274.

Lebron, I., Schaap, M.G., and Suarez, D.L., 1999, Saturated hydraulic conductivity prediction from microscopic pore geometry measurements and neural network analysis. *Water Resources Research*, **35**: 3149–3158.

Leeson, J., Edwards, A., Smith, J.W.N., and Potter, H.A.B., 2003, *Hydrogeological Risk Assessments for Landfills and the Derivation of Groundwater Control and Trigger Levels*. Report number LFTGN01. Environment Agency (England and Wales), Bristol.

Lehr, J., 1988, An irreverent view of contaminant dispersion. *Ground Water Monitoring Review*, **8**: 4–6.

Lerner, D.N., 1986, Leaking pipes recharge groundwater. *Ground Water*, **24**: 654–662.

Lerner, D.N., 2002, Identifying and quantifying urban recharge: a review. *Hydrogeology Journal*, **10**: 143–152.

Lerner, D.N. and Walton, N.R.G. (eds), 1998, Contaminated land and groundwater: future directions. *Geological Society Engineering Geology Special Publication No. 14*. Geological Society, London.

Lerner, D.N., Issar, A.S., and Simmers, I., 1990, Groundwater Recharge. A guide to understanding and estimating natural recharge. *International Contributions to Hydrogeology*, Vol. 8. International Association of Hydrogeologists/Verlag Heinz Heise, Hannover.

Lewis, M.A., Lilly, A., and Bell, J.S., 2000, Groundwater vulnerability mapping in Scotland: modifications to classification used in England and Wales. In: Robins, N.S. and Misstear, B.D.R. (eds) Groundwater in the Celtic Regions: studies in hard rock and Quaternary hydrogeology. *Geological Society, London, Special Publications*, **182**: 71–79.

Liedl, R., Sauter, M., Hückinghaus, D., Clemens, T., and Teutsch, G., 2003, Simulation of the development of karst aquifers using a coupled continuum pipe flow model. *Water Resources Research*, **39**: article no. 1057.

Lindholm, G.F. and Vaccaro, J.J., 1988, Region 2, Columbia Lava Plateau. In: Back, W., Rosenshein, J.S., and Seaber, P.R. (eds), *Hydrogeology. The Geology of North America*, Vol. O-2. Geological Society of America, Boulder, CO, pp. 37–50.

Llamas, M.R., 1988, Conflicts between wetland conservation and groundwater exploitation: two case studies. *Environmental Geology*, **11**: 241–251.

Llamas, M.R., 2003, Lessons learnt from the impact of the neglected role of groundwater in Spain's water policy. In: AlSharhan, A.S. and Wood, W.W. (eds) *Water Resources Perspectives: evaluation, management and policy*. Elsevier Science, Amsterdam, pp. 63–81.

Lloyd, J.W. and Heathcote, J.A., 1985, *Natural Inorganic Hydrochemistry in Relation to Groundwater. An introduction*. Clarendon Press, Oxford.

Loaiciga, H.A., Maidment, D.R., and Valdes, J.B., 2000, Climate-change impacts in a regional karst aquifer, Texas, USA. *Journal of Hydrology*, **227**: 173–194.

Lofts, S. and Tipping, E., 2000, Solid-solution metal partitioning in the Humber rivers: application of WHAM and SCAMP. *Science of The Total Environment*, **251**: 381–399.

Lohman, S.W., (ed.), 1972, Definition of ground-water terms: revisions and conceptual refinements. *US Geological Survey Water Supply Paper 1988*.

López Chicano, M., Calvache, M.L., Martín-Rosales, W., and Gisbert, J., 2002, Conditioning factors in flooding of karstic poljes – the case of Zafarraya polje (South Spain). *Catena*, **49**: 331–352.

Loredo, J., Ordóñez, A., and Pendás, F., 2002, Hydrogeological and geochemical interactions of adjoining mercury and coal mine spoils in the Morgao catchment (Mieres, NW Spain). In: Younger, P.L. and Robins, N.S. (eds), Mine Water Hydrogeology and Geochemistry. *Geological Society, London, Special Publications*, **198**: 327–336.

Lu, S., Molz, F.J., Fogg, G.E., and Castle, J.W., 2002, Combining stochastic facies and fractal models for representing natural heterogeneity. *Hydrogeology Journal*, **10**: 475–482.

Mackay, D.M., 1998, Is cleanup of VOC contaminated groundwater feasible? In: Lerner, D.N. and Walton, N.R.G. (eds) *Contaminated Land and Groundwater: future directions*. Geological Society Engineering Geology Special Publication No. 14. Geological Society, London, pp. 3–11.

Maddock, I.P., Petts, G.E., Evans, E.C., and Greenwood, M.T., 1995, Assessing river–aquifer interactions within the hyporheic zone. In: Brown, A.G. (ed.), *Geomorphology and Groundwater*. Wiley, Chichester, pp. 53–74.

Malard, F., Plenet, S., and Gibert, J., 1996, The use of invertebrates in ground water monitoring: a rising research field. *Groundwater Monitoring and Remediation*, **16**: 103–113.

Malcolm, R. and Soulsby, C., 2000, Modelling the potential impact of climate change on a shallow coastal aquifer in northern Scotland. In: Robins, N.S. and Misstear, B.D.R. (eds), Groundwater in the Celtic Regions: studies in hard rock and Quaternary hydrogeology. *Geological Society, London, Special Publications*, **182**: 191–204.

Malcolm. I.A., Soulsby, C., Youngson, A.F., and Petry, J., 2002, Heterogeneity in ground water–surface water interactions in the hyporheic zone of a salmonid spawning stream. *Hydrological Processes*, **17**: 601–617.

Malcolm, I.A., Youngson, A.F., and Soulsby, C., 2003, Survival of salmonid eggs in a degraded gravel-bed stream: effects of groundwater–surface water interactions. *River Research and Applications*, **19**: 303–316.

Malcolm, I.A., Soulsby, C., Youngson, A.F., and Hannah, D.M., 2005, Catchment-scale controls on groundwater–surface water interactions in an upland Scottish catchment: implications for salmon embryo survival. *Rivers Research and Applications*, **21**: 977–989.

Malmström, M., Destouni, G., Banwart, S.A., and Strömberg, B., 2000, Resolving the scale dependence of mineral weathering rates. *Environmental Science and Technology*, **34**: 1375–1377.

Manning, A.H. and Solomon, D.K., 2003, Using noble gases to investigate mountain-front recharge. *Journal of Hydrology*, **275**: 194–207.

Marsily, G. de, 1986, *Quantitative Hydrogeology: groundwater hydrology for engineers*. Academic Press, Orlando, Florida.

Martinez-Alier, J., 2002, *The Environmentalism of the Poor: a study of ecological conflicts and valuation*. Edward Elgar Publishing, Cheltenham.

Marty, B., Dewonck, S., and France-Lanord, C., 2003, Geochemical evidence for efficient aquifer isolation over geological timeframes. *Nature*, **425**: 55–58.

Mastin, L.G., 1997, The roles of magma and groundwater in the phreatic eruptions at Inyo Craters, Long Valley Caldera, California. *Bulletin of Volcanology*, **53**: 579–596.

Mather, J.D., 2001, Joseph Lucas and the term "hydrogeology". *Hydrogeology Journal*, **9**: 413–415.

Maund, J.G. and Eddleston, M., (eds), 1998, *Geohazards in Engineering Geology*. Geological Society, London, Engineering Geology Special Publications no. 15. Geological Society, Bath. 448pp.

Maxey, G.B., 1964, Hydrostratigraphic units. *Journal of Hydrology*, **2**: 124–129.

Mazor, E., 1991, *Applied Chemical and Isotopic Groundwater Hydrology*. Wiley, New York.

McCarty, P.L., Goltz, M.N., Hopkins, G.D., et al., 1998, Full scale evaluation of *in situ* cometabolic degradation of trichloroethylene in groundwater through toluene injection. *Environmental Science and Technology*, **32**, 88–100.

McDonald, M.C. and Harbaugh, A.W., 1988, A modular three-dimensional finite-difference ground-water flow model. *US Geological Survey Techniques of Water-Resources Investigations*, book 6, chap. A1. United States Geological Survey, Washington, DC.

McGuire, V.L., 2003, *Water-level Changes in the High Plains Aquifer, Predevelopment to 2001, 1999 to 2000, and 2000 to 2001*. US Geological Survey Fact Sheet FS–078–03. United States Geological Survey, Lincoln, NE.

McKendry, I.G., 2003, Applied climatology. *Progress in Physical Geography*, **27**: 597–606.

McLean, I. and Johnes, M., 2000, *Aberfan. Government and Disasters*. Welsh Academic Press, Cardiff.

Meinzer, O.E., 1923a, The occurrence of ground water in the United States, with a discussion of principles. *US Geological Survey Water-Supply Paper 489*.

Meinzer, O.E., 1923b, Outline of ground-water hydrology, with definitions. *US Geological Survey Water-Supply Paper 494*.

Misstear, B.D.R. and Daly, D., 2000, Groundwater protection in a Celtic region: the Irish example. In: Robins, N.S. and Misstear, B.D.R. (eds), Groundwater in the Celtic Regions: studies in hard rock and Quaternary hydrogeology. *Geological Society, London, Special Publications*, **182**: 53–65.

Mitsch, W.J. and Gosselink, J.G., 2000, *Wetlands*, 3rd edn. John Wiley, New York.

Molins, S., Carrera, J., Ayora, C., and Saaltink, M.W., 2004, A formulation for decoupling components in reactive transport problems, *Water Resources Research*, **40**, paper no. W10301.

Moore, G.W. and Sullivan, G.N., 1997, *Speleology: caves and the cave environment*. Cave Books, St Louis, MO.

Morris, B.L., Lawrence, A.R.L., Chilton, P.J.C., Adams, B., Calow, R.C., and Klinck, B.A., 2003, *Groundwater and its Susceptibility to Degradation: a global assessment of the problem and options for management*. Early Warning and Assessment Report Series **03-3**. United Nations Environment Programme, Nairobi, Kenya.

Mühlherr, I.H., Hiscock, K.M., Dennis, P.F., and Feast, N.A., 1998, Changes in groundwater chemistry due to rising groundwater levels in the London Basin between 1963 and 1994. In: Robins, N.S. (ed.), Groundwater pollution, aquifer recharge and vulnerability. *Geological Society, London, Special Publications*, **130**: 47–62.

Mul, M.L., Savenije, H.H.G., Luxemburg, W.M.J., and Weng, P., 2003, Groundwater induced floods, Somme River (March 2001). *Geophysical Research Abstracts*, **5**: abstract 01412.

National Research Council, 1997, *Valuing Ground Water: economic concepts and approaches*. National Academy Press, Washington DC.

National Research Council, 2000, *Natural Attenuation for Groundwater Remediation*. National Academy Press, Washington DC.

Nordstrom, D.K., 2003, Modeling low-temperature geochemical processes. In: Drever, J.I. (ed.), Surface and Ground Water, Weathering and Soils, *Treatise on Geochemistry*, Vol. 5. Elsevier, Amsterdam, pp. 37–72.

Nordstrom D.K., Alpers C.N., Ptacek C.J., and Blowes D.W., 2000, Negative pH and extremely acidic mine waters from Iron Mountain, California. *Environmental Science and Technology*, **34**: 254–258.

NRA, 1992, *Policy and Practice for the Protection of Groundwater*. National Rivers Authority (England and Wales), Bristol (UK).

Nuttall, C.A. and Younger, P.L., 1999, Reconnaissance hydrogeochemical evaluation of an abandoned Pb-Zn orefield, Nent Valley, Cumbria, UK. *Proceedings of the Yorkshire Geological Society*, **52**: 395–405.

Nuttall, C.A., Adams, R., and Younger, P.L., 2002, Integrated hydraulic–hydrogeochemical assessment of flooded deep mine voids by test pumping at Deerplay (Lancashire) and Frances (Fife) Collieries. In: Younger, P.L. and Robins, N.S. (eds), Mine Water Hydrogeology and Geochemistry. *Geological Society, London, Special Publications*, **198**: 316–326.

Nyer, E.K., 1992, *Groundwater Treatment Technology*, 2nd edn. John Wiley and Sons. New York.

Oldroyd, D.R., 1996, *Thinking about the Earth: a history of ideas in geology*. The Athlone Press, London.

Oliver, J.E., 1998, Evapotranspiration. In: Herschy, R.W. and Fairbridge, R.W. (eds), *Encyclopaedia of Hydrology and Water Resources*. Kluwer Academic, Boston, pp. 266–271.

Olsthoorn, T.N., 1985, The power of the electronic worksheet – modeling without special programs. *Ground Water*, **23**: 381–390.

Orchard, R.J., 1975, Working under bodies of water. *The Mining Engineer*, **170**: 261–270.

Orghidan, T., 1959, Ein neuer Lebenraum des unterrirdischen Wassers: Der hyporheische Biotop. *Archiv für Hydrobiologie*, **55**: 392–414.

Owen, M., Headworth, H.G., and Morgan-Jones, M., 1991, Groundwater in basin management. In: Downing, R.A. and Wilkinson, W.B. (eds) *Applied Groundwater Hydrology*. Clarendon Press, Oxford, pp. 16–34.

Parkhurst, D.L. and Appelo, C.A.J., 1999, *User's guide to PHREEQC (Version 2) – a computer program for speciation, batch-reaction, one-dimensional transport, and inverse geochemical calculations*. Water-Resources Investigations Report 99–4259. US Geological Survey, Denver, CO.

Penman, H.L., 1948, Natural evaporation from open water, bare soil and grass. *Proceedings of the Royal Society, London, Series A*, **193**: 120–145.

Penman, H.L., 1949, The dependence of transpiration on weather and soil conditions. *Journal of Soil Science*, **1**: 74–89.

Pentecost, A., 1996, The Quaternary travertine deposits of Europe and Asia Minor. *Quaternary Science Review*, **14**: 1005–1028.

Petts, G.E., Bickerton, M.A., Crawford, C., Lerner, D.N., and Evans, D., 1999, Flow management to sustain groundwater-dominated stream ecosystems. *Hydrological Processes*, **13**: 497–513.

Pettyjohn, W.A., 1985a, Ground water–surface water relationship. In: *US EPA Protection of Public Water Supplies from Ground-water Contamination*. United States Environmental Protection Agency, Technology Transfer Seminar Publication **EPA/625/4-85/016**: 83–105.

Pettyjohn, W.A., 1985b, Regional approach to ground water investigations. In: Ward, C.H., Giger, W. and McCarty, P.L., (eds) *Ground Water Quality*. John Wiley and Sons, New York, pp. 402–417.

Pewe, T.L., 1990, Land subsidence and earth-fissure formation caused by groundwater withdrawal in Arizona; a review. In: Higgins, C.G. and Coates, D.R. (eds), Groundwater geomorphology; the role of subsurface water in Earth-surface processes and landforms. *Geological Society of America Special Paper*, **252**: 218–233.

Pickford, J., 1991, *The Worth of Water. Technical briefs on health, water and sanitation*. Intermediate Technology Publications Ltd, London.

Pinder, G.F. and Sauer, S.P., 1971, Numerical simulation of flood wave modification due to bank storage effects. *Water Resources Research*, **7**: 63–70.

Piper, A.M., 1944, A graphic procedure in the geochemical interpretation of water analyses. *American Geophysical Union Transactions*, **25**: 914–923.

PIRAMID Consortium, 2003, *Engineering Guidelines for the Passive Remediation of Acidic and/or Metalliferous Mine Drainage and Similar Wastewaters*. European Commission 5th Framework RTD Project no. EVK1-CT-1999-000021 "Passive in-situ remediation of acidic mine/industrial drainage" (PIRAMID). (*www.piramid.org*). University of Newcastle Upon Tyne, Newcastle Upon Tyne, UK.

Plemper, B., 2004, *A System Dynamics Interpretation of Groundwater Movement*. Unpublished PhD thesis, University of Sunderland, UK.

Poland, J.F. (ed.), 1984, *Guidebook to Studies of Land Subsidence due to Ground-water Withdrawal. UNESCO Studies and Reports in Hydrology*, vol. 40. United Nations Educational, Scientific and Cultural Organization, Paris.

Poulson, T.L. and White, W.B., 1969, The cave environment. *Science*, **165**: 971–81.

Power, G., Brown, R.S., and Imhof, J.G., 1999, Groundwater and fish – insights from North America. *Hydrological Processes*, **13**: 401–422.

Powers, J.P., 1992, *Construction Dewatering: new methods and applications*, 2nd edn. John Wiley and Sons, New York.

Powrie, W., 2004, *Soil Mechanics: concepts and applications*, 2nd edn. Spon Press, London.

Preene, M., Roberts, T.O.L., Powrie, W., and Dyer, M.R., 2000, *Groundwater Control – design and practice*. CIRIA Publication no C515. Construction Industry Research and Information Association (CIRIA), London.

Price, M., 1996, *Introducing groundwater*, 2nd edn. Chapman and Hall, London.

Price, M., 1998, Water storage and climate change in Great Britain – the role of groundwater. *Proceedings of the Institution of Civil Engineers – Water, Maritime and Energy*, 130: 42–50.

Price, M., Morris, B., and Robertson, A., 1982, A study of intergranular and fissure permeability in Chalk and Permian aquifers, using double-packer injection testing. *Journal of Hydrology*, **54**: 401–423.

Prickett, T.A. and Lonnquist, C.G., 1971, *Selected Digital Computer Techniques for Groundwater Resource Evaluation*. Illinois State Water Survey Bulletin No. 55. Illinois State Water Survey, Urbana, IL.

Prickett, T.A., Naymik, T.G., and Lonnquist, C.G., 1981, *A "Random-walk" Solute Transport Model for Selected Groundwater Quality Evaluations*. Illinois State Water Survey Bulletin No. 65. Illinois State Water Survey, Champaign, IL.

Pryce, W., 1778, *Mineralogia Cornubiensis; a treatise on minerals, mines and mining: containing the theory and natural history of strata, fissures, and lodes, with the methods of discovering and working of tin, copper and lead mines, and of cleansing and metalizing their products; showing each particular process for dressing, assaying, and smelting of ores. To which is added, an explanation of the terms and idioms of miners*. Privately Published, London. (Facsimile reprint published in 1972 by D Bradford Barton Ltd, Truro, Cornwall, UK.).

Pyne, R.G.D., 1995, *Groundwater Recharge and Wells: a guide to aquifer storage recovery*. CRC Press, Boca Raton, FL.

Rawson, P.F., Allen, P.M., Brenchley, P.J., et al., 2002, *Stratigraphical Procedure*. Geological Society Professional Handbook. Geological Society, London.

Rees, S.B., Bowell, R.J., and Wiseman, I., 2002, Influence of mine hydrogeology on mine water discharge chemistry. In: Younger, P.L. and Robins, N.S. (eds), Mine Water Hydrogeology and Geochemistry. *Geological Society, London, Special Publications*, **198**: 379–390.

Reynolds, J., 1992, *Handpumps: toward a sustainable technology. Research and development during the Water Supply and Sanitation Decade*. UNDP–World Bank Water and Sanitation Programme, Washington DC.

Rhoades, J.D., Kandiah, A., and Mashali, A.M., 1992, *The Use of Saline Waters for Crop Production*. FAO Irrigation and Drainage Paper 48. Food and Agriculture Organization of the United Nations, Rome.

Richards, L.A., 1931, Capillary conduction through porous medium. *Physics*, **1**: 318–333.

Rittmann, B.E., 2004, Definition, objectives, and evaluation of natural attenuation. *Biodegradation*, **15**: 349–357.

Robins, N.S., 1996, *Hydrogeology of Northern Ireland*. British Geological Survey/HMSO, London.

Robins, N.S. (ed.), 1998, Groundwater pollution, aquifer recharge and vulnerability. *Geological Society, London, Special Publications*, **130**.

Robins, N.S., Jones, H.K., and Ellis, J., 1999, An aquifer management case study – the chalk of the English South Downs. *Water Resources Management*, **13**: 205–218.

Rosenthal, E., Weinberger, G., Berkowitz, B., Flexer, A., and Kronfeld, J., 1992, The Nubian Sandstone Aquifer in the central and northern Negev, Israel – delineation of the hydrogeological model under conditions of scarce data. *Journal of Hydrology*, **132**: 107–135.

Rowe, P.W., 1986, The potentially latent dominance of groundwater in ground engineering. In: Cripps, J.C., Bell, F.G., and Culshaw, M.G. (eds), *Groundwater in Engineering Geology*. Engineering Geology Special Publication no. 3. Geological Society, London.

Rudolph, M., 2001, Urban hazards – sinking of a titanic city. *Geotimes*, **46**: 10.

Rushton, K.R., 2003, *Groundwater Hydrology. Conceptual and computational models*. John Wiley and Sons, Chichester.

Rushton, K.R. and Al-Othman, A.A.R., 1994, Control of rising groundwater in Riyadh, Saudi Arabia. In: Wilkinson, W.B. (ed.) *Groundwater Problems in Urban Areas*. Thomas Telford, London, pp. 299–309.

Rushton, K.R. and Redshaw, S.C., 1979, *Seepage and Groundwater Flow: numerical analysis by analog and digital methods*. John Wiley and Sons, Chichester.

Sachs, H.M., 2002, *Geology and Drilling Methods for Ground-source Heat Pump Installations: an introduction for engineers*. American Society of Heating, Refrigerating and Air-Conditioning Engineers, Atlanta, GA.

Salem, O.M., 1992, The Great Manmade River Project. *Water Resources Development*, **8**: 270–278.

Sammis, T.W., Evans, D.D., and Warwick, A.W., 1982, Comparison of methods to estimate deep percolation rates. *Water Resources Bulletin*, **18**: 465–470.

Sandstrom, K., 1995, Modeling the effects of rainfall variability on groundwater recharge in semi-arid Tanzania. *Nordic Hydrology*, **26**: 313–330.

Sanner, B., Karytsas, C., Mendrinos, D., and Rybach, L., 2003, Current status of ground source heat pumps and underground thermal energy storage in Europe. *Geothermics*, **32**: 579–588.

Scanlon, B.R., Healy, R.W., and Cook, P.G., 2002, Choosing appropriate techniques for quantifying groundwater recharge. *Hydrogeology Journal*, **10**: 18–39.

Schüring, J., Schulz, H.D., Fischer, W.R., Böttcher, J., and Duijnisveld, W.H.M., 2000, *Redox. Fundamentals, processes and applications*. Springer-Verlag, Berlin.

Seaber, P.R., 1988, Hydrostratigraphic units. In: Back, W., Rosenshein, J.S., and Seaber, P.R. (eds), *Hydrogeology. The Geology of North America*, Vol. O-2. Geological Society of America, Boulder, CO, pp. 9–14.

Sear, D.A., Armitage, P.D., and Dawson, F.H., 1999, Groundwater dominated rivers. *Hydrological Processes*, **13**: 255–276.

Selroos, J.O., Walker, D.D., Strom, A., Gylling, B., and Follin, S., 2002, Comparison of alternative modelling approaches for groundwater flow in fractured rock. *Journal of Hydrology*, **257**: 174–188.

Senarath, D.C.H. and Rushton, K.R., 1984, A routing technique for estimating groundwater recharge. *Ground Water*, **22**: 142–147.

Sharp, J.M., Jr, 1988, Alluvial aquifers along major rivers. In: Back, W., Rosenshein, J.S., and Seaber, P.R. (eds), *Hydrogeology. The Geology of North America*, Vol. O-2. Geological Society of America, Boulder, CO, pp. 273–282.

Shepherd, R.G., 1989, Correlations of permeability and grain size. *Ground Water*, **27**: 633–638.

Sherif, M.M. and Singh, V.P., 1999, Effect of climate change on sea water intrusion in coastal aquifers. *Hydrological Processes*, **13**: 1277–1287.

Simmers, I., Villarroya, F., and Rebollo, L.F. (eds), 1992, Selected papers on aquifer overexploitation. Papers presented at the 23rd Congress of IAH, 15–19 April 1991, Puerto de la Cruz, Tenerife (Spain). International Association of Hydrogeologists/Verlag Heinz Heise, Hannover.

Simmons, C.T., 2003, Happy 200th birthday Mr Darcy and our thanks for your law! A tribute editorial celebrating the life and times of the father of our science, Henry Darcy (1803–1858). *Hydrogeology Journal*, **11**: 611–614.

Sinclair, W.C., 1982, Sinkhole development resulting from ground-water withdrawal in the Tampa area, Florida. *United States Geological Survey Water Resources Investigations*, **81–50**.

Sklash, M.G. and Farvolden, R.N., 1979, The role of groundwater in storm runoff. *Journal of Hydrology*, **43**: 45–65.

Sloan, C.E. and van Everdingen, R.O., 1988, Region 28, Permafrost region. In: Back, W., Rosenshein, J.S., and Seaber, P.R. (eds), *Hydrogeology. The Geology of North America*, Vol. O-2. Geological Society of America, Boulder, CO, pp. 263–270.

Smakhtin, V.Y. and Toulouse, M., 1998, Relationships between low-flow characteristics of South African streams. *Water S.A.*, **24**: 107–112.

Smith, A.H., Lingas, E.O., and Rahman, M., 2000, Contamination of drinking-water by arsenic in Bangladesh: a public health emergency. *Bulletin of the World Health Organization*, **78**: 1093–1103.

Soltanpour, P.N. and Raley, W.L., 1999, *Livestock Drinking Water Quality*. Colorado State University Cooperative Extension, Livestock Series (Management), No. 4908.

Soulsby, C., Chen, M., Ferrier, R.C., Helliwell, R.C., Jenkins, A., and Harriman, R., 1998, Hydrogeochemistry of shallow groundwater in an upland Scottish catchment. *Hydrological Processes*, **12**: 1111–1127.

Soulsby, C., Rodgers, P., Smart, R., Dawson, J., and Dunn, S., 2003, A tracer-based assessment of hydrological pathways at different spatial scales in a mesoscale Scottish catchment. *Hydrological Processes*, **17**: 759–777.

Soulsby, C., Youngson, A.F., Moir, H.J., and Malcolm, I.A., 2001, Fine sediment influence on salmonid spawning habitat in a lowland agricultural stream: a preliminary assessment. *Science of the Total Environment*, **265**: 295–307.

Stiff, H.A., 1951, The interpretation of chemical water analysis by means of patterns. *Journal of Petroleum Technology*, **3**: 15–17.

Stone, W.J., 1999, *Hydrogeology in Practice: a guide to characterizing ground-water systems*. Prentice Hall, New Jersey.

Stumm, W. and Morgan, J.J., 1996, *Aquatic Chemistry. Chemical equilibria and rates in natural waters*, 3rd edn. John Wiley and Sons, New York.

Theis, C.V., 1935, The relation between the lowering of the piezometric surface and the rate and duration of discharge of a well using ground water storage. *Transactions of the American Geophysical Union*, **16**: 519–524.

Theis, C.V., 1940, The source of water derived from wells: essential factors controlling the response of an aquifer to development, *Civil Engineering*, **10**: 277–280.

Thornthwaite, C.W., and Holzman, B., 1942, *Measurements of Evaporation from Land and Water Surfaces*. US Department of Agriculture Technical Bulletin No. 817. USDA, Washington DC.

Tipping, E., 1994, WHAM – A chemical-equilibrium model and computer code for waters, sediments, and soils incorporating a discrete site electrostatic model of ion-binding by humic substances. *Computers and Geosciences*, **20**: 973–1023.

Todd, D.K., 1980, *Groundwater Hydrology*. John Wiley and Sons, New York.

Tolman, C.F., 1937, *Ground Water*. McGraw-Hill Book Company, New York.

Torrens, H.S., 2004, The water-related work of William Smith (1769–1839). In: Mather, J.D. (ed.), 200 years of British hydrogeology. *Geological Society, London, Special Publications*, **225**: 15–30.

Trottier, J., 1999, *Hydropolitics in the West Bank and Gaza Strip*. Palestinian Academic Society for the Study of International Affairs, Jerusalem.

Tufenkji, N., Ryan, J.N., and Elimelech, M., 2002, The promise of bank filtration. *Environmental Science and Technology*, **36**: 422A–428A.

Turner, B.R., Younger, P.L., and Fordham, C.E., 1993, Fell Sandstone Group lithostratigraphy southwest of Berwick-Upon-Tweed: implications for the regional development of the Fell Sandstone. *Proceedings of the Yorkshire Geological Society*, **49**: 269–281.

Turton, P., 1998, Water use. In: Herschy, R.W. and Fairbridge, R.W., (eds), *Encyclopaedia of Hydrology and Water Resources*. Kluwer Academic Publishers, Boston, pp. 770–775.

Uma, K.O., Egboka, B.C.E., and Onuoha, K.M., 1989, New statistical grain-size method for evaluating the hydraulic conductivity of sandy aquifers. *Journal of Hydrology*, **108**: 343–366.

US EPA, 1995, *Manual: ground-water and leachate treatment systems*. EPA Publication no. EPA/625/R-94/005. Center for Environmental Research Information, Cincinatti, OH.

US EPA, 1996, *Pump-and-treat ground-water Remediation: a guide for decision makers and practitioners*. EPA Publication no. EPA/625/R-95/005. United States Environmental Protection Agency. Office of Research and Development, Washington DC.

Vaccaro, J.J., 1992, Sensitivity of groundwater recharge estimates to climate variability and change, Columbia Plateau, Washington. *Journal of Geophysical Research–Atmospheres*, **97**: 2821–2833.

Vrba, J. and Zaporozec, A. (eds), 1994, *Guidebook on Mapping Groundwater Vulnerability*. International Contributions to Hydrogeology Series, Vol. 16. International Association of Hydrogeologists/Verlag Heinz Heise, Hannover.

Wagner, E.G. and Lanoix, J.N., 1959, *Water Supply for Rural Areas and Small Communities*. Monograph No. 42. World Health Organisation, Geneva.

Walter, A.L., Frind, E.O., Blowes, D.W., Ptacek, C.J., and Molson, J.W., 1994a, Modelling of multicomponent reactive transport in groundwater: 1. Model development and evaluation. *Water Resources Research*, **30**: 3137–3148.

Walter, A.L., Frind, E.O., Blowes, D.W., Ptacek, C.J., and Molson, J.W., 1994b, Modelling of multicomponent reactive transport in groundwater: 2. Metal mobility in aquifers impacted by acidic mine tailings discharge. *Water Resources Research*, **30**: 3149–3158.

Wardrop, D.R., Leake, C.C., and Abra, J., 2001, Practical techniques that minimize the impact of quarries on the water environment. *Transactions of the Institution of Mining and Metallurgy (Section B: Applied Earth Sciences)*, **110**: B5–B14.

Watt, S.B. and Wood, W.E., 1979, *Hand Dug Wells and their Construction*, 2nd edn. Intermediate Technology Publications, London.

WHAT, 2000, *Governance for a Sustainable Future*. World Humanity Action Trust, London.

WHO, 2004, *Guidelines for Drinking-water Quality*, 3rd edn. World Health Organization, Geneva. (In three volumes: 1–Recommendations; 2–Health criteria and other supporting information; 3–Surveillance and control of community supplies.) (On-line at: *http://www.who.int/water_sanitation_health/dwq/gdwq3/en/*)

Wilkens, H., Culver, D.C., and Humphreys, W.F., (eds), 2000, *Subterranean Ecosystems. Ecosystems of the World*, Vol. 30. Elsevier, New York.

Wilkinson, W.B. and Cooper, D.M., 1993, The response of idealized aquifer river systems to climate-change. *Hydrological Sciences Journal*, **38**: 379–390.

Williams, J.R., 1970, *Ground Water in the Permafrost Regions of Alaska*. United States Geological Survey Professional Paper 696. US Government Printing Office, Washington DC.

Wilson, M.D. (ed.), 1994, *Reservoir Quality Assessment and Prediction in Clastic Reservoirs*. SEPM Short Course 30. SEPM Society for Sedimentary Geology, Tulsa, OK.

Wilson, J., McNabb, J., Balkwill, D., and Ghiorse, W., 1983, Enumeration and characterization of bacteria indigenous to a shallow water-table aquifer. *Groundwater*, **21**: 134–142.

Winchester, S., 2001, *The Map that Changed the World. The tale of William Smith and the birth of a science*. Viking, London.

Winter, T.C., Harvey, J.W., Franke, O.L., and Alley, W.M., 1998, Ground water and surface water: a single resource. *United States Geological Survey Circular*, **1139**.

Wood, W.W. and Fernández, L.A., 1988, Volcanic rocks. In: Back, W., Rosenshein, J.S., and Seaber, P.R., (eds), *Hydrogeology. The Geology of North America*, Vol. O-2. Geological Society of America, Boulder, CO, pp. 353–365.

World Commission on Environment and Development, 1987, *Our Common Future. (The Brundtland Report)*. Oxford University Press, New York.

WRI-UNEP, 1998, *World Resources 1998–99: Environmental change and human health*. A joint publication by the World Resources Institute (WRI), the United Nations Environment Programme (UNEP), the United Nations Development Programme, and The World Bank.

WWF, 2004, *Freshwater Biodiversity and Sustainable Development*. Factsheet. Worldwide Fund for Nature (WWF), London.

Young, B. and Culshaw, M.G., 2001, *Fissuring and Related Ground Movements in the Magnesian Limestone and Coal Measures of the Houghton-le-Spring Area, City of Sunderland*. British Geological Survey, Technical Report WA/01/04.

Younger, P.L., 1989, Devensian periglacial influences on the development of spatially variable permeability in the chalk of South East England. *Quarterly Journal of Engineering Geology*, **22**: 343–354.

Younger, P.L., 1992, The hydrogeological use of thin sections: inexpensive estimates of groundwater flow and transport parameters. *Quarterly Journal of Engineering Geology*, **25**: 159–164.

Younger, P.L., 1993, Simple generalised methods for estimating aquifer storage parameters. *Quarterly Journal of Engineering Geology*, **26**: 127–135.

Younger, P.L., 1996, Submarine groundwater discharge. *Nature*, **382**: 121–122.

Younger, P.L., 1998, Long term sustainability of groundwater abstraction in north Northumberland. In: Wheater, H. and Kirby, C. (eds.), *Hydrology in a Changing Environment. Proceedings of the International Symposium organised by the British Hydrological Society*, Vol. II. Exeter, UK, 6–10 July 1998. Wiley, Chichester, pp. 213–227.

Younger, P.L., 2002, Deep mine hydrogeology after closure: insights from the UK. In: Merkel, B.J., Planer-Friedrich, B., and Wolkersdorfer, C., (eds) *Uranium in the Aquatic Environment. Proceedings of the International Conference Uranium Mining and Hydrogeology III and the International Mine Water Association Symposium*. Freiberg, Germany, 15–21 September 2002. Springer-Verlag, Berlin. pp. 25–40.

Younger, P.L., 2004a, Environmental impacts of coal mining and associated wastes: a geochemical perspective. In: Gieré, R. and Stille, P. (eds) Energy, waste, and the environment: a geochemical perspective. *Geological Society, London, Special Publications*, **236**: 169–209.

Younger, P.L., 2004b, "Making water": the hydrogeological adventures of Britain's early mining engineers. In: Mather, J.D. (ed.) 200 years of British hydrogeology. *Geological Society, London, Special Publications*, **225**: 121–157.

Younger, P.L. and Bradley, K.M., 1994, Application of geochemical mineral exploration techniques to the cataloguing of problematic discharges from abandoned mines in north-east England. In: *Proceedings of the 5th International Minewater Congress*. Nottingham, UK, Sept 1994, **2**: 857–871.

Younger, P.L. and McHugh, M., 1995, Peat development, sand cones and palaeohydrogeology of a spring fed mire in East Yorkshire. *The Holocene*, **5**: 59–67.

Younger, P.L. and Milne, C.A., 1997, Hydrostratigraphy and hydrogeochemistry of the Vale of Eden, Cumbria, UK. *Proceedings of the Yorkshire Geological Society* **51**, 349–366.

Younger, P.L. and Stunell, J.M., 1995, Karst and pseudokarst: an artificial distinction? In: Brown, A.G. (ed.), *Geomorphology and Groundwater*. Wiley, Chichester, pp. 121–142.

Younger, P.L., Donovan, R.N., and Hounslow, A.W., 1986, Barite Travertine from Zodletone Mountain in the Slick Hills, South Western Oklahoma. In: Donovan, R.N. (ed.), The Slick Hills of South Western Oklahoma–Fragments of an Aulacogen? *Oklahoma Geological Survey Guidebook*, **24**: 75–81.

Younger, P.L., Mackay, R., and Connorton, B.J., 1993, Streambed sediment as a barrier to groundwater pollution: insights from fieldwork and modelling in the River Thames Basin. *Journal of the Institution of Water and Environmental Management*, **7**: 577–585.

Younger, P.L., Banwart, S.A., and Hedin, R.S., 2002a, *Mine Water: Hydrology, Pollution, Remediation*. Kluwer Academic Publishers, Dordrecht.

Younger, P.L., Teutsch, G., Custodio, E., Elliot, T., Manzano, M., and Sauter, M., 2002b, Assessments of the sensitivity to climate change of flow and natural water quality in four major carbonate aquifers of Europe. In: Hiscock, K.M., Rivett, M.O., and Davison, R.M. (eds), Sustainable groundwater development. *Geological Society Special Publications*, **193**: 303–323.

Yusoff, I., Hiscock, K.M., and Conway, D., 2002, Simulation of the impacts of climate change on groundwater resources in eastern England. In: Hiscock, K.M., Rivett, M.O., and Davison, R.M., (eds), Sustainable groundwater development. *Geological Society Special Publications*, **193**: 325–344.

Zeckster, I.S. and Loaiciga, H.A., 1993, Groundwater fluxes in the global hydrological cycle – past, present and future. *Journal of Hydrology*, **144**: 405–427.

Zeckster, I.S., Loaiciga, H.A., and Wolf, J.T., 2005, Environmental impacts of groundwater overdraft: selected case studies in the southwestern United States. *Environmental Geology*, **47**: 396–404.

Zhang, B. and Lerner, D.N., 2000, Modeling of ground water flow to adits. *Ground Water*, **38**: 99–105.

Zhu, Y., Fox, P.J., and Morris, J.P., 1999, A pore-scale numerical model for flow through porous media. *International Journal for Numerical and Analytical Methods in Geomechanics*, **23**: 881–904.

Glossary

All of the terms highlighted in bold when first introduced in this book are listed here. Note that the definitions relate only to the use of these terms in hydrogeology, and are practical (as opposed to highly formal) in nature.

Note: q.v. stands for *quod vide* (which see), indicating that a term used in a definition is itself defined elsewhere in the glossary.

acidic Waters with a *p*H below 6.5.

acidity The ability of a water to neutralize strong alkali to a given end-point *p*H (usually 8.5). In most groundwaters, acidity measures the dissolved concentrations of hydroxide-forming metals (Fe, Al, etc.) as well as of protons (q.v.).

acid-sulfate soils Soils which originally contained large concentrations of the mineral pyrite (FeS_2), at least some of which has since been oxidized to form strongly acidic sulfate salts of iron (and in some cases aluminum).

activated carbon Particulate carbon which has been artificially subjected to high temperatures, giving it highly chemically reactive surfaces. (Activated carbon is widely used in the treatment of groundwaters contaminated with organic compounds.)

activity That fraction of the total dissolved molar concentration (q.v.) of a given solute (q.v.) which is "active" in solution (i.e. available to participate in speciation and mineral precipitation/dissolution reactions etc.).

actual evapotranspiration That portion of the potential evapotranspiration (q.v.) which actually occurs over a specified period of time, given the limited rate of supply of moisture to the soil surface and/or to transpiring plants. Under most circumstances, actual evapotranspiration is only a small fraction of the potential evapotranspiration rate.

adit dewatering The use of one or more long, near-horizontal tunnels (adits) to provide an artificial high-permeability drainage pathway which achieves dewatering (q.v.) of overlying ground.

adsorption Adhesion of a solute to a solid surface by electrostatic attraction.

advection The movement of solutes with the bulk flow of groundwater.

advection-dispersion equation The partial differential equation which describes solute transport as the sum of advection (q.v.) and dispersion (q.v.) processes.

air sparging A form of bioremediation (q.v.) in which air is pumped down wells into contaminated groundwater to stimulate the beneficial activity of aerobic bacteria.

air stripping Cascading polluted groundwater through an aeration device to encourage the release of volatile pollutant compounds to the air.

air-lifting The use of a current of compressed air within a well to displace water (and any entrained solids) to surface. (Often a key activity during well development (q.v.).)

alkaline Waters with a *p*H above 8.5.

alkalinity The ability of a water to neutralize strong acid to a given end-point *p*H (usually 4.5). In most groundwaters, alkalinity is primarily a reflection of dissolved bicarbonate (HCO_3^-).

analytical model A mathematical model (q.v.) which is solved using an analytical solution (q.v.).

analytical solution An exact solution to a given mathematical model (q.v.), applicable at any space-time point within a specified model domain, which is derived using the methods of advanced calculus and allied mathematical techniques.

anions Negatively charged ions (q.v.).

anisotropic Exhibiting anisotropy (q.v.).

anisotropy The condition in which the value of a physical property of a body of some specified material (e.g. an aquifer (q.v.)) varies significantly depending on the orientation of any measurement. For instance, the measured hydraulic conductivity (q.v.) of many aquifers (q.v.) is much less in the vertical direction than in the horizontal plane; such aquifers are said to display anisotropy.

anoxia The complete absence of oxygen from water, such that aquatic organisms cannot perform aerobic respiration.

antiform An up-fold in rock strata.

aquifer A body of saturated rock that both stores and transmits important quantities of groundwater.

aquifer overexploitation The pumping of such great quantities of groundwater from an aquifer that a number of undesirable impacts develop.

aquifer storage and recovery Artificial cyclical storage of water in an aquifer, achieved by using artificial recharge (q.v.) to build up a store fresh water within an aquifer, which is later pumped out again during periods of high demand.

aquifer susceptibility The likelihood that a given aquifer will develop declining water levels, ecologically damaging decreases in outflows, land subsidence and/or groundwater quality degradation as a consequence of high rates of abstraction.

aquitard A saturated body of rock that impedes the movement of groundwater.

artificial recharge The deliberate introduction of water into the subsurface.

atomic absorption spectrophotometry An instrumental technique for measuring dissolved concentrations of cations, using the spectral ranges of light absorbed by metals which have been restored to their uncharged, elemental forms by exposure to high temperatures.

bail test A technique for measuring the hydraulic conductivity (q.v.) of an aquifer (q.v.) in which a large proportion of the water which was originally standing in the well is suddenly withdrawn (by bailing), and the rise of water levels back to their original state is monitored.

bank filtration An improvement in water quality observed during the process of induced recharge (q.v.).

bank storage The temporary storage of surface water in riverside aquifer materials during periods of high stage (q.v.).

barrier boundaries Geological features which form an essentially impermeable seal along one or more edges of a given aquifer (q.v.). (Typical barrier boundaries include fault planes lined with low-permeability materials, and the contact surface between an aquifer (q.v.) and an aquitard (q.v.).)

baseflow The background level of stream flow during dry periods (which in many cases will be due solely to groundwater discharge).

baseflow index The ratio of baseflow to total flow for a given watercourse over a given period of time.

baseflow recession curves Graphs of stream flow rate versus time displaying baseflow recessions (q.v.).

baseflow recessions The natural patterns of decrease in baseflow rates during extended dry periods.

biodegradation Degradation (q.v.) of organic compounds as a result of biochemical processes.

biodiversity The range of species present in a given ecosystem.

biomonitors A biological measurement which can be used to infer previous patterns of water quality/flow rates in a freshwater ecosystem.

bioremediation The removal of contaminants from soil or groundwater by the use of biochemical processes (most commonly those involving microbial activity).

black box model A type of mathematical model (q.v.) in which output values are calculated from input values using a mathematical formula which need not have any obvious physical meaning (e.g. a statistical correlation).

block-centred Applications of finite difference methods (q.v.) in which computational nodes (q.v.) are located in the centres of finite difference cells ("blocks").

borehole catchment That portion of an aquifer (q.v.) which feeds water to a particular well.

boundary condition The mathematical expression of the relationship between processes occurring inside a defined mathematical model domain (q.v.), and those occurring outside it.

brackish water Water containing between 1000 mg/L and 10,000 mg/L total dissolved solids (q.v.).

brine Water containing more than 100,000 mg/L total dissolved solids (q.v.). (Synonym: hypersaline water.)

BTEX A collective term for the contaminants benzene, toluene, ethylbenzene and xylene.

buoyant support The support offered by groundwater to the roof of a cave (or other flooded underground void).

capillary fringe A thin mantle immediately overlying the water table (q.v.) in which the pores are entirely filled with water, but the water pressure is less than atmospheric pressure.

capillary fringe conversion Process by which the water table rises very rapidly by a height equal to the prior thickness of the capillary fringe, as tension-saturation changes to pressure-saturation.

capture zone A synonym for borehole catchment (q.v.).

casing A well lining with solid walls, typically fixed in place with an impermeable grout. (Casing is usually installed above the water table, but can also be used below the water table to seal off undesired horizons from the well.)

catchment The entire surface area feeding runoff to a given point on a surface water drainage system.

cation-anion balance The difference between the sum of concentrations of all cations and the sum of all anions (both expressed in milliequivalents per liter), normalized by dividing with the sum of both.

cation-exchange capacity A measure of the propensity of minerals or solid organic matter to participate in sorption/ion exchange reactions.

cations Positively charged ions (q.v.).

chemical constituents Those chemical substances (elements, compounds, ions) which are present within a given water.

chemical potentials The energy contents of given dissolved substances, as functions of temperature, pressure and composition.

chlorinated solvents A category of DNAPLs (q.v.) which contain chlorine in their molecular structures, and which are widely used as industrial solvents.

circum-neutral Displaying a *p*H in the range 6.5–8.5.

climate The totality of all types of weather (q.v.) experienced in a given place over a specified period of years, decades or centuries.

collapse doline An approximately circular, steep-sided depression in the Earth's surface, formed by collapse of an underground cave followed by migration of the resultant void to surface.

collective parameters Measurements of water quality which reflect the influence of more than one dissolved chemical constituent. (e.g. hardness (q.v.) represents the dissolved concentrations of both calcium and magnesium.)

colloids Minute particles suspended (rather than dissolved) in water.

computer model Any analytical model (q.v.) or numerical model (q.v.) which is solved using a digital computer.

conceptual model An assemblage of justifiable assumptions which simplify a real-world system in a manner which makes it amenable to analysis.

conductivity (As a physicochemical property of a water.) The ability of a water to conduct electricity, which is directly proportional to (and therefore useful as a measurement of) the total concentration of ions (q.v.) dissolved in it.

cone of depression A conical depression in the water table (or piezometric surface) (q.v.) which develops around a pumping well, as a consequence of the fact that drawdown (q.v.) increases with proximity to the well.

confined aquifer An aquifer (q.v.) lying below an aquitard (q.v.), such that there is no unsaturated zone (q.v.) between the base of the aquitard and the groundwater within the aquifer.

congruent dissolution The dissolution of minerals which dissociate completely in water without depositing any new solid phases.

connate water Groundwater (q.v.) which is believed to have been present in the pores of a sedimentary rock ever since it was deposited.

constant-head tests A means of measuring the permeability of a sample of aquifer material in the laboratory, by recording the amount of water which needs to be fed into a cylinder containing the sample in order to maintain a predetermined head gradient across the cylinder. This technique is best suited for high permeability materials (hydraulic conductivity > 0.1 m/day).

constant-rate test pumping Test pumping (q.v.) in which the rate of pumping is held at a single constant rate for the duration of the test.

consumptive use A use of water which effectively removes it permanently from the local natural environment (e.g. by evaporation of cooling waters, or export as moisture in fruit grown with irrigation waters).

contact spring A spring (q.v.) which arises at the lowest-lying point of outcrop of the stratigraphic contact between an aquifer (q.v.) and an underlying aquitard (q.v.).

contaminated land Land which has been left in a contaminated condition as a result of former use.

converged The state reached by a solver (q.v.) when the changes in values between successive iterations become negligible.

Darcy's Law The basic law of laminar groundwater flow, which states that flow rate (Q) is equal to the product of the hydraulic conductivity (K), the hydraulic gradient (i) and the area (A) of aquifer material (perpendicular to the direction of flow) through which flow is taking place. Darcy's Law is most commonly written: $Q = K \cdot i \cdot A$.

deep well (in context of dewatering q.v.) A well penetrating to an aquifer below the sole of the excavation associated with a given construction site.

deformable elements A type of computational element used in certain variants of finite element methods (q.v.).

degradation Diminution of the concentration of a specified solute or colloid due to its *in situ* disintegration into its constituent parts. (Degradation processes may be radioactive or biological, depending on the nature of the solute/colloid in question.)

deployable output The output of a commissioned water supply source (or group of such sources) as constrained by legal regulations; water quality; environmental issues; water treatment system capacity; the spare capacity of raw water mains and/or aqueducts; and limitations of pumping plant.

depression spring A spring (q.v.) formed through the intersection of the land surface by the water table.

desorption Release into solution of ions (q.v.) which were previously subject to adsorption (q.v.).

deterministic modeling An approach to the application of a mathematical model in which a single set of input values is used and a single set of values is obtained. (Deterministic modeling is based on a (often deliberately naïve) "cause-and-effect" premise.)

dewatering Removal of groundwater from the vicinity of an excavation to facilitate mining/construction.

diagenesis The array of geochemical and mineralogical changes which affect sedimentary deposits after their initial deposition, typically resulting in changes in porosity (either increases due to dissolution of minerals, or decreases due to precipitation of mineral cements in original pore space).

diffuse sources of pollution Extensive areas within which a given type of pollutant solute might have entered the ground surface, potentially at millions of individual points.

dilute-and-disperse The outmoded concept that the dilution capacity of natural aquifers will be sufficient to disperse polluted leachates (q.v.) without causing any problems.

direct recharge Recharge (q.v.) which occurs by rainfall soaking downwards immediately below its point of impact, passing beyond the root-suction base (q.v.) and continuing all the way to the water table (q.v.).

discharge zone A zone in which groundwater (q.v.) is flowing out from an aquifer (q.v.) onto the Earth's surface (or into a surface water body). (Head (q.v.) tends to decrease towards the ground surface in discharge zones.)

dispersion The displacement of solutes within aquifers beyond the pathways one would anticipate from the operation of advection (q.v.) alone. Dispersion is the sum of the effects of molecular diffusion (q.v.) and mechanical mixing (q.v.). Dispersion effectively leads to the spreading out of solutes from areas of high concentration into surrounding areas of lower concentration.

dispersion coefficient A quasi-Fickian (q.v.) mathematical description of dispersion (q.v.), which is assumed to equal the product of the rate of advection (q.v.) and a constant known as the dispersivity (q.v.).

dispersivity A "characteristic length" which is taken to typify the propensity of a given portion of an aquifer to generate mechanical mixing effects. (In many ways this is a convenient fiction ("fudge factor") rather than a true physical property of aquifers; it essentially accounts for variations in the magnitude and azimuth of advective fluxes at fine scales below the resolution of our measurements or models.)

disposition The mode of use of abstracted water (whether this is consumptive use (q.v.), or leads to return flows (q.v.)).

dissolved organic carbon The total concentration of dissolved compounds of organic carbon present in a water (i.e. excluding colloids).

dissolved oxygen Oxygen gas dissolved in water.

DNAPLs Dense nonaqueous phase liquids: synthetic organic compounds which form liquids at ambient temperatures and pressures, which are more dense than water and therefore tend to sink through the water column.

doline Closed depression in the ground surface, particularly common in areas of limestone or gypsum karst terrain.

doline spring A spring (q.v.) associated with a doline (q.v.).

DRASTIC Acronym for an aquifer vulnerability classification system first developed in the USA, referring to the following classification criteria: *D*epth to groundwater, *R*echarge rates, *A*quifer media, *S*oil media, *T*opography, *I*mpact of unsaturated zone, and *C*onductivity (hydraulic) of the saturated zone.

drawdown The difference between the initial water level in a given well and the observed water level at any specific time during a period of pumping.

dry deposition The settling of airborne particles from the atmosphere on plant and/or soil surfaces, whence they may be dissolved or entrained to become chemical constituents (q.v.) of surface runoff and/or groundwaters.

Dupuit–Forchheimer Assumption The assumption that regional groundwater flow is predominantly horizontal in orientation.

ecology The scientific study of the interactions between living organisms, and between living organisms and their environment.

ecosystem Any specific assemblage of organisms and their natural environmental surroundings.

ecotone A transitional zone in which one ecosystem gives way to its neighbor; typically this will be a zone in which elements of both ecosystems are identifiable.

effective porosity The ratio of the volume of interconnected pores to the total volume of a given body of rock.

Eh A measure of the status of redox (q.v.) reactions in a given water. (Closely related to the redox potential (q.v.).)

ejectors Wells used in certain construction dewatering applications, in which decompression of introduced water at the base of a borehole locally lowers air pressure, causing water to flow into the well, whence it is removed together with the introduced water.

elastic storage Storage of groundwater which is achieved by compression of the water and dilation of the pores. (The predominant form of storage in a confined aquifer (q.v.).)

electron acceptors Substances that receive electrons during redox (q.v.) reactions. (Synonymous with reductants (q.v.).)

electron donors Substances that lose electrons during redox (q.v.) reactions. (Synonymous with oxidants (q.v.).)

endemic (Referring to a species of animal or plant.) Displaying endemism (q.v.).

endemism The tendency of certain species of animals or plants to be restricted entirely to small geographical areas.

end-member mixing analysis (EMMA) Calculation of the relative proportions of different waters mixing during hydrological events, in which chemical analyses allow distinction between two "end-members," the mixing of which is deduced by the concentrations of particular solutes in the mixture.

enhanced natural attenuation Interventions in natural attenuation (q.v.) which increase the rates of key biogeochemical reactions. (Usually undertaken if monitored natural attenuation (q.v.) indicates that outcomes are not satisfactory from an environmental protection perspective.)

environmental isotopes Stable isotopes (q.v.) which are present naturally in the environment.

ephemeral Adjective describing any hydrological feature (e.g. a river or a spring) which flows for only short periods of time in any one year (usually only during, and shortly after, storms).

estevelle A type of depression spring (q.v.) which discharges groundwater when the water table is high, but which can also drain water back into the subsurface when the water table is low.

evaporation Vaporization from an open water surface.

evapotranspiration The transfer of water vapor to the atmosphere by the combined effects of transpiration (q.v.) and evaporation (q.v.).

Expanded Durov Diagram A hydrochemical classification diagram comprising six diamonds disposed in two clusters of three about a square central plotting field, which is useful in classifying groundwaters and inferring the geochemical reactions they have undergone.

external dewatering A form of dewatering (q.v.) in which one or more pumping wells located outside the current zone of excavation are used to intercept groundwater (q.v.) which would otherwise have flowed into the active working area.

falling head tests A means of measuring the permeability (q.v.) of a sample of aquifer (q.v.) material in the laboratory by recording the decline in head (q.v.) over time as water moves through a cylinder containing the sample. This technique is best suited for low permeability materials (hydraulic conductivity < 0.1 m/day).

fault spring A spring (q.v.) which arises where a fault brings an aquifer (q.v.) into contact with an aquitard (q.v.) at the ground surface.

feedback A situation in which one attribute of a system (X) affects another (Y), whilst Y in turn affects X.

ferric iron Ions of iron carrying a net electrical charge of 3+. Usual symbol: Fe^{3+}.

ferrous iron Ions of iron carrying a net electrical charge of 2+. Usual symbol: Fe^{2+}.

Fick's Law A mathematical formula which describes molecular diffusion (q.v.).

Fickian Any model which follows the mathematical form of Fick's Law (q.v.), whether or not it is used to model molecular diffusion (q.v.).

field capacity When an unsaturated soil contains its full specific retention (q.v.) of water, so that sufficient moisture is available to meet all demands from plants.

filter pack Sand and/or gravel packed around a screen (q.v.) to strain fine sediment present in the aquifer (q.v.) from groundwater (q.v.) entering a well.

finite difference methods The most widely used mathematical technique used for obtaining numerical solutions (q.v.) to the partial differential equations (q.v.) which describe groundwater (q.v.) processes.

finite element methods The second-most popular mathematical technique (after finite difference methods (q.v.)) used to obtain numerical solutions (q.v.) to the partial differential equations (q.v.) which describe groundwater (q.v.) processes.

finite volume methods A less common mathematical technique used to obtain numerical solutions (q.v.).

flow duration curve A cumulative frequency curve showing stream flow rates versus the percentage of time the indicated flow rates were equalled or exceeded at a given measurement point.

flow lines Lines representing the likely pathways of groundwater (q.v.) flowing through an aquifer. (Usually constructed on contour maps of groundwater head (q.v.).)

flow net An assemblage of head (q.v.) contours and flow lines (q.v.) delineating groundwater (q.v.) flow patterns in a particular system.

flow tube The area enclosed between two adjacent flow lines (q.v.).

fluviotrophic An adjective describing wetlands (q.v.) which receive their water mainly from inflows of surface water.

forward modeling Using a mathematical model to predict the likely outcome from known starting conditions.

fracture porosity Porosity (q.v.) arising from the presence of fractures (joints and/or faults) in rock.

free product Single-phase gasoline or other light nonaqueous phase liquid (LNAPL) (q.v.).

fresh water Water containing less than 1000 mg/L total dissolved solids (q.v.).

freshwater ecosystem Ecosystem in which the habitat is dominated by water with a total dissolved solids content of less than about 10,000 mg/L. (Note: discrepancy in definition of "fresh" between water resource engineers and ecologists.)

gaining streams Those which receive water from an adjoining aquifer.

geochemical mass balance Calculation of the net changes in dissolved and solid phases to account for observed geochemical changes.

geographical information systems (GIS) Computer program designed to store, manipulate and combine spatially-referenced data sets.

geostatistics The statistics of spatially-correlated data; the application of stochastic process theory and statistical inference to geographically distributed phenomena.

geothermal gradient The rise in groundwater temperature with increasing depth.

GOD Acronym for an aquifer vulnerability classification system, referring to the classification criteria: Groundwater occurrence, Overall lithology of aquifer and Depth to water.

graphical user interface (GUI) A computer program which uses clear graphical displays to simplify the process of developing and scrutinizing the inputs and outputs of mathematical models (q.v.).

gravity drainage The movement of moisture downwards through a porous medium due to the force of gravity.

ground-source heat resources Ubiquitous shallow groundwaters or soil atmospheres, with temperatures close to the local mean annual air temperature, which are potential sources of thermal energy only if processed using electrically or mechanically actuated heat pumps.

groundwater Subsurface water (q.v.) below the water table (q.v.).

groundwater ecology The study of ecosystems in aquifers (q.v.) and hyporheic zones (q.v.).

groundwater mining The unremitting depletion of irreplaceable groundwater (q.v.) storage by excessive pumping.

groundwater rebound A widespread rise in groundwater (q.v.) levels following a cessation of former pumping.

habitat Natural environmental surroundings for specific organisms.

half-life The time which it takes for an original mass of a given substance to be reduced by half due to radioactive decay (q.v.) or biodegradation (q.v.).

hardness (In the context of water quality.) A collective parameter (q.v.), the magnitude of which is determined by the dissolved concentrations of calcium and magnesium ions (q.v.).

head A measure of the energy content of water at any given point in the subsurface, equalling the sum of: (i) the water pressure measured at that point, and (ii) the elevation of the point of measurement relative to a specified datum (usually sea level).

head-dependent flux boundary A boundary condition (q.v.) in which the relationship between processes occurring within and outside of a mathematical model (q.v.) domain can be summarized as a water flux rate (into or out of the domain) whose direction and magnitude are determined by the head (q.v.) difference across the boundary and its permeability (q.v.).

heterogeneity The condition in which the value of a physical property of a body of some specified material (e.g. an aquifer (q.v.)) varies significantly depending on location within the body. For instance, if the value of transmissivity (q.v.) is greater at one point than at another, the aquifer is displaying heterogeneity.

heterogeneous Exhibiting heterogeneity (q.v.). (In hydrogeology, this adjective is most often applied to aquifers within which *K* and *S* are known to vary significantly from one place to another.)

high-enthalpy resources Natural steam and super-heated groundwaters (at temperatures > 150°C) useful for geothermal power generation.

history-matching A form of inverse modeling (q.v.), especially used in modeling of groundwater (q.v.) flows, in which observed changes in hydraulic head (q.v.) and/or flow rates from springs/wells etc. are compared with model output, and the model parameters are adjusted until a match is obtained between the observed and modeled changes.

homogeneity The condition in which the value of a physical property of a body of some specified material (e.g. an aquifer (q.v.)) does not vary at all, at any location within that body. For instance, if the value of transmissivity (q.v.) is found to be virtually the same at all points in an aquifer (q.v.), then the aquifer is displaying homogeneity.

homogeneous Displaying homogeneity (q.v.). (In hydrogeology, this adjective is most often applied to aquifers within which *K* and *S* scarcely vary from one place to another.)

horizontal flow treatment wells Pairs of dual-screened wells fitted with rotating mixers, which force the water downwards in one well and upwards in the other, so that a closed loop of groundwater circulation develops between the two wells, within which bioremediation (q.v.) can be promoted by the addition of microbial metabolites.

Hortonian overland flow A synonym for infiltration-excess overland flow (q.v.).

Hot Dry Rock (HDR) A category of geothermal energy in which rocks of originally low permeability are artificially fractured and injected with cool waters, which then heat in contact with the rock. (Sometimes termed "Enhanced Geothermal Systems" in mainland Europe.)

Hounslow Diagram A hydrochemical classification diagram for discriminating between saline waters of different origins, in which molar ratios of various major ions (sodium, chloride, calcium and sulfate) are cross-plotted.

humic and fulvic substances Naturally occurring, large, carbon-bearing molecules, commonly present in shallow groundwaters.

hydraulic conductivity The coefficient of proportionality between flow rate and hydraulic gradient in Darcy's Law (q.v.). Essentially a measure of the permeability (q.v.) of a given body of rock with respect to *fresh water* (q.v.). Usually denoted by the letter *K* (upper case).

hydraulic retention time The time taken for natural inflow and outflow to completely replace the water present in a given hydrological feature (e.g. a lake or wetland).

hydrochemical facies Distinctive bodies of water within groundwater systems characterized by particular combinations of major and minor ions.

hydrogeochemical modeling Mathematical modeling of geochemical reactions which affect groundwater quality.

hydro-seral succession The natural process of change whereby many wetlands (q.v.) tend naturally to evolve into dry lands over time.

hydrostratigraphy The classification of sequences of strata according to their ability to store and transmit groundwater. (It essentially involves identifying, naming and specifying the extents and properties of the aquifers (q.v.) and aquitards (q.v.) in a given area.)

hyper-saline water Synonym of brine (q.v.).

hypogean fauna Animals which live in the subsurface (many of them in groundwater (q.v.) systems).

hyporheic flow Flow of surface water or a mixture of groundwater (q.v.) and surface water through the bed sediments of a surface watercourse or wetland (q.v.).

hyporheic zone Zone of bed sediment in which hyporheic flow (q.v.) occurs.

incongruent dissolution Mineral dissolution which is accompanied by simultaneous precipitation of new solid phases.

indirect recharge Recharge (q.v.) which occurs where rainfall *fails* to soak into the soil surface on which it first lands, but instead forms surface runoff, which later enters the subsurface at some distance from its point of initial impact, usually via macropores (q.v.).

induced recharge Influx of water from a river to an aquifer (q.v.) as a consequence of artificial lowering of groundwater head (q.v.) by pumping.

inductively coupled plasma Instrumental chemical analysis technique in which an incandescent cloud of gas, within the force field of a powerful electromagnet, is subjected to emission detection to measure the concentrations of dissolved constituents in water samples.

infiltration-excess overland flow Surface runoff which arises when a soil is sufficiently wetted that infiltration is occurring at the maximum possible rate for that soil, so that any further rainfall landing on the soil surface will be unable to enter the subsurface and will thus become overland flow. Also known as "Hortonian Overland Flow."

integrated finite difference methods An infrequently used mathematical technique for obtaining numerical solutions (q.v.) to the partial differential equations (q.v.) which describe groundwater processes.

intermediate enthalpy resources Groundwaters with temperatures between 100 and 150°C which are useful for a range of direct and indirect geothermal applications.

intermittent Adjective describing any hydrological feature (e.g. a river or a spring) which flows for only specific seasons of the year.

intrinsic permeability A synonym for permeability (q.v.) sometimes used by hydrogeologists to emphasize the distinction between hydraulic conductivity (q.v.), which is a function of both rock properties and water viscosity/density, and permeability (q.v.), which is a function of the intrinsic properties of the rock alone.

intrinsic vulnerability Vulnerability (q.v.) quantified solely in terms of generic hydrogeological properties of a groundwater system, without reference to specific contaminants.

inverse modeling Using a mathematical model to reconstruct the chain of events which led to a known outcome.

ion chromatography Analytical technique that separates the various ions according to their relative affinity for a static adsorbent material lining the walls of a long tube.

ion exchange Coupled process of adsorption (q.v.) of one type of ion with desorption (q.v.) of another, so that the sorbing ion is exchanged for the desorbed ion on the mineral surface.

ionization The process of electron addition/loss which results in atoms becoming charged (i.e. becoming ions (q.v.)).

ions Charged chemical constituents (q.v.) dissolved in water.

isotope An atom of a particular element which contains a different number of neutrons in its nucleus (q.v.) than another atom of the same element.

isotropic Displaying isotropy (q.v.).

isotropy The condition in which the value of a physical property of a body of some specified material (e.g. an aquifer (q.v.)) does not vary with the orientation of any measurement. For instance, if the measured hydraulic conductivity (q.v.) of many aquifers (q.v.) is the same in the vertical and all horizontal directions, then it is said to display isotropy.

iteration A computational sweep of the entire array of nodes which make up a model grid.

Jacob Method A widely-used graphical technique for interpreting time versus drawdown (q.v.) data arising from test pumping (q.v.) exercises.

Jacob plot A plot of time (logarithmic axis) versus drawdown (arithmetic axis) on semi-logarithmic paper, as used in the Jacob Method (q.v.).

jetting A means of well installation in which water is pumped at high pressure down a well casing, which erodes its way into the subsurface.

juvenile waters Waters which have not previously participated in the hydrological cycle during the entire history of our planet.

laminar flow The characteristically gentle nature of most groundwater flow, which occurs smoothly with little mixing, as if the groundwater were a stack of separate layers ("laminae").

landfills Holes in the ground (often former quarries/surface mines) filled with solid wastes.

landslides The rapid movement of a mass of rock, earth or debris down a slope.

leachate A contaminated liquid (usually a brackish water (q.v.)) derived from leaching of waste materials.

light nonaqueous phase liquids (LNAPLs) Synthetic organic compounds forming liquids at ambient temperatures and pressures which are less dense than water, and therefore tend to float on water.

liquefaction The complete loss of strength by a soil, due to high pressure groundwater forcing grains apart, so that the soil begins to behave like a liquid.

local equilibrium assumption An assumption made in order to simplify certain tasks in reactive transport modeling (q.v.), which states that the groundwater achieves geochemical equilibrium with the enclosing aquifer materials at every point along its flow path.

longitudinal dispersion Dispersion in the same direction as the advection of solutes/colloids.

losing streams Those which lose water to an underlying aquifer.

low-enthalpy resources Aquifers at reasonably shallow depths (<3 km) containing groundwater at temperatures in the range 25–100°C, potentially useful for direct geothermal applications.

lumped-parameter model A synonym of "black box model" (q.v.).

macrofauna Large animals (fish, mammals etc.).

macropores Large pores (typically > 1 mm minimum diameter) in soils, of diverse geometries and origins, including desiccation cracks, animal burrows, root casts (i.e. voids formed where a plant root has decayed) and the interfaces between woody plant roots and the surrounding soil.

major ions Ions (q.v.) which are usually present at concentrations in excess of 1 mg/L.

mathematical model The translation of a conceptual model (q.v.) into mathematical form, i.e. as a set of equations of state (those which express relationships between state variables and system parameters) and relevant parameter values, which can then be solved to answer various questions concerning the system represented by the conceptual model.

mechanical mixing The mixing of highly and weakly concentrated groundwaters due to variations in hydraulic conductivity and effective porosity at scales below that at which these parameters are resolved in the calculation of advection (q.v.).

meiofauna Medium-sized animals (e.g. worms, beetles etc.).

membrane filtration The production of brines (q.v.) by the selective movement of water through a low-permeability material under high pressure, such that the solutes (q.v.) are left behind.

method of characteristics (MOC) A less common mathematical technique used to obtain numerical solutions (q.v.) to the equations describing pollutant transport in groundwater systems.

micrograms per liter (μg/L) Common units of expression used for concentrations of dissolved trace constituents in groundwaters.

migration of fines Movement of fine grained sediment from aquifer material into a well (which has been poorly completed and developed).

milliequivalents per liter (meq/L) Units of concentration especially useful in graphical display and mass-balance interpretation of groundwater compositions, obtained by dividing the concentration in millimoles per liter (q.v.) by the valence (q.v.) of the ion (q.v.) in question.

milligrams per liter (mg/L) Most common units of expression for concentrations of dissolved species in groundwaters.

millimoles per liter (mmol/L) The number of millimoles (an expression of the total number of molecules) of a given chemical constituent (q.v.) in a liter of water.

mineralization (Of a given groundwater.) An informal term referring to the relative concentration of total dissolved solids (q.v.).

minor ions Ions (q.v.) which are usually present at concentrations in the range 0.01–1 mg/L, but which are occasionally present at far higher concentrations, such that they can locally be regarded as major ions.

model domain The area contained within the boundaries of a mathematical model (q.v.).

model refinement The progressive adjustment of a conceptual model (q.v.) and/or mathematical model (q.v.) in the light of testing of the agreement between observed and predicted values of key variables.

molar concentration A unit of dissolved concentration in which the mass of a given substance is expressed in terms of the number of moles of it which are present in a specified volume of water.

molecular diffusion The tendency for molecules to spread themselves out evenly within a solution over time (due to the effects of Brownian Motion). (One of the two principal components of dispersion (q.v.).)

moles per liter (mol/L) The number of moles (an expression of the total number of molecules) of a given chemical constituent (q.v.) in a liter of water.

monitored natural attenuation Natural attenuation (q.v.) which is carefully monitored to ensure that the outcomes are acceptable from an environmental protection perspective.

Monte Carlo modeling A form of probabilistic modeling (q.v.) in which a restricted number of input values are selected randomly from the input probability distributions.

mountain front recharge A form of indirect recharge (q.v.) occurring where water leaks from the beds of mountain stream channels into underlying permeable alluvial fan deposits.

MTBE (methyl tertiary butyl ether) A petroleum-associated contaminant which is both rather soluble and problematic at very low concentrations.

multi-phase flow The simultaneous movement of gas, water and/or other liquids through the same system of interconnected pores.

natural attenuation An assemblage of physical, chemical, or biological processes that, under favorable conditions, act without human intervention to reduce the mass, toxicity, mobility, volume, or concentration of contaminants in soil or groundwater.

net gain In a river augmentation (q.v.) strategy, the ratio of the volume of pumped groundwater to the sum of the volumes of decreased natural outflow plus induced recharge from the river.

neutral *p*H *p*H = 7. (Denotes a water which is neither acidic nor alkaline.)

nitrate Nitrogen tri-oxide (NO_3^-), a natural constituent of many groundwaters, which is particularly common in areas of intensive arable agriculture.

nodes The points in a grid in a model domain (q.v.) for which a solver (q.v.) will obtain a numerical solution (q.v.) of the governing equations.

nonionized Adjective describing an atom which has no net electrical charge.

nonuniqueness A problem afflicting deterministic modeling (q.v.), in which it is formally impossible to conclude that the set of input values used, and output values obtained, represent the only possible solution to a given mathematical model.

nucleus That part of an atom in which both protons and neutrons reside.

numerical dispersion Errors arising during the solution of the advection-dispersion equation (q.v.) by finite difference (q.v.) or finite element methods (q.v.), which manifest themselves in a spurious increase in the effects of dispersion (q.v.) on solute concentration distributions.

numerical model A mathematical model (q.v.) solved using a numerical solution (q.v.).

numerical solution An *approximate* solution to a given mathematical model, obtained by replacing terms containing derivatives (i.e. differential expressions such as $\delta h/\delta x$) by simple algebraic expressions relating to subdivisions of finite portions of the entire model domain (q.v.).

oasis A form of depression spring (q.v.) in an arid region, from which all natural groundwater loss occurs by evapotranspiration (q.v.), without surface outflow.

object-oriented programming An approach to computer programming which defines both the nature of a data structure and the types of operations/functions that can be applied to the data structure. The data structure is thus defined as an object that includes both data and functions. Programming is used to create relationships between a range of objects, which can inherit characteristics from other objects.

ombrotrophic An adjective describing wetlands (q.v.) which receive all of their water from rainfall.

orbital An energy shell within an atom in which electrons reside.

oxidants Substances which are doing the oxidizing (and in the process being reduced) in redox reactions. Synonymous with electron donors (q.v.).

PAH See polycyclic aromatic hydrocarbon.

parameterization The process of providing values for all of the parameters used in a given mathematical model (q.v.).

partial differential equation A type of equation common in mathematical models (q.v.) of groundwater flow, and which are typically solved using numerical solutions (q.v.).

particle tracking A well-established technique for modeling pollutant transport in groundwater systems, which generally employs the random walk method (q.v.) to represent hydrodynamic dispersion (q.v.) of solutes (q.v.).

parts per billion (ppb) Units of expression used for concentrations of dissolved trace constituents in groundwaters in terms of mass of solute per mass of solution, closely equivalent to microgrammes per liter (q.v.) in nonsaline groundwaters.

parts per million (ppm) Units of expression for concentrations of dissolved species in groundwaters (mass per unit mass), closely equivalent to milligrammes per liter (q.v.) in nonsaline groundwaters.

passive treatment processes Water treatment technologies which use only naturally available energy sources, in gravity-flow treatment systems which are designed to require only infrequent (albeit regular) maintenance to operate successfully over their design lives.

pathogens Disease-causing microbes.

pe A measure of the status of redox (q.v.) reactions in a given water. Closely related to Eh (q.v.), though expressed on a different scale of units.

perennial Adjective describing any hydrological feature (e.g. a river or a spring) which flows all year round.

permanent hardness Hardness (q.v.) which remains even after boiling.

permeability A measure of the ability of a given rock to transmit water. (Note: permeability, strictly speaking, is a property of the rock alone, whereas hydraulic conductivity (q.v.) reflects both rock properties and the viscosity and density of water.)

permeable reactive barrier (PRB) A form of passive treatment (q.v.) in which a permeable medium of geochemically appropriate material is placed in an aquifer (q.v.) (in the form of a "barrier" perpendicular to the flow path) to treat contaminated groundwater *in situ*.

pesticides Synthetic organic compounds used to kill unwanted plants or insects.

pH The negative log to the base 10 of the activity (q.v.) of hydrogen ions in solution. As such, it is a common measure of the acidity/alkalinity balance of a given solution.

phreatic eruption A highly explosive volcanic eruption caused by sudden contact between rising magma and groundwater.

phreatotrophic An adjective describing wetlands (q.v.) which receive receive their water mainly from groundwater discharge.

physically based Adjective used to describe certain types of mathematical model (q.v.), signifying that the model is based upon realistic mathematical representations of the true physics of a natural process.

physico-chemical parameters Properties which define the physical status of water (e.g. temperature), and which also provide indications of chemical conditions (e.g. conductivity (q.v.), redox potential (q.v.)).

piezometer A small-diameter well specially constructed to measure the head (q.v.) at a specific depth within an aquifer (q.v.).

piezometric surface An imaginary surface, defining the levels to which water in a confined aquifer would rise were it everywhere pierced with wells.

piezometry The measurement of head (q.v.)

piping The development of conduits and caverns by seepage erosion (q.v.). (Most commonly occurs in sandstones, unconsolidated sands, loess, peat and other soft sediments.)

point sources of pollution Pollutant sources which are potentially identifiable at individual locations on the Earth's surface.

polychlorinated biphenyls Large synthetic organic molecules containing chlorine which are widespread in certain types of electrical equipment.

polycyclic aromatic hydrocarbons (PAH) Coal-tar by-products commonly found at former sites of town gasworks and cokeworks.

porosity The proportion of a given volume of rock that is occupied by pores.

potential evapotranspiration The rate at which evapotranspiration (q.v.) would occur, given the ambient conditions of atmospheric temperature humidity and solar radiation, if there were no limit to the supply of water to the soil surface and/or to plants.

potential recharge Infiltrating waters passing below the root-suction base (q.v.).

potential yield The yield of a commissioned source or group of water supply sources as constrained only by well and/or aquifer properties for specified conditions and demands.

potentiometric surface A synonym of "piezometric surface" (q.v.).

precipitation Rain, hail, sleet or snow.

principle of electroneutrality The phenomenon that means water cannot carry a net electrical charge (positive or negative), but must always be electrically neutral.

principle of superposition The principle that drawdown (q.v.) due to a number of wells pumping in consort equals the sum of the drawdowns each well would cause were it pumping alone.

probabilistic modeling An approach to the application of a mathematical model in which all inputs and outputs are probability distributions, rather than single parameter values.

protons Hydrogen ions (H^+).

pump-and-treat Remediation of contaminated groundwater (q.v.) by pumping the water to surface and using conventional water treatment technologies to remove contaminants from the water. Subsequently, the clean groundwater is either re-injected, discharged to a surface water course, or dispatched for use elsewhere.

Q_{95} The flow rate in a stream which is equaled or exceeded for 95% of the time.

Qanat Ancient systems of horizontal wells/tunnels used to extract groundwater from mountain-front aquifers in the Persian and Arabic worlds.

quicksand A body of saturated sand in which the movement of water lifts the grains away from one another, resulting in the sediment having far less load-bearing capacity than would normally be expected.

radioactive decay A form of degradation (q.v.) exhibited by certain elements, in which partial disintegration of atoms, accompanied by the release of ionizing radiation, results in a change of elemental identity.

radioactive isotope An isotope (q.v.) which is prone to radioactive decay (q.v.).

radius of influence The distance from the outer limit of the cone of depression (q.v.) to the pumping well.

Ramsar Convention An international agreement for the protection of birdlife (principally through the protection of wetland (q.v.) habitats (q.v.)), originally adopted in 1971.

Ramsar Sites The world's most important avian wetland habitats

random walk method A method of representing hydrodynamic dispersion of solutes in groundwater systems in which random numbers are used to represent the variability of dispersive displacements.

reactive transport modeling Mathematical modeling of the simultaneous movement of solutes and their geochemical reactions within groundwater systems. Essentially, a combination of solute transport modeling (q.v.) and hydrogeochemical modeling (q.v.).

realization One of many potential solutions to a mathematical model (q.v.), obtained in the course of probabilistic modeling (q.v.), especially using the Monte Carlo modeling approach (q.v.).

rebound (1) An increase in contaminant concentrations in groundwater after the cessation of a pump-and-treat (q.v.) operation. (2) The recovery of piezometric/water table levels in an aquifer following the cessation of pumping (especially in mining districts).

recharge The entry of water into the saturated zone (q.v.).

recharge area Any area in which recharge (q.v.) to an aquifer (q.v.) commences its journey into the subsurface. Head (q.v.) tends to decrease with depth in recharge areas.

recharge boundaries Geographical/geological features (e.g. rivers/lakes with permeable beds; other aquifers) which can act as sources of further recharge (q.v.) to an aquifer (q.v.) under particular head (q.v.) conditions, especially those relating to drawdown (q.v.).

recirculating wells A technology for pump-and-treat (q.v.) in which abstraction and reinjection of water occurs within the same wells.

redds Cavities created by salmonid fish within the gravel beds of streams, within which they bury their eggs for later hatching.

redox Simultaneous oxidation and reduction, involving the gain and loss of electrons by reacting ions.

redox potential A measure of the status of redox (q.v.) reactions in a given water. Closely related to Eh (q.v.), though sometimes expressed on a different scale of units.

reductants Substances which are doing the reducing (and in the process being oxidized) in redox reactions. Synonymous with electron acceptors (q.v.).

refugia Small zones within ecosystems (q.v.) where conditions suitable to survival of particular species are maintained even when they are lost elsewhere (e.g. during a drought).

retardation Slowing down of the rate of movement of a given solute or colloid normally due to sorption (q.v.), relative to that which would be expected from the affects of advection (q.v.) and dispersion (q.v.).

return flows The return of used water to the same part of the water environment from which it was originally abstracted.

Richards Equation A partial differential equation (q.v.) which fully describes the distribution of moisture and head (q.v.) in, and quantifies rates of water movement through, variably saturated soils.

river augmentation Seasonal pumping of groundwater into a nearby river to maintain flows in the latter during periods of drought/high demand.

river basin Synonym of catchment (q.v.) provided that the "given point" in the definition of "catchment" is the mouth of the river.

root-suction base The maximum depth from which water can be removed by the suction exerted by plant roots.

saline intrusion Movement of sea water into an aquifer which previously contained fresh groundwater, usually due to artificial lowering of head (q.v.) by pumping.

saline water Water containing between 10,000 mg/L and 100,000 mg/L total dissolved solids (q.v.).

Salinity High concentrations (>10,000 mg/L) of total dissolved solids (q.v.), i.e. elevated mineralization (q.v.).

salinization The accumulation of salt in the soil, often due to excessive irrigation with groundwaters rich in sodium and chloride.

sand pumping Production of sand along with water from an inadequately completed and developed well.

sapping The development of valleys by seepage erosion (q.v.), often due to collapse of conduits originally formed by piping (q.v.).

saturated (1) Completely filled with water. (2) The geochemical condition in which the ionic activities of specified dissolved species exactly equal the thermodynamic threshold value above which the water would be likely to begin precipitating a given mineral. A saturated water is at thermodynamic equilibrium with respect to the minerals in question.

saturated thickness The thickness of the saturated zone (q.v.) at a given place in a given aquifer (q.v.). In an unconfined aquifer (q.v.) it is calculated by the difference in elevation between the water table and the base of the aquifer material. In a confined aquifer (q.v.), it equals the total thickness of aquifer material.

saturated zone (Also known as *phreatic* zone.) Those parts of the subsurface in which pores are completely filled with water which has a pore pressure greater than atmospheric pressure.

saturation-excess overland flow Surface runoff which occurs whenever the water table rises so rapidly during a storm event that groundwater discharges at the ground surface, and any further rainfall cannot infiltrate the saturated soil and thus also joins the overland flow.

screen A well lining with slotted walls, typically installed below the water table, to facilitate entry of groundwater into the well.

seepage erosion Erosion of soils or rocks due to the entrainment of grains by discharging groundwater.

seepage face An area of ground (often steeply sloping) through which groundwater discharges in a diffuse manner.

silica Silicon dioxide (SiO_2). A common natural component of many minerals and a common uncharged solute in groundwaters.

slug test A technique for measuring the hydraulic conductivity (q.v.) of an aquifer (q.v.) by monitoring of the decline of water levels in a well back to their original state, following the sudden addition of a known volume ("slug") of water to the top of the water column in a well.

soakaways Chambers (1–3 m deep) deliberately intended to discharge water into the subsurface, used for disposal of surface drainage and/or wastewaters.

soil moisture Subsurface water (q.v.) above the water table (q.v.).

soil moisture deficit The amount of water which would need to be added to a given body of dried soil in order to bring its moisture content up to field capacity (q.v.).

soil zone The uppermost layer of unconsolidated earth materials, which will normally support plant life if sufficient moisture is present.

solute A dissolved chemical constituent (q.v.) in a water.

solute transport modeling Mathematical modeling of the movement of solutes within groundwater systems.

solution doline A type of doline (q.v.) formed by direct dissolution of the bedrock surface (either exposed or beneath a thin soil cover) by recharge (q.v.) waters.

solver A numerical technique which inverts the matrices arising from the approximation of equations of state in a mathematical model, yielding resultant values of key variables (e.g. head (q.v.), pollutant concentration etc.).

sorption A collective term including both adsorption (q.v.) and desorption (q.v.). (Almost synonymous with surface processes (q.v.).)

source elimination The removal of a contaminant source to prevent ongoing release of pollutants to the subsurface.

speciation The distribution of dissolved components among free ions, ion pairs, and complexes.

species richness A measure of biodiversity (q.v.), computed as the number of species in a given area.

specific electrical conductance Synonym of conductivity (q.v.), of which this is the more formal term.

specific gain Value obtained if the transmission gain between two streamflow measurement stations is divided by the distance between the two stations. (Typically expressed in units of m^3/s per km length of stream channel).

specific retention The water retained in the unsaturated zone (q.v.) after it has been allowed to fully drain under the influence of gravity. (Typically expressed as the ratio of the volume of water retained in the soil after drainage to the total volume of the soil.)

specific vulnerability Vulnerability (q.v.) of a groundwater system to a specific contaminant or group of contaminants (e.g. chlorinated pesticides).

specific yield The amount of water which drains freely from a unit volume of initially saturated rock per unit decline in water table elevation. (Typically expressed as the ratio of the volume of water draining from the soil to the total volume of the soil.)

specified flux boundary A boundary condition (q.v.) in which the relationship between processes occurring within and outside of a model domain (q.v.) can be summarized as a specific value of water flux rate (into or out of the domain).

specified head boundary A boundary condition (q.v.) in which the relationship between processes occurring within and outside of a model domain (q.v.) can be summarized as specified values of head, which do not vary as the heads within the domain vary.

speleogenesis The formation of caves and cave systems.

spring A discrete opening in the Earth's surface from which groundwater emerges and flows in an open channel.

stable isotope An isotope (q.v.) which will not radioactively decay.

stage The water level in a stream (river) at any one point along its length and at any specified time.

standard operating procedures (SOPs) Formally agreed methods for undertaking work in the field or laboratory, typically specified as part of a quality assurance plan.

steady-state model A mathematical model (q.v.) of a situation in which there are no temporal changes in the key variables calculated by the model.

step-drawdown test A particular form of test pumping (q.v.) which is used to assess the performance efficiency of abstraction wells. It involves pumping the well in a number of steps of increasing flow rate and measuring the resultant increases in drawdown.

step-test Synonym of step-drawdown test (q.v.).

stochastic modeling A synonym for "probabilistic modeling" (q.v.).

stoichiometry Branch of chemistry which deals with the molar concentrations of given substances participating in particular chemical reactions. Widely used in the interpretation of hydrogeochemical reactions occurring in groundwaters.

storativity The amount of water which will be released from a unit volume of confined aquifer (q.v.) per unit decline in head (q.v.) within that aquifer.

subsidence The localized lowering of the ground surface, usually forming closed depressions.

subsidence doline A broad, shallow depression formed due to down-warping of overlying rocks as an underlying soluble rock layer is gradually dissolved.

sub-soil zone The zone between the root-suction base (q.v.) and the water table (q.v.), comprising unsaturated soils and/or rocks, within which the soil moisture (q.v.) is slowly seeping downwards.

subsurface water All natural water beneath the ground surface, amounting to the sum of soil moisture (q.v.) and groundwater (q.v.).

suffosion doline A crater left behind as loose sediments fall/are washed into voids in underlying bedrock.

sump dewatering A method of dewatering (q.v.) in which any water which enters the working area is diverted to one or more storage ponds ("sumps"), whence it is pumped out of the excavation.

supersaturated The geochemical condition in which the ionic activities of specified dissolved species exceed the thermodynamic thresholds above which the water is likely to begin precipitating a given mineral.

surface processes Electrostatic attraction processes which result in adsorption (q.v.) or desorption (q.v.). (Almost synonymous with sorption (q.v.).)

synform A down-fold in rock strata.

system dynamics A technique for developing and solving mathematical models (q.v.) of systems in which complex feedbacks (q.v.) occur.

temporary hardness Hardness (q.v.) which can be removed by boiling, due to reaction of calcium and magnesium with bicarbonate alkalinity (q.v.) to precipitate carbonate scale.

test pumping A means of measuring transmissivity (q.v.) (and sometimes also storativity (q.v.)), by recording the changes in drawdown (q.v.) in and around a well which is being pumped at a known rate.

total dissolved solids (TDS) The sum total of all dissolved chemical constituents (q.v.) in a given water.

total organic carbon (TOC) The total amount of organic carbon compounds present in a water, as colloids and/or dissolved.

toxins Substances which are potentially harmful to humans, animals and/or plants.

trace elements Any dissolved species present at concentrations below 0.01 mg/L.

trace ions Ions which are present at concentrations less than 0.01 mg/L.

trace metals Dissolved metals present at concentrations below 0.01 mg/L.

transfer function A mathematical formula used to calculate output values from input values in a black box model (q.v.).

transient Varying over time.

transmission gain The increase in flow rate between successive gaging points down a river.

transmission loss The decrease in river flow rate between two successive flow gaging stations along the same channel.

transmissivity The integration of the values of (horizontal) hydraulic conductivity between the base and top of the aquifer. (In practice, a simplification is often used which equates transmissivity with the product of the saturated thickness (q.v.) and the average horizontal hydraulic conductivity (q.v.).)

transpiration The release of water vapor to the atmosphere from the leaves and stems of plants.

transverse dispersion Dispersion perpendicular to the direction of advection (q.v.).

travertine Encrustations of minerals deposited by discharging groundwater.

travertine spring A spring (q.v.) which precipitates travertine (q.v.).

tributyl tin oxide (TBTO) An artificial organic compound of tin, long used as a component of anti-fouling paints in the maritime industry.

trophic chains Predation hierarchies in ecosystems with one animal being eaten by another, which is in turn eaten by others etc.

unconfined aquifer An aquifer (q.v.) in which the upper limit of saturation (q.v.) (neglecting the capillary fringe (q.v.)) is the water table (q.v.), such that unsaturated soil or sub-soil lies between the upper boundary of the aquifer and the ground surface.

underground thermal energy storage (UTES) The introduction of heat into the subsurface (either with down-hole heat exchanges or by injection of warmed water), whence it can later be removed during a period of high heat demand.

undersaturated The geochemical condition in which the ionic activities of specified dissolved species do not exceed the thermodynamic thresholds above which the water would be likely to begin precipitating a given mineral.

uniformitarianism The principle that observation of present-day processes is the key to understanding past events, as evident in the geological record.

unsaturated zone Those parts of the subsurface in which pores are only partly filled with water, and thus contain a mixture of air and soil moisture (q.v.).

urban heat island An urban area which has a warmer climate than surrounding rural areas, due to the disposal of waste heat from multiple sources.

vadose Obsolete term meaning "unsaturated" (as in "unsaturated zone" (q.v.)).

valence The magnitude of the charge on a given ion (q.v.).

variable source areas Runoff-generating areas of ground that wax and wane substantially in surface area and water depth during and after storms.

virtual reality A computer-based technology for simulating the visual, auditory, and other sensory aspects of complex environments.

volatile organic compounds (VOCs) A class of organic compounds (most of them pollutants) of high volatility, which are prone to exsolve from water if it is exposed to the atmosphere.

vulnerability The readiness with which a given aquifer (q.v.), or portion thereof, is likely to succumb to pollution.

water table The upper surface of the saturated zone (q.v.), corresponding to the base of the capillary fringe (q.v.) and the surface upon which pore water pressure is exactly equal to atmospheric pressure. (In practice, the water table corresponds to the level to which water will settle in a well dug into an unconfined aquifer (q.v.).)

watershed US usage: a synonym of catchment (q.v.). UK usage: the crest of a ridge dividing one catchment (q.v.) from another.

weather Whatever is happening outdoors at a given place and given time with regard to precipitation, temperature, wind conditions and barometric pressure.

well completion Installing pipe work and other materials (e.g. gravel pack, cement, etc.) in a borehole to prevent the collapse of the hole and leave it in a suitable condition for its future intended use.

well development Agitating the water within a well by mechanical means or by air-lifting (q.v.), so that any fine sediments within the screen (q.v.), filter pack (q.v.) and nearby parts of the aquifer are dislodged, and can be removed from the well before it is commissioned for long-term pumping.

well loss The difference between the real water level observed during pumping and that which one would anticipate, taking into account the aquifer transmissivity (q.v.) and storativity (q.v.) only.

well-bore storage effect The response of a well to pumping during the first few minutes after starting the pump, during which time the water which was previously standing within the well casing/screen is removed. (It results in a linear relationship between drawdown (q.v.) and elapsed time of pumping, as distinct to the curvilinear response characteristic of the subsequent depletion of aquifer storage.)

wellfield A group of pumping wells located in relatively close proximity to one another.

wellpoints Numerous shallow, narrow-bore wells installed around a construction site, which are pumped together by suction from a surface pump, to which they are all joined by a common header pipe.

wetland A body of surface water which is nowhere deeper than 6 m.

willow carr An area of wet ground (often a groundwater discharge zone) thoroughly colonized by willow trees.

Younger Diagram A hydrochemical classification diagram for discriminating between mine waters of different origins in terms of the balance of acidity (q.v.) and alkalinity (q.v.) and the relative proportions of chloride and sulfate.

zero flux boundary A boundary condition (q.v.) in which the relation ship between processes occurring within and outside of a mathematical model (q.v.) domain can be summarized by stating that no water enters or leaves the domain on the boundary in question.

Index

Figures in *Italic*, Tables in **Bold**, information in Boxes B, information in notes n